WIND AT OUR BACKS

LEADING DURING THE COLD WAR

WIND AT OUR BACKS

LEADING DURING THE COLD WAR

James B. Morin
Rear Admiral, U. S. Navy-Retired

John H. Bradley (Contributing Author & Editor)

Reutel Press
Houston

Title Page Photograph aboard USS *Coral Sea,* May 1965.

Back Row: CDR J. Morin, CPT G. Hart, LT W. Fidelibus, LCDR W. Rivers, LTJG J. Hughes
Front Row: LT R. Floreani, LTJG K. Holthaus, ENS G. Blosser

First edition published in April 2011.

Cover designed by John H. Bradley.

U. S. Navy photograph

Printed in Texas by Americas-Press.

Published by Reutel Press
14027 Memorial Drive, 189
Houston, TX 77079

ISBN 978-0-578-08449-7

1. Naval Aviation 2. Cold War 3. Vietnam War 4. Middle East 5. Mediterranean

DEDICATED TO

BETTY

My children

My shipmates and squadron mates,

especially

the Silver Foxes of 1965 and 1966

Thanks for your love and support

and

all the memories

May the road rise up to meet you.
May the wind always be at your back.
May the sun shine warm upon your face,
and rains fall soft upon your fields.
And until we meet again,
May God hold you in the palm of His hand.

CONTENTS

SPECIAL NOTES

Additions

Besides editing and formatting Admiral Morin's narrative, I wrote the following additions to his autobiography.

The "Prologue" and "Epilogue."
The chapters entitled, "The Gathering," "Dangerous Times," and "The Ceremony."
The Cold War narratives and comments in most chapters.
The Chronologies that appear in most chapters.
Some additional information woven into the Morin narrative.
Several special sidebars.
The footnotes.
The Appendices.

Two Styles

To differentiate authors, I have used TWO STYLES.

Morin Narratives

Admiral Morin's narratives have indented paragraphs and expanded line spacing such as this paragraph. I have used the symbol **Ψ** to identify his narratives and to separate paragraphs in them.

Bradley Additions

My chapters, narratives, and comments have block paragraphs with tight spacing such as this paragraph. I have used the symbol **Δ** to separate paragraphs in my narratives.

John H. Bradley

Acknowledgments

Many people helped with the development of this book.

Jim Mattly initially proposed the project several years ago. He supported my travel and my work, gave me many useful documents and photos to work with, provided his Flight Log, and brought together many of his VA-155 squadron mates to help with it. Diane Mattly loaned me a collection of photos from the 2000 reunion and provided information about the reunion.

Bill Kish provided extensive original papers, documents and news clippings, reviewed the manuscript, and provided many useful additions and comments.

Ed and Marilyn Hiebert added important information and provided photos.

Bill, Ed, and Marilyn flew into Houston to discuss the project with Jim and Diane Mattly and me.

George Blosser, Ken Holthaus, and Chris Morss wrote new papers and provided valuable information for the story.

Dick Floreani's earlier paper proved most useful.

Albie Hyde sent me excellent information, and he provided his Flight Log for his time flying against North Vietnam.

Two young sailors from the USS *La Salle* - Rex Harrison and John Tomaio - e-mailed me recollections of their service with CAPT Morin.

Steve Lockwood, a young officer on the USS *Franklin D. Roosevelt* wrote detailed information about his service under CAPT Morin on that ship.

One of CDR Morin's young sailors on the *Coral Sea,* Bill Maynor, wrote a splendid e-mail to Admiral Morin that I have enclosed as an appendix.

Admiral Morin helped greatly. He talked to me for several days at his home in Tallahassee, provided me papers and photographs,answered many questions, and reviewed and commented on the manuscript.

Dr. Ann Bradley edited the manuscript as did Penny Leas. Both provided many valuable comments. Desiree Terry helped me learn Quark so that I could prepare the manuscript in that software.

My thanks to all.

John H. Bradley

Glossary of Special Terms

AAA	Antiaircraft artillery
ABM	Anti Ballistic Missile
ARVN	Army of South Vietnam
CAG	Commander, Air Group or Air Wing
CAT 4	Lowest mental category of enlistees/draftees
CENTO	Central Treaty Organization
CG	Commanding General (Air Force, Army, Marine Corps)
Charlie	Viet Cong, short for VICTOR CHARLIE
CHARLIE	Approval to land on carrier
CO	Commanding Officer (All services)
CNO	Chief of Naval Operations, U. S. Navy
CON	Officer has control of the ship
CV/CVA	Aircraft carrier
Davey Jones	Bottom of the sea - ruler of the sea
Feet Wet	Over water
Flight	Two or three aircraft
Frag/to frag	Hand grenade thrown at an American officer or non-commissioned officer by an American soldier
HMS	Her/His Majesty's Ship
ICBM	Intercontinental Ballistic Missile
Indian Country	Enemy area
JCS	Joint Chiefs of Staff
KIA	Killed in Action
MED	Mediterranean Sea
MIA	Missing in Action
MIRV	Multiple Independently Targeted Re-Entry Vehicle
NAS	Naval Air Station
NVA	North Vietnamese Army
NVN	North Vietnam
OJT	On the job training
OPS	Operations
Petty Officers	Non-commissioned officers in the Navy
Pickle off	Release/drop bombs
Polliwog	A sailor who had not crossed the equator
Recce	Reconnaissance
ROG/ROGER	Understand
SALT	Strategic Arms Limitation Treaty
SAM	Surface to Air Missile
SAR	Search and Rescue/Recovery
Section	Two flights of aircraft
Shellback	A old or veteran sailor - one who has crossed the equator
Silver Fox	Member of VA-155 squadron
Squadron	Formal air organization - generally twelve aircraft
SVN	South Vietnam
UCMJ	Uniform Code of Military Justice
USS	United States Ship
XO	Executive Officer
Viet Cong/VC	Communist guerrillas in South Vietnam
WIA	Wounded in Action
Wing	Formal air organization - generally three or four squadrons
Wingie	Wingman - aviator flying off the wing of another aviator

Ranks/Pay Grades

CWO		Chief Warrant Officer
ENS	O1	Ensign
2LT	O1	2nd Lieutenant, U. S. Air Force, U. S. Army, U. S. Marine Corps
LTJG	O2	Lieutenant, junior grade
1LT	O2	1st Lieutenant, U. S. Air Force, U. S. Army, U. S. Marine Corps
LT	O3	Lieutenant
CPT	O3	Captain, U. S. Air Force, U. S. Army, U. S. Marine Corps
LCDR	O4	Lieutenant Commander
MAJ	O4	Major, U. S. Air Force, U. S. Army, U. S. Marine Corps
CDR	O5	Commander
LTC	O5	Lt. Colonel, U. S. Air Force, U. S. Army, U. S. Marine Corps
CAPT	O6	Captain, U. S. Navy
COL	O6	Colonel, U. S. Air Force, U. S. Army, U. S. Marine Corps
RADM	O7	Rear Admiral, Lower Half
BGEN	O7	Brigadier General, U. S. Air Force, U. S. Army, U. S. Marine Corps
RADM	O8	Rear Admiral, Upper Half
MG	O8	Major General, U. S. Air Force, U. S. Army, U. S. Marine Corps
VADM	O9	Vice Admiral
LTG	O9	Lt General, U. S. Air Force, U. S. Army, U. S. Marine Corps
ADM	O10	Admiral
GEN	O10	General, U. S. Air Force, U. S. Army, U. S. Marine Corps
FA	O11	Fleet Admiral
GOA	O11	General of the Army

Note
The rank for a Navy captain is abbreviated CAPT in the text to distinguish it from Air Force, Army, and Marine Corps captains [CPT].

Prologue

James Brendon Morin *is one hell of a guy.* I have spent a lot of hours talking with soldiers, sailors, and airmen, and generally have found that most have remarkable stories to tell. I have also found that no one knows much about most of the veterans because they never were in the right place at the right time (or the wrong place at the wrong time) to gain public recognition, or they never achieved high enough positions in their services to gain public attention. But when I have talked with people who knew them, particularly those who worked for them, I have gained a deep respect for their skills, contributions, and their impact.

Δ

This is a story about such a man. Born in Cohoes, NY on 27 October 1927, Morin was an only son. His father, Paul Joseph Morin, of French-Canadian parentage, was born in Lunenberg, MA; his mother, who immigrated to the United States in 1921, was born Nora Theresa Wynne in Kilfynn, County Kerry, Ireland.

A Catholic, Morin gained success in a Protestant-dominated Navy. A charming man with a great sense of humor, he is Irish to his core, he was a tough taskmaster when needed. Very bright, he, however, was a player not a by-stander. Red-headed, he had the inner fire associated with his look-a-likes.

Morin would have made a fabulous salesman or he could have become a political boss in his home town. He could have been an out-and-out rogue with his good looks and charm and touch with people. But, he was driven to fly and left college to become a Naval Aviator. Once commissioned, Morin became a skilled pilot and an outstanding naval leader. He served well in all his assignments, but the command of an attack squadron in Vietnam and the command of an aircraft carrier - the apogees of his career - tested all his skills and professionalism.

Morin did not graduate from the United States Naval Academy at Annapolis, but he still made flag rank in the days when most line admirals graduated from Annapolis. His achievement of high rank was impressive considering that he had moral courage equal to his physical courage to stand his ground or take the correct action when others would have waffled or compromised in order to meet the desires of their superiors and get ahead. And he earned his flag in spite of being threatened with two court-martials in mid-career.

Perhaps his greatest moral test came when he was on the brink of higher rank and greater success, and he left the Navy he loved to be with the wife he loved who had been stricken with cancer.

Morin performed as a high-impact leader during challenging times and stood out among the many exceptional war-tested leaders in the Navy. If you spent a couple of hours talking with him, you grasped that immediately. He talks about his men and what they meant to him. He gives them credit for much of his success. Morin admires them: "I was blessed with a squadron of great officers and men." He talks about senior officers who taught him how to lead. He talks about training and its importance. He talks about setting the example.

Δ

When you talk with Morin's subordinates, their stories about his leadership jump out at you. I would have loved to have worked for him. He would have probably tested me in ways that I never expected, and on some days I would have thought him to be crazy or heavy handed or impossible to work for. But on the day that I needed him or my mates needed him, he would have been there backing us up, giving us everything he had, and if necessary he would have been out front pulling us along.

He would have put himself, his career, and his life on the line for us, and those are the guts of leadership and those traits build trust and respect that are never forgotten.

Leadership, trust, and respect marked Jim Morin, an honored leader.

John H. Bradley
Houston, Texas
4 April 2011

One

THE GATHERING

SILVER FOXES 1965

LEFT TO RIGHT,
Top row: Hiebert, Hughes, Floreani, Mattly, Hart
Middle Row: Hyde, Starling, Holthaus, Kish, Blosser
Front row (flight suits): Fidelibus, Rivers, Chasko, Harris, Morin, Mares
In Khaki: Shaw, Tillis, Renning (baseball cap), Morss

VA-155

January 1965		October 1965
	COMMAND	
CDR Jack Harris	CO	CDR James Morin
CDR James Morin	XO	
	Pilots	
LCDR Wendy Rivers	Operations Officer	
LCDR Ernest Mares	Administrative Officer	LCDR Ernest Mares
LT William Fidelibus	Maintenance Officer	LT William Fidelibus
LT Edwin Hiebert	Safety Officer	LT Edwin Hiebert
LT Edward Dickson		
		LCDR Paul Moore
LT Richard Floreani		LT Richard Floreani
CPT George Hart, USMC		CPT George Hart, USMC
LTJG Gerald Chasko		LT Gerald Chasko
LTJG Kenneth Holthaus		LTJG Kenneth Holthaus
LTJG James Hughes		LTJG James Hughes
LTJG Albert Hyde		LTJG Albert Hyde
LTJG William Kish		LTJG William Kish
LTJG James Mattly		LTJG James Mattly
ENS George Blosser		LTJG George Blosser
	SUPPORT	
LTJG Chris Morss	Air Intelligence Officer	LTJG Chris Morss
LT James Renning	Asst Maintenance Officer	LT James Renning
ENS Roy Shaw	Asst Maintenance Officer	ENS Roy Shaw
CWO James Tillis	Aviation/Ordnance Officer	CWO James Tillis

A Special Group

They called themselves Silver Foxes. In their fifties and sixties, most still remained active in their chosen work as they traveled to Pensacola Naval Air Station for another reunion with their skipper, Rear Admiral James B. Morin, United States Navy-Retired. Veterans of the Navy's Attack Squadron 155 (VA-155), they had flown beautiful and agile A4E Skyhawks from the USS *Coral Sea* on Yankee Station over North and South Vietnam and Laos in 1965, and they completed more than 2,000 sorties between February and October 1965. Each surviving pilot flew an average of 115 missions during his deployment and earned a parcel of Air Medals for his combat service.[1] One Silver Fox died in action; one became a POW. Overall, the enemy shot down twenty-nine of the sixty plus aircraft from the *Coral Sea*, killed a total of eight pilots, and captured two in 1965.[2]

AWE

"THOSE OF US WHO DID NOT FLY CAN ONLY STAND IN AWE OF THEIR COURAGEOUS ACCOMPLISHMENTS.

I SHALL ALWAYS HAVE A PROFOUND RESPECT FOR THEIR CONTINUING FRIENDSHIP THAT HAS LASTED 45 YEARS."

CHRIS MORSS, 2010

The Silver Foxes considered themselves to be a special group and considered that they had a special bond. Richard Floreani expressed it well:

> *Seldom in one's lifetime will one be in a group where the mixture of many diverse personalities results in a group chemistry that was as unique, as uplifting as was ours. Together we lived through the most stressful, exciting, and intense time in our lives, nothing else has or will come close. We are all so different individually; it is really surprising how collectively we blended into that "perfect fit". There was a special aura in that Silver Fox squadron of the 1964-65 combat cruise. We had developed a feeling of brotherhood that has never left us. Indeed, we had become very close.*[3]

They even had their own calling card.

A Special Leader

Albie Hyde added: "I think that VA-155 does deserve to be remembered not so much as an outstanding navy squadron, which I think we were, but more for the idea of what a group of guys can accomplish with really inspiring leadership."[4]

"I THINK THAT VA-155 DOES DESERVE TO BE REMEMBERED NOT SO MUCH AS AN OUTSTANDING NAVY SQUADRON, WHICH I THINK WE WERE, BUT MORE FOR THE IDEA OF WHAT A GROUP OF GUYS CAN ACCOMPLISH WITH REALLY INSPIRING LEADERSHIP."

AL HYDE

Another Silver Fox expanded on that thought: "I left the Navy after my initial tour of duty in 1967 and to this day I believe the guidance and leadership Jim Morin provided has made me a better man. Just what are the lessons I have carried and felt so enriched by? Is it friendship forged out of difficult times? Is it comradeship bonded by respect? Is it the sense of fairness and loyalty shared? Is it a sense of honor that was brought into existence and fostered by this man? Is it a sense of humor that allows a smile even in the worst of circumstances? Is it mutual respect? Is it the ability to maintain your integrity and sense of what is right when career considerations would suggest another course? Is it decisiveness when others seem irresolute?"

"No matter what direction any of our lives had taken, when we got together we reverted to who we were in 1965."

He continued: "I know that since leaving the Navy, Jim Morin has never been far from my consciousness...Every time I procrastinated doing a project, I recall paint blistering on the bulkheads of his room on the Coral Sea. When I talked with my sons during their growing years, I tried to share with them the sense of integrity and honor I felt I had learned from him. When I dealt with co-workers in my chosen profession, I tried to remember little things like offering encouragement when things looked grim. I tried to be light-hearted and not overwhelmed when tragedy was at hand.

For some reason, I always wanted the "Skipper" to be proud of me. I was a professional pilot and throughout my career when I did instructing I always attempted to impart more information than the syllabus provided. I felt, as I think any "Morin-trained man" did, that you could never know too much. Something you had learned and shared could very possibly help someone out in a way you could not imagine. You should never be ashamed about demanding excellence from both yourself and from those whom you may be leading. Accept mistakes only as part of a larger learning process."

The old Fox wrote about the VA-155 reunions: "No matter what direction any of our lives had taken, when we got together we reverted to who we were in 1965. The laughter and comradeship came so easily. The many years in the interim had no impact at all on the close comradeship we shared. I know many pilots from other squadrons and this kinship does not exist with them." He added, "I suppose one could attribute this to the mix of people who "happened" to get assigned to VA-155. I think it is more than that; I think it is the leadership effect of Jim Morin."[5]

Δ

"He knew what he wanted to do with the squadron and was quick to let us know his desires."

George Blosser

George Blosser recalled in 2003: "By the time Skipper Morin assumed command of VA-155, he had already spent over a year as the squadron's executive officer. We all enjoyed and respected him as an officer and a pilot. His piloting skills ranked among the best, but never in a braggey manner. He was the consummate professional Naval Aviator."

Then he wrote: "When Skipper Morin assumed command..., he immediately set a new course for our squadron. He knew what he wanted to do with the squadron and was quick to let us know his desires. During our first "all pilots meeting," the Skipper delivered a very concise outline of how things were going to be in our squadron. He was not demanding or overbearing, he simply wanted to let us know what he expected and what he would do for us in return. Anyone present that day knew good and well how our "new Skipper" intended to run the squadron. As he closed the meeting, he said, "I want you to know I can only do so much; I need your help."

Blosser emphasized that Morin's "one simple statement said it all about the inner workings of our group. We respected the Skipper, and we wanted to help. The last thing any of us wanted was to embarrass the squadron in front of the entire air group… never mind how it would make the Skipper feel." He concluded,

> *When Skipper Morin first assumed command, we were a mediocre unit. However, by the end of the cruise, we were as good as anyone in the air group. ... a characteristic we were all proud to accept.*

The JG left the Navy and joined National Air Lines as a co-pilot on a Boeing 727 in January 1968. He ended his "new career" in May 2003.

Reflecting on his long civilian career, Blosser wrote: "Looking back, there is absolutely no question that the influences of VA-155 and the direct leadership of Jim Morin had a profound effect on my successful career. The highlight of my airline career was being selected to run the B-727 Flight Training program for Delta. In some sense, this was similar to commanding a squadron in the Navy. Skipper Morin's philosophy was that he "never asked a squadron mate to do anything he would not do himself." As I presided over the 727 group for over three years, I recounted my total flight time on the B-727 and the hours spent instructing in the cockpit and simulator. The numbers are not important… What is important is the fact that I never "dented" an aircraft or injured a passenger. Is this the result of superior airmanship… certainly not? It is rather the product of excellent instruction by those that are in a position of authority and who are charged with "teaching" the young folks how it should be done."

Blosser summed up the reasons for his own success: "Jim Morin was as good as anyone instilling these traits in his junior pilots. Without question the success I enjoyed in my airline career can be traced directly to those early days in VA-155."[6]

Δ

Ed "Hoob" Hiebert wrote the following about Jim Morin:

> *The main difference I saw was that of the four skippers I had in VA-155, Morin was the only one that could be one of the guys but still maintain the authority position. Most of realized we were in a sh____ war but there was*

TICKET PUNCHING

TICKET PUNCHING AND CAREERISM BEGAN TO INFECT THE ARMED SERVICES ABOUT THIS TIME.

TICKET PUNCHERS TENDED TO SPEND MINIMUM TIME IN RISKY COMMAND POSITIONS WHERE SUBORDINATES COULD MAKE OPERATIONAL MISTAKES AND BLEMISH THEIR RECORDS. MOST TICKET PUNCHERS SOUGHT MAXIMUM TIME IN SAFE STAFF POSITIONS WHERE THEY COULD NOT BE BLAMED FORTHE MISTAKES OF OTHERS.

TICKET PUNCHERS AND CAREERISTS AVOIDED TAKING CHANCES BECAUSE MISTAKES COULD BLEMISH THEIR RECORDS AND JEOPARDIZE THEIR CAREERS. EVEN WORSE, FEW TOOK CARE OF OR STOOD BY THEIR SUBORDINATES BECAUSE THEY FOCUSED ON TAKING CARE OF THEMSELVES.

MOST TICKET PUNCHERS WOULD NEVER HAVE TOLD THEIR OFFICERS AND MEN, "I WANT YOU TO KNOW I CAN ONLY DO SO MUCH; I NEED YOUR HELP," AS MORIN DID.

JOHN BRADLEY

> *no way out and at least he was standing up for us when dealing with guys like Fighting Hank [CDR Henry Glindeman]. I appreciated that and wanted to make him look as good as I could."*[7]

"I, FOR SOME REASON ALWAYS FELT THAT HE WAS LOOKING OUT FOR OUR INTEREST, EVEN IF DOING SO MEANT JEOPARDIZING HIS OWN CAREER..."

ED HEIBERT

Hiebert added an interesting and perceptive comment: "I, for some reason always felt that he was looking out for our interest, even if doing so meant jeopardizing his own career. I never felt that he was making decisions which were designed to further his climb up the ladder."[8]

Δ

Ken Holthaus spoke well of his new XO, Jim Morin: "It didn't take long for all of us to recognize that our new XO was a rare individual. One that was demanding, but had common sense."

He added:

> *As the cruise went on and Jim Morin replaced Jack Harris as CO, the squadron took on a new sense of pride. Most of this was due to Skipper Morin getting the group of diverse individuals to operate with a professionalism that separated us from most squadrons.*

And then talking about harmonizing the reserve officers who were serving to fulfill their obligations and the regular officers whom he perceived to be concerned about doing or saying anything that could damage their careers, Holthaus wrote:

> *As CO, Commander Morin found a way to treat both with a sense of appreciation for putting our asses on the line. As a result none of us wanted to disappoint him because he had our respect.*[9]

"...OUR NEW XO WAS A RARE INDIVIDUAL. ONE THAT WAS DEMANDING, BUT HAD COMMON SENSE."

KEN HOLTHAUS

Δ

Looking back to his service with CDR Morin, Albie Hyde wrote:

> *There are three qualities that are required in a naval flight officer and most have them in greater or lesser degrees: professional ability, the ability to do the day to day job of being a naval officer and carrying out your assigned job in the chain of command; flying ability, the skill to control your aircraft in a manner that allows you to complete your mission; and leadership ability, the ability to instill in your men a desire to follow you into any situation.*[10]

After he joined his new squadron, Hyde said that he had not met an officer who had all these qualities in abundance, but "that changed when CDR[11] Morin became the XO.[12] As XO he was not the squadron leader but after working with him for a short time most of the pilots knew he was the kind of officer that you should try to emulate." He added: "Later after taking over from CDR Jack Harris there was no question. No one tried to push the limits with the Skipper, but he was not stiff or over controlling. When the time allowed he could be just one of the boys. During the time that he was Skipper, CDR Morin became like a father figure to the squadron."[13]

Δ

LCDR Ernie Mares did not become a fan of Morin. The reasons are not clear. Perhaps the bitterness stemmed from Morin not using him as his Operations Officer or XO.[14] Or perhaps, Morin gave him a lackluster efficiency report or the acidic Mares just clashed with the good humored Irishman. Mares' approach to things showed up in his critical commentary about John McCain's actions in Vietnam. In it, Mares focused on McCain's flying skills which he did not think were particularly good, and McCain's actions during combat, which he felt resulted in his being shot down unlike other members of his flight.

Years later, Morin took good care of Mares when he could have just ignored him. Mares never responded.

Morin's Foxes

The Silver Foxes of VA-155 who served under Jim Morin on the USS *Coral Sea* in 1965 were a diverse group. Some were Regular Navy officers; some were Reserve officers.

The Silver Foxes of VA-155 who served under Jim Morin on the USS Coral Sea in 1965 were a diverse group. Some were Regular Navy officers; some were Reserve officers.

Four of them had graduated from the U. S. Naval Academy at Annapolis, MD; most had graduated from other universities such as the California Polytechnic Institute, Harvard, and the University of California. One was an aeronautical engineer; one came from the U. S. Marine Corps. Most were pilots. None, however, had ever flown in combat before. All who flew when Morin commanded VA-155 aboard the USS *Coral Sea* survived the war. After leaving the Navy, many joined major American airlines. Several reflected on their time with VA-155 and with Morin, and Morin, in turn, reflected on them and their service.

Δ

ENS George Blosser was the most junior pilot in the squadron in February 1965. Morin noted that "With red hair, blue eyes and freckles, he looked like a character in a Norman Rockwell painting. He was what every CO wanted in an ensign, eager, bright, interested, and a fast learner."[15] And, according to a peer, "professional and fun."[16] Blosser was the son of Mr. and Mrs. Elbert G. Blosser of Inglewood, CA and was a graduate of Inglewood High School.

ENS George Blosser initially was the most junior pilot in the squadron in February 1965. Morin noted that "With red hair, blue eyes and freckles, he looked like a character in a Norman Rockwell painting."

"George the Red" worked for LT Bill Fidelibus, the squadron Maintenance Officer. He enjoyed his shipboard job, saying, "My job was the aircraft division officer, having three shops under my wing... Air Frames, Power Plant, and the Para Loft. I treasured the job and looked forward every day to working with the men in the shops."[16]

Chasko earned a Distinguished Flying Cross, eight Air Medals , and two Navy Commendation Medals for his service during the war.

By July 1965, Blosser had completed more than seventy combat missions, the majority of which he flew over North Vietnam and "made over 100 successful day and night arrested landings" and joined the "selected ranks of Coral Sea *Centurions*." That month, Morin presented the new JG his first Air Medal and two Gold Stars for second and third awards.[18]

In August, he had an exciting day flying with Bill Fidelibus.

Blosser deployed to the MED after Vietnam and served in VA - 34 on the USS *Saratoga* with Bill Kish and Jim Mattly. The three Coral Sea vets earned a Golden Tailhook Award for their successful landings.

Δ

LT Jerry Chasko, who also hailed from California, flew many missions with Jim Morin beginning in February 1965.[19] Morin recommended Chasko for a Distinguished Flying Cross on 2 May 1965 for his performance against a radar antenna on Bach Long VI Island in the Gulf of Tonkin between Hainan Island and Haiphong. Morin's citation read:

Hanoi-Haiphong-Hainan

> *The island was very heavily defended and the anti-aircraft fire was fierce and devastatingly effective. Lieutenant CHASKO was cognizant of the fact that several aircraft had been hit immediately preceding his attack; however, he did not forget the importance of his objective. With total disregard for his personal safety, he continually circled the target area at low altitude until he was able to locate and identify the radar antenna and then aligned his aircraft for a low level attack oblivious of the heavy fire along his flight path. With calm professionalism, he delivered his load of bombs directly on the radar antenna causing its complete destruction. His aircraft sustained numerous hits resulting in hydraulic failure and serious structural damage; however, through superior display of airmanship, he successfully guided his crippled aircraft to safe recovery aboard the USS CORAL SEA.*[20]

After he completed his tour with VA-155, a tour in the MED with VA-46 with Bill Kish, and then one ashore, Chasko joined a heavy reconnaissance squadron that flew North American RA5Cs, Vigilantes. Morin reported that "The RA5C was a difficult aircraft coming aboard, and experienced carrier pilots were needed to fly them."

Chasko earned a Distinguished Flying Cross, eight Air Medals, and two Navy Commendation Medals for his service during the war.[22] The 1961 Annapolis graduate retired as a commander in 1981.[23]

Δ

Bill Fidelibus, the Aircraft Maintenance Officer during the 1965 cruise of the *Coral Sea*, proved a stout pilot and a good officer with a sense of humor. In the squadron's newsletter,[24] he ribbed his XO: "Well, it looks like a long cruise. Except for the elderly types like the X. O. and the Leading Chief most of us have never spent such a long at sea period without a break." And, then, he commended his maintenance people:

> *I am quite proud of the performance of the personnel in the maintenance department during this long at sea period. You have been asked to work long, tedious, often frustrating hours and in many cases the sense of accomplishment may be lost in an unexpected change of plans or an interrupted holiday routine. To the maintenance department I say – your response to these circumstances has been admirable. As a squadron we showed what we could do when we were called upon to execute "Flaming Dart" 1 and 2 in the strikes on North Viet Nam. Everyone in the maintenance department can take pride in their part in these strikes.*

On 12 August 1965, Fidelibus flew his most memorable mission to search for SAM sites in North Vietnam. George Blosser flew as his wingman.[25]

After searching for forty-five minutes and finding nothing, Fidelibus gave Blosser the hand signal for FEET WET, and the two headed back to the coast. They descended to 200 feet and flew out across numerous rice paddies at 400 knots. Then bad things happened. Blosser recalled the situation:

> *I remember seeing a dike coming into view below Bill's aircraft. The next thing we knew the whole world opened up and it was like the 4th of July. Unknowingly, we had flown directly over a 57mm [millimeter] anti-aircraft site that was positioned out of view on the far side of the dike. The North Vietnamese had watched us fly toward them and simply waited until we were at close range to open fire... much like a hunter waits for birds to fly by his position during the annual dove hunt. I vividly remember two large shells going by my canopy, one on each side of the aircraft. The noise and fireballs were unimaginable. For reasons that to this day I do not understand, my aircraft did not receive a scratch.*
>
> *Bill, however, was not quite so fortunate. In a matter of seconds, it became apparent that his aircraft had been severely damaged."*

The Fidelibus A4

"The starboard wing received at least two hits and the horizontal and vertical stabilizers received at least one hit. It appears that the empennage damage incurred when sections of the wing separated and struck the starboard side of the horizontal stabilizer and elevator. The external stores were jettisoned shortly after being hit, While climbing and heading toward the Gulf of Tonkin, the rudder section caught fire but extinguished itself at approximately 16,000 feet. By the time the pilot reached 18,000 feet both the utility and flight control hydraulic systems had failed.

The wingman reported that the starboard flap, 2/3 of the starboard elevator, 1/4 of the starboard wing forward of the flap, 2/3 of the horizontal stabilizer on the starboard side and the entire starboard elevator were missing. Three ten inch diameter holes were also observed in the starboard side of the rudder."

J. B. Morin

THE RESCUE HELICOPTER WAS RIGHT THERE AS HE ENTERED THE WATER. FIDELIBUS REPORTED THAT IT "LOWERED THE SLING, THE CREWMAN ENTERED THE WATER AND HELPED ME ENTER THE SLING. I HAD NO DIFFICULTY ENTERING THE COPTER AND THE RIDE TO THE SHIP WAS UNEVENTFUL."

Blosser spent the remaining time during the mission working with his flight leader to help him get back to the *Coral Sea*. After Fidelibus climbed to a higher altitude and put out his fires and then executed a difficult maneuver to re-fuel from a tanker, the pair headed back to their carrier. En route, Fidelibus found that he could not fly below 240 knots, and therefore he decided that he had to eject from his A4 ahead of the *Coral Sea* where the rescue helicopter could pick him up. Blosser flew to the right of and behind his leader.

As the two pilots approached the carrier, the rescue helicopter launched. Then, the Squadron LSO, Ed Hiebert, "...gave Bill some final words of encouragement and a reminder to *beat the seat* during the sequence." Completely exhausted by this time, Bill Fidelibus recalled his ejection:

> *I transmitted my intentions to eject to CORAL SEA and was directed to eject to the starboard bow of one of the destroyers. I made an easy turn to the designated area and positioned myself for ejection. I disconnected my zero-delay lanyard, put my feet on the rudders and pulled the face curtain. The ejection was violent. I felt a severe tumbling sensation and the face curtain was jerked from my hands at seat separation. The opening shock of the chute caused my head to jerk and gave me a stiff neck.*[26]

The rescue helicopter was right there as he entered the water. Fidelibus reported that it "lowered the sling, the crewman entered the water and helped me enter the sling. I had no difficulty entering the copter and the ride to the ship was uneventful."[27]

A Navy reporter added:

> *The helo... was flying plane guard when word of [Fidelibus'] ejection was received. [The pilot] immediately proceeded to the downed pilot, hovered five feet above the water and rescued Lt. Fidelibus, 50 seconds after he hit the water.*
>
> *Rescue crewman, M. J. Austin, ADJ3, who had entered the water four times previously to rescue downed personnel, was lowered into the moderately choppy seas to help Lt. Fidelibus into the rescue sling.*
>
> *Helo pilot Lieutenant (j.g.) B. G. Brandow had this to say about the rescue. "The pilot handled himself like a real pro. We couldn't ask for any more cooperation or better set up. We regret that his watch wasn't waterproof."*[28]

When he landed on the carrier Fidelibus greeted Blosser with, "thanks a lot George; I could not have done it without you."

Blosser commented, "Of course I thought nothing of what I had done as my actions were simply those of a good wingman. Once again, doing what we had been trained to do by Skipper Morin .... Jim Morin's leadership and confidence in his pilots led me to take charge of this event and assist Bill Fidelibus in any manner necessary at the moment. At no time did I ever question what should be done; I simply did it... because that is what we had been taught."[29]

"[BILL] KNEW IT WAS GRIM TO FALL BEHIND IN MISSIONS SO AS SOON AS HE COULD, SORE NECK AND ALL, HE WAS FLYING AGAIN."

Morin added to the story: "[Fidelibus] was circling the ship one day in a damaged aircraft. The aircraft finally became uncontrollable - lateral control was limited due to wing damage, requiring the use of full left aileron (both hands) to hold the wing level - and Bill had to punch out about a mile and a half from the carrier. He was rescued quickly [fifty seconds after he hit the water] but had been hurt in the ejection."[30]

According to Jerry Chasko[31] this was the first at-sea rescue by a Navy helicopter," although *Sea Breeze* reported differently.[32]

Morin concluded: "[Bill] knew it was grim to fall behind in missions so as soon as he could, sore neck and all, he was flying again."[33]

Δ

Richard Floreani had been one of the four full Navy lieutenants in the squadron. Called "Da Vinch" because of his Italian heritage, he had deployed with VA-155 in December 1964. When he first arrived in Vietnam and began using real bombs at assumed uninhabited areas, he remarked: "most likely we were not killing anybody and more importantly, nobody seemed to be shooting back, trying to kill us. We were big boys and big boys love this kind of stuff. This was fun." Soon after he arrived in the war zone, he flew reconnaissance missions over South Vietnam and then flew bombing missions against the Ho Chi Minh Trail. When things got tighter, Floreani flew in the three-carrier strike into southern North Vietnam. He recalled: "On that first mission over the north the Silver Foxes experienced our first of many losses. Our mate, Ed Dickson was shot down and killed." The fun stopped.[34]

"THE PACE OF THE AIR WAR QUICKENED. WE FLEW SCORES OF SORTIES, MOST UNDER WITHERING ANTI AIRCRAFT FIRE. OUR TARGETS WERE ROADS, BRIDGES, BARRACKS, AIR FIELDS, RAIL YARDS, WAREHOUSES, AND SAM SITES."

DICK FLOREANI

After returning for an unexpected extended tour on Yankee Station in 1965, he reported that: "The pace of the air war quickened. We flew scores of sorties, most under withering anti aircraft fire. Our targets were roads, bridges, barracks, air fields, rail yards, warehouses, and SAM sites. There were about 30 A4 pilots on the entire ship, the other pilots supported us guys but it seemed that we A4 drivers were carrying the air war on our backs. Yea, the brass in Washington could assign the targets but we thirty guys were called on repeatedly to fly into harm's way to deliver the goods".[35] He added: "The pilots were really getting stressed. We were tired. There were always more missions to fly and the targets became increasingly dangerous ... We exhibited our rage at new high risk, low value missions by having periodic ready room riots with hat and pillow fights and ceaseless witty banter. But, to a man, we always flew our assigned missions. We never let each other down. Every man flew the task assigned to him. It was not passed off to another."[35]

"...We always flew our assigned missions. We never let each other down. Every man flew the task assigned to him. It was not passed off to another."
Dick Floreani

Morin summed up Floreani's combat service:

> *Dick saw more FLAK than the rest of us on any mission. Where we were concentrating in an attack on pipper control on the target, Dick was looking out the side of canopy at the FLAK sights. He made craters in the damndest places in NVN. On a road recce mission one day, Dick caught an Army truck out on a road going like hell. He had already dropped his bombs but decided to use his 20mm cannons. The A4 had notoriously unreliable guns. When Dick had the truck in his sights, he squeezed the trigger. One or two rounds were fired before the guns jammed. When Dick got back to the ship, he was still fuming as he filled out the maintenance discrepancy sheet. He wrote: "THE GUNS WENT RAT and THEY'RE SUPPOSED TO GO RAT-A-TAT-TAT.*[36]

After completing eleven months on line in 1965, Floreani returned to the United States for a few months, and then he joined with CDR Morin and a couple of other *Coral Sea* comrades for a second combat cruise on the USS *Constellation.*

The "unusually gentle, reverent, and unassuming pilot" who lacked "the bravado of some of the others" and was "much liked by all," flew 167 combat missions.[37]

The Navy released Floreani from service in June 1966. He left with seven Air Medals and two Navy Commendation Medals. "Da Vinch" went on to fly with an airline, stating, "The new flying was so wonderfully safe and boring."[38]

Δ

CPT George Hart

CPT George Hart, a career marine from Kansas, served as the Marine Exchange Officer with VA-155 in 1965. All the squadron officers respected Hart for his flying skills and dedication and they liked his great sense of humor. The "tall and imposing dark-haired officer"[39] seemed to volunteer for more than his fair share of missions but did it in an inconspicuous way.[40] Morin really liked Hart. He reported that,

> *Typical of Marine Aviators, he didn't have much carrier experience when he came to us, but you'd never know it. He was a quiet person, always in the ready room ready to volunteer for any mission.*[41]

Hart, however, had more time in the A4 than anyone else in the squadron except CDR Jack Harris when he joined the squadron in 1964. Previously, he had served as an instructor pilot on shore and had completed one full sea tour with an operational marine squadron. With his rank and experience, Hart flew as one of the senior section leaders in the Silver Fox squadron. After leaving VA-155, he returned to Vietnam later with another marine squadron. Hart retired in the early 1980s as a major after twenty years of service.

GEORGE HART

"GEORGE HART RECALLED THAT AS A TEENAGER FLYING IN A CUB SOMEWHERE IN KANSAS, HIS ENGINE SPUTTERED AND HE AND HIS COMPANION DEADSTICKED THE SMALL PLANE INTO A FIELD. THE TWO GOT OUT, UNSCREWED THE GAS LINE, BLEW HARD INTO IT, DISLODGED WHATEVER WAS BLOCKING THE LINE, SCREWED IT BACK IN, AND TOOK OFF AGAIN."

CHRIS MORSS, 2010

Δ

The squadron's Landing Signal Officer (LSO) and Safety Officer, Ed Hiebert, became another air line captain. Jim Morin judged the CAL POLY graduate to be "smart as a whip" and "great carrier pilot and air-to-ground man," but he added that "his spelling was terrible."[44] Hiebert, in turn, said that the Skipper had one major fault: "He had this phobia about misspelled words. I could never understand his concern about misspelling a word now and then. I would much rather use a big important sounding word and misspelling it a little than using a ten cent word that anyone could spell."[45] Apparently his spelling deficiency did not prevent Hiebert from gaining his position with American Air Lines.

On 11 February 1965, Ed Hiebert flew as a section leader with Ken Holthaus on the second FLAMING DART Mission. After clearing the target. Hiebert found that some of his bombs had hung up underneath his A4, and he could not land back on the carrier. Holthaus recalled:

> *After escorting me back to the ship, Ed proceeded to Danang. Unbeknown to me, Ed had multiple wave offs at Danang due to higher priority emergencies. On his final approach with a low fuel state Ed forgot to put his gear down and made a successful landing on his bombs. Running away from the aircraft Ed look's back to see emergency equipment rolling his aircraft off the runway into a scrap heap.*
>
> *Meanwhile I had made a normal landing and had made it to the Ready Room for a de-briefing. Shortly thereafter, Skipper Morin gets the news that his diverted aircraft in Danang was totalled.*
>
> *Morin says, "How the hell do I lose an aircraft on a divert, and of all people my "Safety Officer." Skipper Morin throws his ball cap at the blackboard. That wasn't the last time the Skipper would be throwing his hat."*[45]

JIM MORIN JUDGED THE CAL POLY GRADUATE TO BE "SMART AS A WHIP" AND "GREAT CARRIER PILOT AND AIR-TO-GROUND MAN," BUT HE ADDED THAT "HIS SPELLING WAS TERRIBLE."

Hiebert provided a lighter moment for Albie Hyde:

> *I remember returning to the ship in a two-man [flight]; don't remember what we were doing over the beach with Hoob, Ed Hiebert. Ed was the squadron safety pilot and a very smooth stick. After we were FEET WET he turned to me and gave me the hand signal to move out*

> *from the normal wing position. After I had moved out a fair distance Ed's A/C started to rock and bounce all over the sky. I couldn't figure what was going on until he smoothed out after a few minutes and motioned me back. Ed turned to me with a big smile and showed me a zip-lock bag filled with yellow fluid. You have to understand that there is no relief-tube on the A-4 and the process of going down thru your survival vest, harness, flight suit, g-suit, and various layers of underwear to fill that bag in the tight confines of an A-4 cockpit without turning upside down or spilling the whole bag was at least worthy of an Air Medal.*[46]

"ED TURNED TO ME WITH A BIG SMILE AND SHOWED ME A ZIP-LOCK BAG FILLED WITH YELLOW FLUID. YOU HAVE TO UNDERSTAND THAT THERE IS NO RELIEF-TUBE ON THE A-4..."

AL HYDE

Morin recently remembered a story about Hoob Hiebert.

> *I had returned from a mission and after the debriefing I went back to our ready room. CAG McDonald came in to tell me that we were wanted in the admiral's ops spaces to be briefed on a mission that the staff wanted flown the next day. There was a radar site northwest of Hanoi that was an integral part of the North Vietnamese missile air defense system. There was no photography. All we had were two ECM cuts that would give us the approximate area. The site was in an area of trees/jungle. Once in the target area, we would have to go in there at low altitude and slow airspeed to search the area visually. Here we would have two aircraft motoring around low and slow in the densest AAA defense area in the whole world.*
>
> *I told the staff that this was a KAMIKAZE MISSION. Odds were we would lose at least one aircraft if not both. I was not happy.*
>
> *CAG and I returned to my ready room. I told him that I couldn't send two of my guys on that ridiculous mission. I was the most experienced aviator in the squadron therefore I would lead it. I would need my most reliable person as my wingman. I decided on Ed Hiebert as he was my oldest and most steadfast junior officer. I told Ed that we had a special mission to be flown the next day. We went down to the intel spaces to break out our charts to do the planning. When I told Ed about the mission, he was incredulous. This was going to be a no fun trip.*
>
> *When the next day's air operations plan came out the mission was not on it. Saner heads on the staff realized that this was a no brainer and cancelled it. THANK YOU LORD!*[47]

"I TOLD THE STAFF THAT THIS WAS A KAMIKAZE MISSION. ODDS WERE WE WOULD LOSE AT LEAST ONE AIRCRAFT IF NOT BOTH. I WAS NOT HAPPY."

J. B. MORIN

Hiebert recalled that "the mission was being considered and the skipper was not happy. I was available on that day and that I was being considered for the mission...I for the life of me cannot remember actually planning the thing. I do remember that my first comment was F___."[48]

Δ

"King Kamehameha" or "The "King," Ken Holthaus, graduated from St. Cloud State College in St. Cloud, MN with a degree in Business Administration and a goal to pursue a business career. With the war in Vietnam going on and the draft still in effect, Holthaus decided to try to get into the Navy's flight program. He wrote about these early years:

> *I was very fortunate to have had great leadership and role models as coaches and teachers. Among them were General Ed Tieman who recruited me and friends to join the National Guard. I was discharged with three years of longevity in the Guard along with college graduation.*
>
> *Due to my military experience in the Guard, I found Pre-Flight rather easy and in Class 25-61, I graduated without a single demerit. I don't know if that had ever been done before or has been done after.*
>
> *After receiving my wings I was ordered to the War College at Newport, RI to attend the Navy Justice School and after that reported to VA-155 as the Legal Officer. Tom Russell was the skipper and Jack Harris the XO. I was the first nugget to join the squadron in some time as VA-155 had just returned from cruise and was a veteran squadron.*
>
> *After awhile, Tom Russell said, "Ken, it's nice to have a hard-charging bachelor in the squadron and I wish I had several more like you." I told him I would do my best to uphold the compliment.*

Ken reported that he liked life in the squadron on shore, in part, because he lived in "a target rich environment with the ladies of the valley."[49]

His squadron mates knew Ken as a "fun-loving bachelor" and a definite party-animal.[50] The swashbuckling, big-boned, and muscular officer started off his career with VA-155 with a serious confrontation with his XO, CDR Morin. According to a friend, "Ken had no interest in the Navy other than flying airplanes and intended to get out of the service at the earliest possible date. He was a very competent pilot, but his performance in his non-flying duties left something to be desired. His primary motivation was to have a grand party with his life and fly during the non-party times." Because of this the XO asked him one day, "Ken, you know that normal squadron working hours are 8-5 if you are not on the flight schedule. How is it that you come in at 10, go to the O'club at 11, return at 1 pm and head for the club again at 2:30?" Holthaus replied: "It's not my fault if I can do in two hours what it takes those other guys eight to accomplish."

READY ROOM RELIEF

"IN JULY OR AUGUST 1965, A LARGE PACKAGE ARRIVED FROM THE UNITED STATES CARRYING A GIFT FOR EACH PILOT.

THE GIFT TURNED OUT TO BE A STUFFED AND DECORATED BRA WHICH THE PILOTS ATTACHED TO THEIR READY ROOM CHAIRS. THEREAFTER, THEY ENJOYED THE "LASCIVIOUS COMFORT OF THE PADDED CUPS."

THE WIVES AT LEMOORE HAD TASKED DOTTIE FIDELIBUS TO BUY THE BRAS. SHE HAD GONE TO A LOCAL LINGERIE SHOP, ASKED FOR THE LARGEST CUP SIZE AVAILABLE, AND THEN BOUGHT TWENTY [WITH 44 C CUPS]."

CHRIS MORSS

The smart-ass answer set up a collision with the XO. That came a few days later. Because he had partied too hard the night before, Holthaus failed to make a 6:30 a.m. meeting for "a navigation and weapons mission" which Ken was scheduled to fly with the Squadron Commander.

"KEN SIMPLY COULD NOT BE BOTHERED WITH COLLATERAL DUTIES. HE TOLD ME AFTER ED DICKSON WAS LOST AND HE WAS DESIGNATED THE WEAPONS OFFICER THAT THE FIRST THING HE INTENDED TO DO WAS CERTIFY ALL OF THE CLASSIFIED DOCUMENTS IN THE WEAPONS SAFE AND THEN THROW THE KEY OVERBOARD BECAUSE THAT WAY HE COULD NOT BE ACCUSED OF LOSING OR MISPLACING ANY CLASSIFIED MATERIAL."

BILL KISH, 2011

When Morin arrived and heard about the incident, he placed Holthaus "in hack" for the Labor Day weekend and "made him the permanent squadron duty officer for the next few weekends." The XO told the rest of us that "he was not making an example of Ken, he just expected him to do his job and if he did not this was the treatment he could expect." Later, Holthaus said "he had not been so well rested in his young life as during this period when he only needed to be concerned about a fire in the hanger and had plenty of time to sleep." Hearing his comment, Morin "laughed and shook his head."[51]

Holthaus, however, became a valued member of the squadron and reportedly "the most senior lieutenant, junior grade in the Navy." Morin wrote that "Ken initially had a problem with rules and regulations. Life was too much fun to take things seriously, but if I were ever in a squadron, I'd want Ken in it." Morin remembered, "in those long, frustrating, tiring line periods, with no end in sight, Ken was there taking his turn."[52]

Silver Fox One highlighted one significant contribution that his JG often made to the squadron. "Ken had an upper arm as big as my thigh. When we'd be at the Cubi Officer's Club [at Subic Bay in the Philippines] between line periods, one of the FOXES would call out "Who wants to arm wrestle our champion for a pitcher of beer?" Morin continued,

> *I remember this Marine Lieutenant took the challenge. His neck started somewhere near his shoulders. Depending on how thirsty we were or how close to empty our pitcher was, Ken would play with his opponent or take him right down. The Marine didn't last long and bought the beer.*[53]

"WHO WANTS TO ARM WRESTLE OUR CHAMPION FOR A PITCHER OF BEER?"

Holthaus also had emphasized his individuality by sporting a three-inch handlebar moustache. By May 1965 while on Yankee Station, he had to shave it off "for want of grooming wax."[54] Ken left the Navy after the Vietnam cruise, having earned eleven Air Medals and three Navy Commendation Medals.

He soon joined Braniff Airlines and spent seventeen years with Braniff. When Braniff shut down, Ken and twelve partners founded Sun Country Airlines, and he flew with that airline for seventeen more years until he retired.

Looking back, the King concluded, "After flying a single engine aircraft off an aircraft carrier in a combat situation - 34 years of commercial aviation seemed like a walk in the park."[55]

Δ

Genial and outgoing LTJG Jim Hughes had joined the squadron as a replacement pilot along with Albie Hyde.[56] Early on, Hughes flew as a wingman to CDR Jack Harris and was flying with his squadron leader the day the North Vietnamese shot Harris down. A commentator[57] wrote that: "Hughes was on a near simultaneous run in a loose-deuce formation and saw lots of tracers. Shortly after being hit, CDR Harris reported that his engine had flamed out. He turned south and converted his speed to altitude to get as far south as he could." Hughes reported that he "told Harris that he was on fire."

Just after joining the squadron, Hughes flew as Morin's wingman and did so for over a year. His boss wrote that "As a new aviator out of the A4 training squadron, Jim had a tendency to do strange things. I asked the Skipper – Jack Harris – for Jim as my wingman so I could watch and train him. Jim was another whose sense of humor and vitality helped get us through the long hard times."[58]

After he resigned from the Navy in September 1968, Hughes joined Eastern Air Lines and became a First Officer or a Captain with Eddie Rickenbacker's famous air line.

Δ

LTJG Albie Hyde, a Regular Naval officer, had joined VA-155 from the A4 Replacement Air Group (RAG), VA-125 just before CDR T. B. Russell turned over the squadron to CDR Jack Harris in July 1964. At the time, Morin served as the Squadron Executive Officer.

Hyde had graduated from the U. S. Naval Academy at Annapolis in 1962, and like the others he had completed his flight training at Pensacola less than two years before he entered combat.

The Navy had awarded him a Distinguished Flying Cross and five Navy Commendation Medals for his combat actions.[59] Morin commented: "There wasn't a braver pilot than Albie." However, he qualified that by saying, "There's a fine line between foolhardiness and bravery and Albie was so aggressive that I made him my wingman, so I could keep an eye on him." Morin added, "I was the first pilot to get over 100 missions over NVN. I believe that Al was the first to get 200 missions."[60] Eventually, many Navy attack pilots got 300, 400, or more missions over NVN.

Hyde, in turn, wrote about his boss: "I remember flying on one Alfa Strike on the Skipper's wing. The target was an airfield northwest of Hanoi. The strike being a large and important one was led by the CAG with course, altitudes, airspeeds, descents, and turn points all planned out to the second. The CAG was a little new and as we approached the beach and started a descent the planned airspeed sort of went out the window. We

Flying Hours Albie Hyde

USS *Coral Sea* Feb-Oct 1965

Beginning Hours	704.7
Ending Hours	1,002
Total Hours	297.3
Total Combat Hours	239.6
Combat Missions	117
Average Missions per Month	13

USS *Constellation* Jun-Nov 1966

Beginning Hours	1,155.8
Ending Hours	1,348.4
Total Hours	192.6
Total Combat Hours	175.2
Combat Missions	100
Average Missions per Month	16-17

All

Total Combat Flying Hours	414.7
Total Combat Missions	227
Average Mission Time	1.9 hours

...WHAT HYDE REMEMBERED MOST ABOUT HIS TWO CRUISES WAS JUST BEING TIRED.

rocketed past the turn point and only the Skipper's "we missed the turn" on the radio kept the strike from turning into complete chaos." He continued, "The fact that the Skipper could keep track of everything while trying to fly formation on about 25 other A/C was impressive but his overall grasp of the situation was brought home to me even more when he made a radio transmission: *#2, me, move it out a little. I can't see you.* I had gotten a little sucked, too far aft, and he was just looking out for me."[61]

Interestingly, what Hyde remembered most about his two cruises was just being tired. "At the end of a line period, I got so tired that I could sleep in my stateroom that was just below the arresting gear while aircraft were landing over my head. If you ever heard an A-3 or Vigilante make an arrested landing you would know that was no easy task."[62] Fatigue nearly proved disastrous one night when Hyde was stacked up waiting to land on the *Coral Sea.* He recalled, "... I fell asleep on an outbound leg and woke up heading for China and much too far from the ship. Luckily for me I had enough time before my CHARLIE TIME to get back and catch up. If I hadn't I might have woken up the only POW in China vice North Viet Nam."[63]

Hyde, who entered the Naval Reserve after leaving active duty, had retired as a lieutenant commander in the Naval Reserve.

Like Hughes, Hyde became an airline Captain, but with Northwest Airlines.

Δ

Bill Kish, another LTJG and one of Jim Mattly's close friends since Vietnam, probably was the youngest looking one in the entire Navy during the time that he served with VA-155 off Vietnam. He had flown over 100 missions from the *Coral Sea.* Bill – who had earned a bachelor's degree in mathematics from the University of California at Berkeley - joined the Navy in August 1962 as an Aviation Officer Candidate. Along with other recent college graduates, he completed the sixteen-week training program under the professional eyes of marine drill sergeants. After receiving his commission as an Ensign in the Naval Reserve, the Navy sent him to pilot training at Pensacola.

Bill completed flight training in March 1964 and joined the replacement air group at Lemoore, CA. There, he completed one hundred hours of training in the A4 and qualified to land on carriers.

In August 1964, when Bill joined VA-155, CDR Jack Harris commanded the squadron, CDR Jim Morin served as the Executive Officer, and Ed Hiebert and Ernie Mares were pilots. Kish found that Harris had fleet experience with the A4, but that few of the others in the squadron - including Morin - had flown the ship. Dubbed a "nugget" because he was a brand new pilot, Bill found his new squadron quite inexperienced compared to sister squadrons and realized that much had to be done before it would become a competent outfit. He also discovered that CDR Harris seemed to wilt when the aggressive commander of the air group to which VA-155 belonged criticized the Silver Foxes, but his XO, Morin, "clearly bristled every time the CAG criticized us." He soon found Morin to be someone to admire.[64]

WHEN THE CAPTAIN ASKED "SON, IS THERE ANYTHING YOU WANT?" KISH REPLIED: "YES SIR, SEND ME HOME TO SELL WAR BONDS."

In turn, Morin said that Kish "was another excellent stick and throttle man and an accurate bomber pilot. He was dependable, reliable, with the ability to lighten up a serious situation to help us get through it."[65]

During his tour, "Razor" received eight Air Medals and two Naval Commendation Medals with V devices for valor for his 115 combat missions. His squadron mates, however, remember him for other actions. First, he married a former Rose Bowl Queen during his tour when the squadron withdrew to Subic Bay in the Philippines for a rest period. Pictures abound: his "friends" smashing cake into his face and throwing him into the swimming pool in his dress whites. Second, Bill made the 10,000th launch off the *Coral Sea* during his tour, and when he returned to the carrier, CAPT George L. Cassell[66] ordered him to the bridge to commemorate the event. When the captain asked "Son, is there anything you want?" Kish replied: "Yes sir, send me home to sell War Bonds."[67]

Bill became an airline pilot with Northwest Airlines after he left the Navy and retired in 2000. During his career, he worked as an instructor in the training department for over twenty years, and spent three years managing the technical flight operations department for the Boeing 757 when the airplane came into service.

Δ

MORIN CONSIDERED MARES TO BE "A GOOD A4 DRIVER AND DIVISION LEADER," AND "A SOLID, RELIABLE COMBAT AVIATOR."

Ernie Mares was an accomplished pilot and initially a full lieutenant when he joined VA-155. During the tour on the *Coral Sea*, Mares became a lieutenant commander - one of two in the squadron - and served as Morin's Administrative Officer. Morin considered Mares to be "A good A4 driver and division leader," and "a solid, reliable combat aviator."[68] Mares retired in grade as a lieutenant commander.

On 29 March 1965, Mares, then the Administrative Officer of VA-155 when Morin was the XO, helped in the rescue of CDR Harris. Harris had been hit making a low level attack on Bach Long Island "against a concentrated anti-aircraft gun emplacement. His aircraft was hit and flamed out immediately."

Mares reported, "I was leading the second division and had Harris in sight. I followed the skipper as he headed south to put some separation between himself and the target. Harris stayed with his aircraft until the last possible moment and ejected at

50 feet above the water. His parachute opened instantly and both he and his aircraft hit the water a couple of seconds later." Mares immediately "set up an orbit over the downed pilot" with Bill Kish and transmitted a MAYDAY message. Mares then saw a submarine quickly surface near Harris who had climbed into his inflated raft. The sub, however, sailed away from Harris so Mares called the submarine and told it to make a 180 degree turn to pick up his squadron commander. The sub did that.[69]

JAMES B. MATTLY DEEPLY RESPECTED JAMES B. MORIN.

A few days later, CDR Harris returned to the *Coral Sea* and received "the royal Silver Fox" welcome, which included a brass band and hundreds of handshakes, along with cake and ice cream."[70]

Bill Kish remained close to Ernie after the war. He often had flown as his wingman when Mares led a division of four planes into combat. Both got into trouble after a bombing incident in Laos. Jim Morin stood up for them; CDR Harris did not.

Δ

LTJG Jim Mattly, nicknamed "Crunch" by Bill Kish, had been an Ensign in the Naval Reserve when he joined the Silver Foxes at Lemoore. Jim had flown over 100 missions from the *Coral Sea,* many at night, and had completed his long tour without a scratch. However, like many of his squadron mates, he found on two occasions that the enemy had hit his aircraft: once a .30 caliber round struck a latch in front of his canopy and shattered.[71] One inch to either side and Jim may have suffered a fatal wound.

He had earned two Navy Commendation Medals with "V" device (for Valor) and ten Air Medals for his combat service.

CRUNCH

"WE WERE OPERATING IN SEVERE PITCHING DECK CONDITIONS IN SEPTEMBER 1964 ON A WORKUP CRUISE AND THE DECK PITCHED UP AS [MATTLY] WAS TOUCHING DOWN...

THE LANDING COULD ONLY BE REFERRED TO AS A "CRUNCH" BECAUSE OF HOW THE GEAR HIT THE DECK AT A MUCH HIGHER RATE OF DESCENT THAN NORMAL BECAUSE OF THE DECK PITCH."

BILL KISH

Jim had always wanted to be a pilot. In fact, he kept a doctored photo of a military pilot in his room when he was in high school so that he could focus on flying while he was having trouble academically in school. Before joining the Navy in 1962, he had earned an Associate Degree in Mathematics from Bakersfield College. Mattly became a fine young officer and pilot. Morin wrote: "He wasn't too loud or one who drew attention to himself, and even though he wasn't too thrilled about the war he'd man up with the rest of us and do the job expected of him."[72] A shipmate considered him to be "calm and courageous" and "modest and laconic by nature.[73] Following Vietnam, he flew off the USS *Saratoga* in the Mediterranean, and during that time he won one Golden Tailhook Award for high quality landings aboard the carrier.

After leaving the Navy, Mattly graduated Phi Beta Kappa with a Bachelor's Degree in Business Administration from the University of California at Berkeley and later earned a MBA from Harvard Business School. Previously a successful senior executive with Butler Aviation, Browning-Ferris Industries, and Loomis

Armored, Inc., Jim was the CEO and President of Loomis, Fargo & Company, a cash-handling service company headquartered in Houston. He and his management team had just completed an overhaul of the company and saw bright days ahead for Loomis, Fargo. Jim and Diane were looking forward to seeing Pensacola again where Jim had taken pilot training in 1962, and where they had visited in 1985 for a Silver Fox reunion that coincided with the celebration of the 75th Anniversary of Naval Aviation.

"STAY OUT AND GIVE ME SOME ROOM BECAUSE THERE'S AN AWFUL LOT OF AIRCRAFT AND AN AWFUL LOT OF SKY."
PAUL MOORE

Jim remains an active aviator, flying his acrobatic aircraft around Houston. Diane is also a licensed pilot.[74]

James B. Mattly deeply respected James B. Morin.[75]

Δ

LCDR Paul Moore joined VA-155 before the North Vietnamese shot Wendy Rivers down. He had very little carrier experience when he arrived, but he quickly distinguished himself as a combat pilot and later on the USS *Constellation* became a leader in VA-155. The Irishman from New Jersey returned to Tonkin Gulf on multiple tours after 1965, and Jim Morin reported that by 1970 Moore had become the most respected attack pilot at Lemoore. Paul Moore retired as a Captain in 1986.[76]

Moore gained some notoriety and a place in the Quotable Quotes' Log when on 23 September 1965 he briefed his section for an attack on the Thanh Hoa Bridge, saying, "Stay out and give me some room because there's an awful lot of aircraft and an awful lot of sky."

Δ

Former LTJG Chris Morss, originally a high school teacher in Massachusetts, had been the squadron's Air Intelligence Officer. Morin wrote, "He was intelligent and a fine young man. How he put up with us aviators, wild men at times, I'll never know."[77] Morss highlighted those times when he wrote:

> *One night in the ready room, we were watching "Twelve O'Clock High" with Gregory Peck, a film about a US bomber crew operating out of England in WW II. One of the crew members goes to the flight surgeon complaining of stress, saying he's had three missions this week. That brought loud guffaws from the Silver Foxes, who were flying one and sometimes two missions a day.*

THE AIR INTELLIGENCE OFFICER HAD GRADUATED FROM HARVARD IN 1962 AND GONE TO PRE-FLIGHT AT PENSACOLA IN OCTOBER.

He added that after reading another book about the bomber war, it made "the comment in the movie a bit more credible."[78]

The Air Intelligence Officer had graduated from Harvard in 1962 and gone to Pre-Flight at Pensacola in October. After he finished in February 1963, he attended the Air Intelligence School in

Wendy

"Wendy Rivers was the bravest man I ever knew..

High strung in some ways, he had premature white hair and was a heavy smoker. He was a product of the American heartland, a Nebraskan and an unshakable patriot whose unwavering belief in our cause enabled him to withstand his seven and a half year imprisonment in North Vietnam.

To some he might have seemed doctrinaire, but this quality allowed him to survive when others might have doubted their own beliefs...

We did not know his fate for some months...It was with a great lump in the throat that I saw him on television, returning in 1973.

The country had changed dramatically in that time. Prisoners returned to find little children now teenagers, sometimes into drugs and often rebellious with long hair that would not have been tolerated in 1965. Wendy's wife had drifted away..."

Chris Morss
2010

Alameda, CA. In June, Morss flew to Cubi Point in the Philippines and joined VA-155 on the *Coral Sea.* He left the squadron in December 1965.

Before leaving the Navy in August 1968, Morss studied Italian for two years at the Defense Language Institute and served two years as the Aide to the Deputy Commander, Strike Force South, a Navy rear admiral, in Naples.

Morss returned to school, earned a Masters in Education, gained a position in the public school in Avon, MA, and taught English there until 1995. Eventually he earned a doctorate in education. In 2011, he remains active with civic work.[79]

Δ

LT Dick Renning, a Mustang, handled maintenance operations for the squadron. The tall and lanky officer had a reputation for swearing exquisitely. "To hear him cuss was a linguistic delight. He had a broad vocabulary and a unique and creative way of stringing together a bunch of curses in combinations that were wholly new and richly satisfying. From anyone else, they would have been crude and objectionable, but not from him."[80]

On 1 April 1965 while on Yankee Station, Renning saw a note on the bulletin board which announced that there would be no movie that night, and after reading it he exclaimed out loud: "No movie again tonight. Goddam a man might have to do work to amuse himself." His comment found its way into the Quotable Quotes Log.[81]

Δ

Wendy Rivers became a special member of the group because he had been shot down in September 1965 during a raid on a rail marshalling yard at Vinh, had been captured, and spent the duration of the war as a POW in the infamous Hanoi Hilton. At the time he went down, Rivers was "Silver Fox 1" and was leading the main element of the squadron. Many thought that the North Vietnamese had shot down his boss, CDR Jim Morin. Rivers, one of the longest serving POWs, spent over seven and a half years in Hanoi. Air Force COL Robinson Risner wrote about Wendy in his memoir. He said:

> *One of the first things that I did was to tap on the walls to see who, if anyone, was next door. On one side was Wendy Rivers. He and I were to become good friends. After I had bored a small hole through the wall with a steel rod pried from the grate I passed all my cigarettes through to him, and it was just a good feeling to be able to look through a hole and see a friendly face. We had made a pact that if anything happened to either of us, the other would visit his family and give them some words that we had for them.*[82]

After taking command of VA-155, Morin tasked Wendy, his Operations Officer, to be his Acting XO because he did not have an assigned Executive Officer.

"BEING SHOT DOWN BEATS BEING KILLED."
WENDY RIVERS

Unlike the other Foxes, Rivers had a severe health problem, to wit: constant surgery on his teeth and gums. "The dentist was cutting him, scraping him and sewing him back up a part at a time." Morin commented,

> *At fifteen pilots, we couldn't afford anyone grounded. [Wendy] felt bad about this and would start flying again before he was really ready. There was no slack. If you fell behind in sorties, you'd get doubled up until you caught up to the average again. We all wanted everyone to have all those gold stars on his Air Medal.*

The North Vietnamese shot Wendy down over Vinh on his ninety-sixth mission before he completed his dental treatment.[83] Could the dental program and the doubling up of missions have taken the edge off Wendy and led to his shoot-down?

Years later, Rivers who with other POWs "waited a day at a time for the war to end" reflected on his lack of medical treatment, poor food, and terrible conditions in Hanoi: "being shot down beats being killed."[84]

The Navy awarded Rivers a Silver Star and a Legion of Merit for his time in Hanoi to go with his Distinguished Flying Cross and numerous Air Medals.[85]

The Silver Foxes had gathered in 1973 to welcome Wendy home. Rivers, who graduated from the U. S. Naval Academy in 1952, retired in January 1977 as a Captain.[86]

Δ

ENS ROY SHAW HAD BEEN AN ENLISTED SAILOR BEFORE GAINING A COMMISSION IN THE NAVY.

ENS Roy Shaw had been an enlisted sailor before gaining a commission in the Navy. Shaw served as an assistant maintenance officer under Bill Fidelibus.

At about forty years of age, Shaw was older than most of the Foxes and was a quiet man. Morin considered him to be a good officer. The ensign was not an aviator.[87]

Δ

LTJG Jay Starling, another quiet man, had earned a Distinguished Flying Cross during the war. After the war, he earned his PhD. CDR Morin wrote: "Jay Starling, our future Ph.D., was the target of a lot of kidding for doing dumb things at times but with enough sense to laugh at himself. Jay was another solid attack pilot."[88]

While a Silver Fox, Starling appeared to be a dreamer and somewhat distracted. His demeanor disguised a "superb intellect." In the air, however, he "displayed mettle and focus."[89]

While a Silver Fox, Starling appeared to be a dreamer and somewhat distracted. His demeanor disguised a "superb intellect." In the air, however, he "displayed mettle and focus."

The Navy called him back to active duty following North Korea's seizure of the USS *Pueblo* on 23 January 1968. He served a year on active duty and an additional eighteen months in the Reserves after his call up. When he finished his active duty, he went back to college to get his degree.

Starling had done well in his undergraduate studies at the University of California at Berkeley, and he stayed at the university and earned a master's degree and a doctorate in Policy Sciences and Organization Theory. He could have studied for his advanced degrees at Harvard, Princeton, or Yale, but elected to remain at Berkeley. Morin noted, "We were all proud of him when we heard he got his doctorate."

The former Silver Fox taught for ten years at Southern Methodist University in Dallas, TX and built SMU's graduate program in Public Administration. He left in 1982 to work for ARCO and then worked for British Petroleum after BP acquired ARCO. Jay stayed with BP until 1999. After leaving BP, he became the Director of Administrative Services for the Florida State Legislature.[90]

Δ

CWO Jim Tillis had been the squadron avionics/ordnance officer. Morin thought the retired warrant officer had been the oldest member of the squadron in 1965 and said that Tillis "put many long days up on the flight deck and nearly worked himself into a serious illness because of it." Morin got a Navy Commendation Medal awarded to his dedicated subordinate.[91]

Tillis was another Mustang, a former enlisted sailor, and a Florida native. He had joined the Navy about the time of Pearl Harbor. "Jim was rather more formal without being stuffy, and he was not given to using sailor's language. He was deferential and respectful, a gentleman who went quietly and highly competently about his work and was much liked in the squadron."[92]

Tillis was another Mustang, a former enlisted sailor, and a Florida native. He had joined the Navy about the time of Pearl Harbor.

Δ

The photo of the squadron officers beside an A4 at the beginning of the chapter shows a jaunty group of young men with some wearing non-regulation bush hats and all looking eager and proud. For most of them, their days with VA-155 were the "most meaningful and memorable days of their lives." One said:

> *We were united by abandoning ourselves to an enterprise that greatly transcended our personal needs and in which we believed passionately. Those who flew in combat together are united by an unbreakable bond of brotherhood.*[93]

Missing

FOR MOST OF THEM, THEIR DAYS WITH VA-155 WERE THE "MOST MEANINGFUL AND MEMORABLE DAYS OF THEIR LIVES."

Two Silver Foxes who served in VA-155 before Jim Morin assumed command of the squadron had died while on Yankee Station: LT Ed Dickson and CDR Jack Harris. The North Vietnamese had shot down and killed Dickson; a terrible fire on the USS *Oriskany* had killed Harris.

Three of Morin's other Foxes who had not been with him on the *Coral Sea* had died during the war: CDR John (not Jack) Harris; CDR Chuck Peters; and LCDR Gene Smith. Harris, who had been assigned as his XO, died of meningitis in December 1965, just after joining the squadron when it returned from the war zone, and Peters, Morin's other assigned XO, perished the day after he had assumed command of the squadron on the USS *Constellation.* Smith who flew with VA-155 on the *Constellation* had died on 27 June 1966 just before Morin turned over VA-155 to Peters.[94]

Four of Morin's 1965 Foxes had died after surviving the war and before the reunion:

- o Bill Fidelibus,
- o George Hart,
- o Paul Moore, and
- o Dick Renning.

Renning died of a heart attack shortly after he left the squadron. He was not yet forty.

Squadron mates had lost track of George Hart, but later they found out that their Marine comrade had died of cancer in the mid-1980s shortly after he had retired. Wendy Rivers reported that Hart and his wife rest in a tiny Kansas cemetery out in the country that Rivers wistfully described as "bleak and lonely."

Bill Fidelibus died of a heart attack in his fifties. Moore, who had flown countless combat missions, also died of a heart attack in July 1996.[95]

Everyone missed their shipmates.

The Gathering

"TIME MARCHES ON...AND OLD FRIENDSHIPS BECOME SWEETER AND DEARER."

Over the years, the Silver Foxes had held reunions hosted by different individuals at different places. Ed and Marilyn Hiebert hosted the first one in Ft. Worth, TX . Wendy Rivers commented on that event: "The high point of these past few years...came in Sept. 1991. It was the first gathering of the Silver Foxes (VA-155, my squadron in Vietnam) in years. So wonderful to see that old gang again." He added: "Time marches on...and old friendships become sweeter and dearer."[97]

This gathering differed from the previous ones because Skipper

Jim Morin ramrodded it, and he tasked Diane Mattly to handle the arrangements. He had called some Silver Foxes and written to all of them and told them to be in Pensacola and bring sports coats.

Most came. George and Joan Blosser arrived in Pensacola from North Carolina. Jerry and Jeanne Chasko joined from Arizona and Richard and Sheena Floreani came in from New Hampshire. Ed and Marilyn Hiebert flew in from Texas and Ken and Joan Holthaus arrived from Minnesota. Iron-man Albie Hyde and his wife Zeb joined the group from Arizona. Irrepressible Bill Kish flew in from Nevada and avid pilots Jim and Diane Mattly flew in from Houston. Wendy and Bette Rivers, Wendy's high school sweetheart, also arrived from Texas, and Jay and Kay Starling drove in from nearby Tallahassee, FL.

THESE AGING SILVER FOXES GATHERED IN MAY 2000 AT PENSACOLA NAVAL AIR STATION TO BE WITH SKIPPER JIM MORIN ONCE AGAIN.

And to the delight of all, Father Evan Greco joined the group as did Jim Tillis and his wife Mae, Dorothy Fidelibus, Bill's widow, and Chris Morss.[98]

Two surviving pilots did not attend: Jim Hughes could not come because of business reasons; Ernie Mares chose not to attend.

These aging Silver Foxes gathered in May 2000 at Pensacola Naval Air Station to be with Skipper Jim Morin once again.

End Notes

[1] Mission numbers found in Bob Good, "Weary, Happy LNAS Pilots Return After 11 Months in Viet Nam," Fresno *Bee* (no date), 1-B and 7-B. Comments about Air Medals from John Bradley (JHB) conversations with pilots,personal knowledge, and related readings.
[2] Richard J. Floreani, *The Silver Fox Squadron of 1964-65*, an unpublished paper, 5. (Possession of JHB). Floreani stated that eleven pilots died, but did not list names. Bill Kish listed eight KIA based upon information in *Sea Breeze*, January 1966. Two enlisted helicopter crews members were killed in an aircraft accident. The current *Coral Sea Website* lists only five pilot losses in 1964.
[3] *Ibid*, 4.
[4] Al Hyde, *Letter to John Bradley*, c. December 2003.
[5] Bill Kish, *Email to John Bradley*, no date. (Possession of JHB).
[6] George Blosser, *Silver Fox Memories*, 28 December 2003, 1, 5 and 6. (Possession of JHB).
[7] Bill Kish, *Email to John Bradley*, 23 June 2010. (Possession of JHB).
[8] Ed Hiebert, *Paper*, 15 July 2010, 2. (Possession of JHB).
[9] Ken Holthaus, *Comments to John Bradley*, December 2010. (Possession of JHB).
[10]Hyde, *Letter to Bradley*. Underlining added.
[11] Current abbreviation for "Commander."
[12] Standard abbreviation for "Executive Officer," the second-in-command in units that do not have a Deputy Commander.
[13] Hyde, *Letter to Bradley*.
[14] Something embittered Mares. Perhaps Morin gave him a good but lukewarm fitness report. Such a report may have reduced his chances of promotion to commander or may have barred any future promotion. Perhaps Mares proved to be a better pilot than a naval officer or perhaps Mares did not have the leadership skills to gain higher rank. Perhaps it was just a personality conflict between an officer with an acid personality and one with a buoyant one. In available commentaries, Mares criticized several naval officers in addition to John McCain. Mares reportedly wrote a 100 page history of his time with VA-155, but he did not want to share it with JHB. One squadron officer did not consider Mares to be either a good officer or a good pilot.
[15] Rear Admiral James B. Morin, *Unpublished Manuscript*, no date (hereafter, Morin, *Manuscript*), 29 (Blosser).
[16] Chris Morss, *Notes for John Bradley*, 15 July 2010.
[17] "Inglewood Resident Awarded Air Medal," (Newspaper clipping, no data. c. July 1965). Clipping provided personal information about Blosser. (Possession of JHB).
[18] Blosser, *Silver Fox Memories*, 2.
[19] Morin, *Manuscript*, 28 (Gerald J. Chasko).
[20] *Letter, Commanding Officer, Attack Squadron ONE HUNDRED FIFTY-FIVE to Commander in Chief U. S. Pacific Fleet, Subject: Distinguished Flying Cross, recommendation for,* 2 May 1965. (Possession of JHB).
[21] Morin, *Manuscript*, 28.
[22] Citations and proposed citations on hand, 3 June 2010. Possession of JHB. James B. Morin, "Belated Honors," listed Chasko having received a DFC during the war. (Possession of JHB).
[23] Gerald Joseph Chasko, Class of 1961, U. S. Naval Academy. U. S. Naval Academy Alumni Association, *Register of Alumni, 1845-1993* (An-

Sidebar Notes

Awe
Chris Morss, *Notes.*

Ticket Punching
John Bradley, 2010.

The Fidelibus A4
Speedletter from Commanding Officer, Attack Squadron 155 to Commanding Officer, U. S. Naval Aviation Safety Center, Subject, "Combat Mishap Report (1)," *nd.*

George Hart
Chris Morss, *Notes.*

Ready Room Relief
Chris Morss, *Notes. Confirmed by Marilyn Hiebert, 23 July 2010.*

Flying Hours, Albie Hyde
Albie Hyde, *Flight Logs.*

Crunch
Bill Kish, *Comments.*

Wendy
Chris Morss, *Notes.*

napolis, MD: 1992), (hereafter, *1993 USNA Register of Alumni*), #30341, 364. There were three other Naval Academy graduates: Jim Hughes, Albie Hyde, and Wendy Rivers.

[24] *THE SILVER FOX, VA 155 NEWS*, Issue #5 - February 1965, 2.

[25] Blosser provided the commentary about the Fidelibus mission. Blosser, *Silver Fox Memories*, 2-4.

[26] *Statement of Lieutenant William T. Fidelibus, 565189/1310, USN*, enclosure to *Speedletter* from Commanding Officer, Attack Squadron 155 to Commanding Officer, U. S. Naval Aviation Safety Center, Subject, "Combat Mishap Report (U)," no date. (Possession of JHB).

[27] *Ibid.*

[28] "Navy Jet Pilot Escapes Injury," *Sea Breeze*, January 1946, 39.

[29] Blosser, *Silver Fox Memories*, 3-4.

[29] *Speedletter* from Commanding Officer, Attack Squadron 155 to Commanding Officer, U. S. Naval Aviation Safety Center, Subject, "Combat Mishap Report (U)," no date and Morin, *Manuscript*, 66 (Fidelibus).

[31] Jerry Chasko, *Notes*, 6 July 2004. (Possession of JHB).

[32] "Navy Jet Pilot Escapes Injury," *Sea Breeze*, January 1946, 39. Article stated that the rescue crewman M. J. Austin "had entered the water four times previously to rescue downed personnel..."

[33] Morin, *Manuscript*, 28 (Fidelibus) and *Speedletter*, "Combat Mishap Report," 1.

[34] Floreani, *The Silver Fox Squadron*, 3.

[35] *Ibid.*

[36] Morin, *Manuscript*, 29 (Richard J. Floreani).

[37] Morss, *Notes*.

[38] Floreani, *The Silver Fox Squadron*, 4.

[39] Morss, *Notes*.

[40] Discussions with Jim Mattly, 23 March 2005.

[41] Morin, *Manuscript*, 28 (George Hart).

[42] Kish, *Email to Bradley*, 23 June 2010 and James Morin, *Email to John Bradley*, 18 May 2009.

[43] Morin, *Manuscript*, 28, (Edwin G. Hiebert).

[44] Hiebert, *Paper*, 15 July 2010.

[45] Ken Holthaus, *Letter to John Bradley*, December 2010.

[46] Hyde, *Letter to Bradley*.

[47] James B. Morin, *Email to John Bradley, 8 September 2010.*

[48] Ed Hiebert, *Email to John Bradley*, 12 September 2010.

[49] Holthaus, *Letter to Bradley*.

[50] Morss, *Notes*, described Holthaus.

[51] Bill Kish, "Jim Morin Project," unpublished paper, c. December 2003, 2.

[52] Morin, *Manuscript*, 28 (Kenneth R. Holthaus).

[53] *Ibid.*

[54] THE SILVER FOX, VA 155 NEWS, 20 May 1965, 4.

[55] Holthaus, *Letter to Bradley*.

[56] Morss, *Notes*. Characterization of Hughes.

[57] Rear Admiral H. P. Glindeman, USN-Retired, "Gulf of Tonkin Submarine SAR," draft article, no date. (Possession of JHB).

[58] Morin, *Manuscript*, 29 (James Hughes).

[59] Albert Raymond Hyde, *1993 USNA Register of Alumni*, #30921, 369. Hyde retired as a LCDR in the U. S. Navy-Reserve.

[60] Morin, *Manuscript*, 29 (Albert R. Hyde). Morin commented on the high number of missions: "I salute those guys. That's too much to ask

of any man."
[61] Hyde, *Letter to Bradley*, 1-2.
[62] *Ibid, 3.*
[63] *Ibid*, 2.
[64] Kish, "Jim Morin Project," 1.
[65] Morin, *Manuscript*, 28 (William J. Kish).
[66] Cassell had been in naval aviation since 1941 and had flown air strikes during World War II against the Philippines, Formosa, French Indo-China, Hong Kong, Amoy, Iwo Jima, and Okinawa. He flew in the first carrier-based strikes on Tokyo. See USS *Coral Sea*, "Fifty Days," *Sea Breeze*, March 1965, 11.
[67] Morin, *Manuscript*, 37.
[68] *Ibid*, 28 (Ernest J. Mares).
[69] RADM H. P. Glindeman, USN-Ret, "Gulf of Tonkin Submarine SAR," (Copy of an article in possession of JHB which Ernie Mares heavily annotated). Glindeman incorrectly identified Mares as the Operations Officer of VA-155. Mares had no use for "Fighting Hank" Glindeman or Admiral E. C. Outlaw.
[70] THE SILVER FOX NEWS, 20 May 1965, 2 provided information about CDR Harris.
[71] "Conversations with John Bradley" at various times provided most of the information about Jim Mattly.
[72] Morin, *Manuscript*, 29 (James B.Mattly).
[73] Morss, *Notes.*
[74]John Bradley, personal knowledge.
[75]John Bradley, personal knowledge.
[76] Bill Kish provided the basic information about Paul Moore. (Paper in possession of JHB). *Quotable Quotes Log* possession of Admiral Morin, Tallahassee, FL, 20 October 2010. Conversations on 20 October 2010 in Tallahassee, FL.
[77] Morin, *Manuscript*, 29 (Morss).
[78] Chris Morss, *Email to Bill Kish*, 23 Jun 2010.
[79] *Ibid.*
[80] *Ibid.*
[81] *Quotable Quotes Log.*
[82] Robinson Risner, *The Passing of the Night, My Seven Years As a Prisoner of War of the North Vietnamese* (Old Saybrook, CT: Konecky & Konecky, 1973), 65.
[83] Morin, *Manuscript*, 27-8 (Wendell B. Rivers).
[84] Beth McPherson, 'Silver Foxes' rewarded, 6 receive Distinguished Flying Cross, Pensacola News Journal.com, 27 May 2000 (James Mattly Papers).
[85] James Morin, *Email to John Bradley*, 18 May 2009. For DFC, see Morin, "Belated Honors" which lists one DFC awarded for VN service.
[86] Wendell B. Rivers, *1993 USNA Register of Alumni*, #22641, 311.
[87] Kish, *Email* (Shaw).
[88] Morin, *Manuscript*, 28 (Jay D. Starling).
[89] Morss, *Notes.*
[90] Kish *Email* (Starling).91 Morin, *Manuscript*, 29 (James Tillis). James
[91] Morin, *Email to John Bradley*, 18 May 2009 reported the award of the Navy Commendation Medal.
[92] Morss, *Notes.*
[93] *Ibid.*
[94] Morin, *Manuscript*, 45 (John Harris), 47 (Peters), 46 (Smith).
[95] Morss, *Notes* (Renning and Hart). Kish, *Papers* (Moore). Hiebert,

Kish, Mattly, *Emails to John Bradley*, 6 November 2010 (Fidelibus).
[96] Ed and Marilyn Hiebert, *Silver Fox Newsletter*, #1, 6/9/91. (Possession of JHB).
[97] Wendy and Bette Rivers, *Holiday Greetings for 1990, 1991, 1992, & 1993.* (Possession of JHB).
[98] *VA-155 Reunion Booklet, Pensacola, Florida,* May 25-28, 2000. (Possession of JHB).

Two

IT BEGAN A LONG TIME AGO

USS Franklin D. Roosevelt
Commissioned 27 October 1945[1]

World War II[2]

Important Events		James Morin
Japan attacked Pearl Harbor.	**7/8 Dec 1941**	Morin at East Boston High School.
U. S. declared war on Japan.	8 Dec 1941	
Germany declared war on U. S.	11 Dec 1941	Morin transferred to St. Peter's High School.
U. S. and Filipino forces surrendered in the Philippines.	May 1942	
Battle of the Coral Sea.	May 1942	
Battle of Midway.	**Jun 1942**	
Allied offensive began in New Guinea. U. S. Marines invaded Guadalcanal.	Aug 1942	
Naval battles of Guadalcanal. Allies invaded North Africa.	Aug-Dec 1942	
	Spring 1943	Morin began flying lessons.
Allies invaded Sicily.	Jul 1943	
Allies invaded Italy.	Sep 1943	
	Oct 1943	Morin soloed in the Fall.
U. S. began Central Pacific Campaign.	Nov 1943	
U. S. invaded Admiralty Islands; opened the door for the return to the Philippines.	Feb 1944	
Allies invaded Normandy.	Jun 1944	
U. S. invaded the Marianas. Battle of the Philippine Sea.	Jun 1944	
U. S. invaded Leyte in the Philippines. Battle of Leyte Gulf.	Oct 1944	
Battle of the Bulge in Europe.	Dec 1944	
U. S. invaded Luzon in the Philippines.	Jan 1945	
U. S. invaded Iwo Jima.	Feb 1945	
U. S. invaded Okinawa.	Apr 1945	
President Franklin D. Roosevelt died. Harry Truman became President.	Apr 1945	
Germany surrendered unconditionally.	**May 1945**	
Allies began the occupation of Germany & Berlin.	May 1945	Morin graduated from High School.
	Jun 1945	Morin entered Holy Cross College.
U. S. dropped atomic bombs on Hiroshima and Nagasaki.	Aug 1945	
U. S. S. R. declared war on Japan.	Aug 1945	
Ho Chi Minh declared Vietnam independent of France.	Aug 1945	
Japan surrendered formally. U. S. began the occupation of Japan.	Sep 1945	**Morin inducted into the U. S. Navy.**
USS *Franklin D. Roosevelt* commissioned.	27 Oct 1945	Morin celebrated 18th Birthday.

Cold War 1945[3]

What the Red Army controlled, Stalin would not give up.

When World War II broke out, Jim Morin was a youngster in Boston and the United States faced a worldwide war to free European, Asian, and Pacific countries from Nazi, Italian, and Japanese despotism. When it ended, he looked forward to flying with the Navy.

Δ

But, when World War II ended in Europe, the Red Army of the Soviet Union occupied most of Eastern Europe, the eastern part of Germany, and Berlin. And, when the war ended in Asia, the Red Army occupied Manchuria, where it soon would supply the Communist Armies of Mao Tse-dung, it occupied the Kuriles and the southern part of Sakhalin Island, and it eventually moved into North Korea to accept the surrender of the Japanese forces there. Fortunately, no Red Navy sailed the seas with a multitude of fast fleet carriers, and the Soviets had no very-long-range bomber such as the American B-29 to strike distant targets. The large, battle-tested Red Army, however, dominated the European and Asian land mass, and it became the rock on which Premier Joseph Stalin based the emerging Cold War with the United States and its Western Allies and it became the brutal instrument to suppress freedom in the countries the Soviets had occupied.

The new American President, Harry Truman, soon found that Stalin was a faux-ally

The new American President, Harry Truman, soon found that Stalin was a faux-ally whom President Franklin Roosevelt had supported to destroy the German Wehrmacht in the East and who, in turn, had used the United States to provide immense quantities of military supplies to aid the Soviet war against Germany. Neither Truman nor Roosevelt seemed to understand that their ally of convenience would not remain one after the end of the war – as Prime Minister Winston Churchill suspected – and would become the nation's new political and military foe. Consequently, when Truman demanded that Poland be granted free elections or the United States would stop economic aid for the U. S. S. R., Stalin gave up the American aid and did not permit a valiant Polish nation to hold free elections.[4] What the Red Army controlled, Stalin would not give up. The Poles who fought so bravely against the Germans at home, during the Battle for Britain, and in mountains of Italy saw their country pass from the brutal control of Adolph Hitler to Joseph Stalin, the man who had conspired with Hitler to invade Poland from the east in September 1939 and whose operatives conducted the heinous Katyn Forrest Massacre of Polish officers and troops soon after.[5]

...Jim Morin would begin his career operating in the new Cold War.

Fortunately, Truman had more success in the Pacific in stopping Soviet expansion. When Stalin demanded to occupy the Hokkaido Island, he told him, "no." And shortly thereafter, when the Chief of the Soviet Mission in Japan told General Douglas MacArthur that he was going to land Soviet troops in Japan to take part in the occupation of that nation, General MacArthur told him if he landed a single Soviet soldier in Japan without MacArthur's permission, he would throw the Soviet general and the entire Soviet Mission in jail.[6] So no Soviets landed in Japan to interrupt the American reconstruction of the country.

Across the sea on the Asian mainland, however, Truman found that the Soviets would not leave the northern part of Korea. Soon a Soviet state emerged there, and even worse, with direct Soviet support, Mao Tse-dung and his route armies would overrun China and make it the most populous Communist nation in the world.

And south of China, Viet Minh guerrillas had been actively working to build a Communist political network throughout Vietnam during World War II, but principally in the north. In December 1944, Vo Nguyen Giap, who would gain international fame as a guerrilla and army leader, founded the Viet-Nam People's Army (VPA). Then after the Japanese surrender on 14 August 1945, the expatriate Communist Ho Chi Minh returned to French Indochina where he proclaimed Vietnam independent from France and established the Democratic Republic of Viet-Nam (DRVN). For more than a year the DRVN would rule Tonkin and northern Annam.[7] Eventually, Giap and Minh with the Viet Minh[8] would begin an armed insurgency against France.

Before long there would be serious confrontations between the Soviet Union and the United States. And, soon Jim Morin would begin his career operating in the new Cold War. In 1945, his future might have appeared to be a cloudless sky, but he could not have guessed that he would spend his entire career involved with the Cold War and would fight a hot war in Southeast Asia.

Ψ

It Began a Long Time Ago[9]

I cannot remember when I first got the desire to fly, but I was pretty young, probably seven or eight years old. Movies about flying were my favorites. It had to be the excitement, the adventure, the glamour of being daring. Aviation in the mid-thirties was still in its infancy and I wanted to be a part of it. The heroes then were the air mail pilots, the pilots setting speed and altitude records, and the civilian and military test pilots. They were replacing the flying heroes of World War I whose exploits I avidly read about.

WE ALL BUILT MODEL AIRPLANES OF BALSA WOOD AND TISSUE PAPER.

Ψ

We all built model airplanes of balsa wood and tissue paper. My father's generation did not have things like that so I did not receive much help in the construction. In fact, early model kits gave you a block of balsa wood from which you had to carve your own propeller! A little kid who did not have a degree in aeronautical engineering would carve some pretty interesting and totally inefficient shapes.

"BRENDON" ABOUT AGE 8.

These kits were called "flying" models, the idea being that, when they were finished, the rubber band running from the prop to the tail could be wound up by turning the prop. Then you would launch this beauty, the result of hours of cutting, pinning, gluing, and papering into the sky and hope for the best. Now if I carved the prop the wrong way, my model upon launching wanted to go backwards. Even without an Aero degree, I knew this was not the way it was supposed to go. With such a prop the model would crash, and I would have to patch it up. If I carved the prop somewhere near the right way, the model would flutter away, the prop *wop-wop-wopping* until it crashed again. Once again I would have to patch it up. What did I know about balance? Never mind the prop not being balanced – the model was not balanced. What did I know about dihedral, angle of attack, and such? And these World War I models, the Spad, the Fokker D VIII, et. al., had more than one wing!

When my latest creation would pull up into a hammerhead stall because it was tail-heavy, then pitch over and reach terminal velocity going straight down with the little prop wop-wopping into overspeed, and then crash with a wing shattering crunch, it was very discouraging. For who knew where the next ten cents would come from to buy another model.

The five and ten cent stores sometimes would have a model

Early one July before World War II broke out, two or three aircraft carriers had docked in Boston harbor. On the Fourth, the combined air groups conducted a fly over. That did it! That is what I wanted to be – a Naval Aviator.

plane expert whom you could ask for helpful hints. This is how I learned how to balance the model and how to make the tissue paper skin tight. And then, praise the Lord, models came out with pre-cut props! With this quantum leap in technology, my models started flying. With success came confidence which meant you really wound the prop up. It is hard to explain the elation watching your creation soaring up and up, turning this way and that, and flying full speed into the second floor of a three-decker house. I could always tell a kid who built models. He would go around making "phtui" sounds as he bit the dried airplane glue off his fingers. And, what fun it was to torch one of your old models as you let it go from your back porch. "Take that - Red Baron!"

Ψ

In the late Thirties, we were living in East Boston, just over the hill in World War I Memorial Park near East Boston Airport, now Logan airport. I spent many hours watching airplanes and peering through the fence at these wonders.

Early one July before World War II broke out, two or three aircraft carriers had docked in Boston harbor. On the Fourth, the combined air groups conducted a fly over of the city. That did it! That is what I wanted to be – a Naval Aviator.

I felt that the mental, physical, and educational requirements were beyond the reach of a skinny Irish kid from East Boston. But still I dreamed.

I read every book and magazine about Naval Aviation I could lay my hands on. Seeing Gary Cooper portraying a pioneer in carrier aviation and Errol Flynn flying SB2Us in movies such as *Divebomber* only whetted my appetite. They looked great in their choker whites with wings of gold, and they had pretty actresses on their arms. I liked the whites and figured the actresses came with the uniform.

I do remember that what I read about Naval Aviation stressed the stiff mental and physical requirements for entry into flight training. The government had opened the Aviation Cadet Program in the late Thirties but candidates had to have a college education to enter it. I felt that the mental, physical, and educational requirements were beyond the reach of a skinny Irish kid from East Boston. But still I dreamed.

Working at the dance hall gave me the money I needed to pay for flying lessons that I started in 1943.

After one half of my freshman year at East Boston High School, my family relocated to Worcester, MA, where I entered St. Peter's High. My best friend, Jackie Hynes, and I worked in his father's dance hall (ballroom, pardon me) where we sold cokes, gum, and cigarettes from a counter. Most of the young men were away in the service so there were not many guys for the girls to dance with. The girls - well these were WOMEN anywhere from seventeen to twenty - would come over and ask me to dance. I was a sophomore then, but I was still kind of short. Dancing a foxtrot with a "woman" about a head taller proved very stimulating. At the time, I would promise myself that I would tell the priest at my next confession about the impure thoughts that I enjoyed.

Working at the dance hall gave me the money I needed to pay for flying lessons that I started in 1943. Leroy McGee, a classmate of mine, and I would drive to Bolton Airport outside of Clinton, MA every Sunday morning to get an hour of instruction. We learned to fly in a Piper J3. Paying $10.50 an hour for instruction did not leave me much mad money, but I was flying. By my sixteenth birthday I had soloed. What a thrill!

"Most Reliable"

"Brendon's avid enthusiasm for aviation excels greatly his other interests which in themselves are many and varied. He already has to his credit several hours in the air. His keen mind, intelligence and general skill are sure to aid him in the field he has chosen."

Profide, 31
St. Peter's High School Yearbook, 1945

One thing World War II did: it created a tremendous demand for pilots. Consequently, the Army and Navy no longer required a college degree to enter flight training. My friend Leroy finished High School in January 1945, a half year ahead of me, and joined the Army Air Corps as a Cadet. Envious, I was eager to finish high school so I could join up. While Leroy was in recruit training, the war ended in Europe, and the Air Corps diverted him to aerial gunners' school. I dropped the idea of going into the Army Air Corps in a hurry.

I had written to my Congressman about applying for the Naval Academy. He had given away his three appointments but said that I could take the exam for West Point. At about this time, Holy Cross notified me that the school had awarded me a scholastic scholarship. As a result, I did not pursue an appoint-

ELIZABETH ANNE KANE

"SHE HAS FRIENDS BECAUSE SHE IS ONE."

"A MERRY INFECTIOUS LAUGH, A WONDERFUL DISPOSITION, A CHARMING MANNER ARE THE TRADE MARKS BY WHICH WE KNOW "BETTE." SHE HAS BEEN A FRIEND TO US ALL AND WE SHALL NOT HESITATE TO ACKNOWLEDGE OUR APPRECIATION.

"BETTE" PLANS TO FURTHER HER EDUCATION AT COLLEGE AND BECOME A BACTERIOLOGIST. MAY FORTUNE BE YOURS, "BETTE," IN SECURING HAPPINESS AND SUCCESS."

***PROFIDE*, 28**
ST. PETER'S HIGH SCHOOL YEARBOOK 1945

ment to West Point.

In the spring of 1945, I noticed a pretty girl who was in many of my classes. I was President of the Senior Prom and I asked this classmate, whom I had been making goo-goo eyes at, to be my date. Betty Anne Kane accepted. We were married in 1949. I continued to make goo-goo eyes at her for the next thirty years. But that is another story.

Ψ

I started at Holy Cross in June 1945. I did not have to work too hard to be a good student in high school, but I was not prepared for college. Being a day student and working nights at the dance hall did not help. All of a sudden I was in classes where everyone was a good student. Halfway through my first semester, the Naval Aviation recruiter came around seeking applicants for the Navy's V-5 program, the Naval Aviation College Program. If accepted, the Navy would order me to active duty as a seaman, send me to a college with an established V-5 program for two years, and then send me to flight training. This program answered all my prayers! When the recruiter guaranteed me that he would get me out of Worcester, I was hooked.

I went to Boston for my mental and physical exams. Except for an overbite that caused me worry all through my training, I had no problems and was accepted and sworn into the Navy on 15 September 1945. After being inducted, the Navy representative told to go home and expect orders to a V-5 unit to start the fall semester.

Many of my friends told me that I was foolish to throw away a four-year college scholarship, particularly now that World War II was over. This did cause me some doubts so I went to see my Advisor at Holy Cross, Father Sullivan, S. J.[10] I asked if the Cross would hold my scholarship until I got out of the Navy. I had never heard a priest swear before. So, I left and have never regretted it."[11]

I HAD NEVER HEARD A PRIEST SWEAR BEFORE. SO, I LEFT AND HAVE NEVER REGRETTED IT.

End Notes

[1] The USS *Franklin D. Roosevelt,* U. S. Navy photo, c. 1947.
[2] World War II dates from Brigadier General Vincent J. Esposito, *The West Point Atlas of American Wars,* Volume II, *1950-1952,* (NY: Praeger, 1959). See maps for selected campaigns and battles.
[3] COMPRESSED TEXT with block paragraphs and with the symbol **Δ** in this chapter and throughout the book were written by John Bradley.
[4] James K. Martin, et. al., *America and Its Peoples, A Mosaic in the Making,* Volume II from 1865 (NY: Pearson Longman, 2004), (hereafter, Martin, *America and Its Peoples,* II), 734-5.
[5] "Katyn Forest is a wooded area near Gneizdovo village, a short distance from Smolensk in Russia where, in 1940 on Stalin's orders, the NKVD shot and killed and then buried over 4,000 Polish service personnel that had been taken prisoner when the Soviet Union invaded Poland in September 1939 in WW2 in support of the Nazis." All told the Soviets massacred over 25,000 Poles. http://www. katyn.org.au/ (29 Dec 2009). In May 2010, the President of Poland and his large party died in an air crash en route to a commemoration of the massacre.
[6] Douglas MacArthur, *Reminiscences* (NY: McGraw-Hill Book Company, 1964), 285.
[7] Bernard B. Fall, "Indochina 1946-54," in Major J. W. Woodmansee, Jr. (Ed), *Revolutionary Warfare, Volume V, French Counterrevolutionary Struggles: Indochina and Algeria* (West Point, NY: Department of History, 1970), 7-8.
[8] *Ibid.* Viet Minh is short for Viet Nam Doc-Lap Dong-Minh Hoi (Vietnamese Revolutionary Independence League) which formed in May 1941 with Chinese help.
[9] OPEN TEXT with indented paragraphs and with the symbol **Ψ** in this chapter and throughout the book were written by RADM James B. Morin and are from his *Manuscript.*
[10] Society of Jesus. A Jesuit priest.
[11] Morin, *Manuscript,* 1-3.

Three

BECOMING A NAVY PILOT

Hellcats, 1943

Cold War 1945-1948

Important Events		**James Morin**
Russian Clerk, Igor Gouzenko, defected from Soviet Embassy in Canada. Named Soviet spies in the U. S. and Canada.	Fall 1945	
Soviet agent, Elizabeth Bentley, defected to the FBI. Named over 150 Americans as Soviet agents, 37 in the U. S. Government.	Nov 1945	
	1946	Morin entered the V-5 Program.
Winston Churchill announced the formation of an "IRON CURTAIN" across europe.	1946	
Civil War began between Chiang Kai-shek's Nationalist Armies and Mao Tse-dung's Communist Red Army for control of CHINA.		
	Feb 1947	Morin became a Naval Aviation Cadet (NAVCAD) at NAS Memphis.
President Harry Truman proclaimed the TRUMAN DOCTRINE.	Mar1947	
	Apr 1947	Morin began Pre-Flight Training at NAS Ottumwa.
	Apr 1947	Morin became an Aviation Midshipman (AVMID).
Secretary of State George Marshall proposed the European Recovery Plan - soon called the Marshall Plan.	Jun 1947	
George Kennan proposed a **CONTAINMENT POLICY**.	Jul 1947	
Department of Defense established. JCS established.	Jul 1947	
	Aug 1947	Morin finished Pre-Flight.
	Sep 1947	Morin began Primary Training.
	May 1948	Morin completed CARQUALS.
Soviet Union blockaded Berlin. Created the 1st Berlin Crisis.	Jun 1948	
BERLIN AIRLIFT began.	**Jun 1948**	
President Truman issued an executive order ending segregation in the armed forces.	Jul 1948	
Whitaker Chambers publicly identified Alger Hiss, a former State Department official, as a Soviet spy.	Aug 1948	
	Oct 1948	**Morin graduated from Flight School; became a Naval Aviator.**
President Truman elected President in his own right.	Nov 1948	

Cold War 1945-1948

Morin took a big risk leaving college for flight school in the post-war period. He obviously had a sure thing at Holy Cross, but completing flight school was not a sure thing. Many aggressive and highly motivated candidates do not earn their wings as pilots in military flight schools. Even those who may be qualified civilian pilots when they enter flight school wash out.

Becoming a fully qualified military pilot is tough. Becoming a Navy pilot capable of flying off of and landing on aircraft carriers — virtual postage stamps on the high seas — is even tougher. Morin faced a challenge.

Δ

Some pilot candidates bust out because of the mental strain, others find they do not like flying, some cannot pass the rigors of the classroom, others do not have the required coordination to fly, some cannot navigate, others cannot land properly, and some become fearful. A few bust their check rides and wash out; some have accidents and wash out. And because flying is hazardous, a handful crash and die young before they can qualify as pilots.

Becoming a fully qualified military pilot is tough. Becoming a Navy pilot capable of flying off of and landing on aircraft carriers - virtual postage stamps on the high seas - is even tougher. Morin faced a challenge.

Moreover, he was stepping into unknown waters as the world recovered from World War II. By the time he received his wings in 1948, the Cold War had blossomed.

In 1945, Igor Gouzenko, a cipher clerk in the Soviet Embassy in Canada, defected and identified a number of Soviet Spies in the United States. The same year, Elizabeth Bentley defected in the United States and told the FBI about a number of Soviet spies in the United States and in the United States Government. She identified Alger Hiss, a former member of the State Department as a Soviet spy.[1] Gouzenko and Bentley proved that the honeymoon with Communist Russia was over.

In 1946, the code breakers in the U. S. Army's Signal Intelligence Branch began deciphering Soviet messages that "stunned American officials" because they showed that "Espionage, not diplomacy, was the subject of the [messages]." One of the first deciphered cables showed that "the Soviet Union has infiltrated America's most secret enterprise, the atomic bomb project." Moreover, "By 1948 the accumulated evidence from other decoded Venona cables showed that the Soviets had recruited spies in virtually every major American government agency of military or diplomatic importance."[2] Venona intercepts verified Bentley's reports and vice versa.

1947
Truman Doctrine

...obviously focused on the Soviet Union. It would become the basis of American foreign policy for years to come.

1947-1948
Marshall Plan

...the United States sent $17 billion dollars of aid to Europe over the next four years, and prevented, in part, the growth of Communist power in the West.

1947
Containment

...a defensive policy that "emphasized the doctrine of limited wars fought for limited goals."

In addition to discovering spies, the United States began dealing with the Soviets and their stand-ins. On 12 March 1947, President Truman appeared before Congress to deal with the situation in the Balkans where Communist rebels threatened the stability of Greece and Turkey.

Truman announced:

> *At the present moment in world history nearly every nation must choose between alternate ways of life...One way of life is based upon the will of the majority, and is distinguished by free institutions, representative government, free elections, guarantees of individual liberty, freedom of speech and religion, and freedom from political oppression. The second way of life is based upon the will of the minority forcibly imposed on the majority. It relies on terror and oppression....*
>
> *I believe that it must be the policy of the United States to support free peoples who are resisting attempted subjugation by armed minorities or outside pressures.*

The **TRUMAN DOCTRINE** obviously focused on the Soviet Union.[3] It would become the basis of American foreign policy for years to come.

In June 1947, because of deteriorating economic, social, and political conditions in Europe, and to combat growing Communist power in Western Europe, Secretary of State George Marshall proposed a massive economic aid program for Europe. Early in 1948, Congress passed the **MARSHALL PLAN** (European Recovery Program) and the United States sent $17 billion dollars of aid to Europe over the next four years, and prevented, in part, the growth of Communist power in the West.

In July 1947, Soviet-expert George Kennan anonymously proposed that "Russia would only expand when allowed to by American weakness. It could be contained to its present borders by a politically, economically, and militarily active United States."

Kennan's ideas became the core of the **CONTAINMENT POLICY** of the United States, a defensive policy that "emphasized the doctrine of limited wars fought for limited goals." Kennan explained that was needed was "the adroit and vigilant application of counter-force at a series of constantly shifting geographical and political points, corresponding to the shifts and maneuvers of the Soviet Union.[4]

Stalin took an aggressive step in June 1948: he stopped all road and rail traffic from entering divided Berlin from the west. If President Truman sent in armed forces to force open the road and railroad, he most probably would have triggered a war between the former Allies. Instead he chose to supply Berlin by air, and the "Berlin Airlift" adequately supplied the Allied sectors of the city with food, coal, clothing, and other essential items for the next year and embarrassed the Soviet leaders.

At home in August 1948, a former Communist operative in the United States, Whittaker Chambers, publicly identified Alger Hiss as a Soviet spy. He had previously reported the fact to an official in the Roosevelt Administration in 1939, but no one took any action against Hiss until after World War II.

With these developments, Jim Morin's cloudless sky darkened as he began his active duty.

American Government Officials Spying for the Soviet Union during World War II

Lauchlin Currie
A White House official

Laurence Duggan
Chief of the Latin American Division in the State Department

Alger Hiss
formerly of the State Department

Harry Dexter White
Assistant Secretary of the Treasury

Ψ

Active Duty

My friends bade me farewell as I waited expectantly for my orders. I was somewhat shocked to receive orders to the V-5 unit at Worcester Polytechnic Institute, about a fifteen minute walk from my house. So much for farewells and my recruiter's guarantee to get me out of Worcester! But I eagerly started the program and one of my first roommates, Joseph Dobronski, was a semester ahead of me, and he helped me a lot. Joe did go on to receive his wings but left the Navy in 1949 or 1950. At this time, the Secretary of Defense, Louis Johnson, imposed some severe manpower reductions on the Navy. As a result, the reductions forced many Naval Aviators such as Joe out of service just before war broke out in Korea. Joe continued flying, became a well-known test pilot, and eventually finished his active flying as the head of Flight Testing with McDonnell-Douglas Aircraft Company in St. Louis, MO.

With World War II over and with returning veterans flocking to colleges, the Navy closed down many V-5 units. After one semester at Worcester Tech, the Navy sent many of us to Stevens Institute of Technology in beautiful, downtown Hoboken, NJ. I was fortunate to attend two excellent engineering schools in the V-5 program. I carried a very heavy course load and attended the summer semester, as did my classmates. Consequently, we completed our remaining year and half of college at Stevens in one year.

At Stevens, I had two excellent roommates: Mike Arnone from New London, CT and Ed Bouffard from Roosevelt, Long Island. Mike was brilliant, but when he got to the first stage of flying he could not make his hands, feet, and head work together. He washed out. Ed, however, completed flight school and then served a very good career in the Navy. He retired as a captain.

As soon as we completed our second year of college, the Navy ordered us to Naval Air Station Memphis in February 1947 and appointed us Naval Aviation Cadets...

As soon as we completed our second year of college, the Navy ordered us to Naval Air Station Memphis in February 1947 and appointed us Naval Aviation Cadets, in the vernacular, NAVCADs. This was the BIG TIME for me. There were sixty in my class, thirty from the Deep South and thirty from the Northeast. Depending on which group you were in, it was the first exposure one had to "Yankees" or "Rebels" and to the funny way "they" talked.

The Navy placed us in "Selective Flight Training," a euphemism for a screening site, where the officials could cull out the real incompetents so that the Navy would not waste more money training them further. We were the last group of NAVCADs to go through Memphis.

Ψ

We began our flight instruction in the NS2, a biplane built by Stearman and called "The Yellow Peril."

We began our flight instruction in the NS2, a biplane built by Stearman and called "The Yellow Peril." The Navy painted its training planes bright colors – generally yellow — so that other pilots could identify student pilots in the air and keep well away from them. I found the Stearman perilous to fly because in a cross-wind-landing the wing tips had a propensity to scrape themselves on the ground. The instructors frowned on that, but it was the airplane's fault, not the fault of the NAVCADs.

These were exciting times. I never realized how cold Memphis could be in February and March, and I never expected to see snow on the ground. The Stearmans had open cockpits so in freezing weather we each dressed in a sheepskin-lined leather coverall, jacket, helmet, and in boots with two-inch soles in order to keep warm. We looked more like today's astronauts than anything recognizable. We certainly did not look like dashing Navy pilots in the tradition of Gary Cooper and Errol Flynn. In addition to looking ungainly, we could not feel how much rudder pressure we were applying while flying because of our thick-soled boots.

Since I never wanted to appear completely stupid, I would nod my head indicating that I had read him loud and clear. Deep down inside, however, I would be nearly hysterical because I did not have the faintest idea what the instructor had said.

In the Stearman, my instructor sat up front and I sat in the rear cockpit. Our aircraft did not have a radio. We communicated between the open cockpits using a device known as a GOSPORT tube. This technology was about one cut below the quality of communications kids used to imagine getting from two coffee cans connected by thirty feet of waxed string. At least that was two-way communications.

In the front cockpit, the instructor wore a canvas mask over his face that attached to his helmet. From the mask a rubber hose led back to and attached to my helmet about where my ears were supposed to be. Over the roar of the engine and the howl of the wind, I was supposed to receive instructions somehow through this garden hose.

Most of what I heard sounded like the grunts and pants of a weight lifter rather than words. I can best approximate the sounds made by putting a hand tightly over my mouth and then trying to talk clearly. The instructor would say: "Okay, Morumph, lets hemf out toomf site sevenedumf." Then, he would look back at me in his mirror seeking my acknowledgment of his instructions. Since I never wanted to appear completely stupid, I would nod my head indicating that I had read him loud and clear. Deep down inside, however, I would be nearly hysterical because I did not have the faintest idea what the instructor had said. This was a one-way communication system, and my classmates and I had no way of saying to our instructors "Say again, sir!" We just nodded our heads. I do not know how we learned anything unless, over time, the umphs and rumfs became associated with real words.

Every training base had numerous outlying fields that we used to practice landings. We had to know where these fields were in relation to the Air Station, so our first instruction flight

...WE HAD TO USE FLASHING LIGHTS, A TETRAHEDRON AND WIND SOCKS TO DETERMINE THE PROPER DIRECTION TO HEAD FOR TAKE-OFF.

was an area checkout. Like the other student pilots, I just went along for the ride while the instructor pointed out Site 9, Site 17, Site 30, etc. In the 1940s, every field I saw in the farming areas of Tennessee and Arkansas looked like a landing field. Somehow, I had to pick out our grass field from every other grass field. I still do not know how we all did it.

Then, we started hands-on flying. I can remember my instructor explaining to me that the aircraft will assume a certain "attitude" at a given air speed and holding a constant altitude. Since "attitude flying" is the whole key to landing aboard ship, we learned about it early on. In the air one day, my instructor explained that I could recognize this correct attitude by noting where the horizon crossed the engine's number one cylinder. But I had a problem: hunkered down in the open cockpit to get away from the freezing cold air, I no more saw my instructor's back in front of me, never mind the horizon or cylinder head. "Do you understamf, Morumph?," he asked, looking at me in his mirror. I peeked up like a groundhog in Pennsylvania and nodded, yes.

As I mentioned, we had no radios, therefore, we could not communicate with the tower. Instead of talking with an air controller in a tower to guide our take-offs, we had to use flashing lights, a tetrahedron, and wind socks to determine the proper direction to head for take-off. During this training, we did not take off and land our Stearmans on the available runways. Instead, we used the "MAT" - a huge asphalted area in the middle of the field - to take off and land. Landing on runways is easy because it forces everyone to follow standard entry and approach procedures. Using the MAT, however, made things interesting. Returning to the Air Station, each cadet had to first determine the wind line by his visual cues and set up his pattern accordingly. The cheese would start to get binding when ten of us would return at about the same time and would set up our landing approaches on ten different but converging headings. Somehow, it all worked itself out, usually by some colorful, descriptive, and profane prompting by our instructors.

Ψ

When we were not flying, the instructors allowed us to go out and watch the other cadets returning from their instructional flights. We enjoyed hooting and hollering at our friends' mistakes as their wing tips would come close to and even strike the asphalt. Those were fun days.

We enjoyed hooting and hollering at our friends' mistakes as their wing tips would come close to and even strike the asphalt.

Memphis introduced me to Southern customs. A bunch of us had gone to Memphis for liberty one Saturday. Returning that evening, we boarded the bus, and I smartly walked to the back and sat down because in high school, when we rode a trolley car or a bus to and from school, the boys always sat in the back. It was "the boys' area." My Southern friends tried to get me to come forward. They seemed embarrassed for me, but I did not know why. Moreover, I saw no need to move. Then, the big, red-necked bus driver looked at me and said: "Get up front, boy, or this bus ain't moving." I moved up front where my friends explained the rules to me.[5]

After eight or nine hours of flight instruction, the instructors gave each student a check-ride. If the student pilot got an "UP," he qualified to solo. Each evening, those who soloed during the day received their informal reward: other cadets would clip their neckties and throw them into the swimming pool. We all looked forward to that proud day. Selective Flight weeded out a few of our classmates, but most of us got through and prepared for the next step.

Each evening, those who soloed during the day received their informal reward: other cadets would clip their neckties and throw them into the swimming pool.

Ψ

Before continuing my story, I must tell you about Frank Price. Frank trained in the class ahead of us, but he failed to move forward with his class because he had "earned" too many demerits. Frank and I lived on the second deck (floor) in a regular barracks at Memphis. One night, Frank decided to play one of his tricks after Taps sounded at 10:00 p.m., and we had turned off our lights. He soaked a tennis ball in lighter fluid,

BEING CAREER MOTIVATED, IT WAS A SIMPLE CHOICE: I CHANGED FROM BEING A NAVCAD TO BEING AN AVIATION MIDSHIPMAN, AVMID.

lighted it, and rolled it the length of the barracks' floor. In the dark, the blazing ball made an impressive sight as it rolled past all of our bunks, then reached the steps, and finally thumped down each step in a blaze of glory. Unfortunately, the Duty Officer was coming up the steps this night to make an inspection. The flaming ball did not please the Duty Officer, and, needless to say, Frank found himself in trouble again. The Navy did not hold Frank back again, however, because the Navy started closing Memphis as a flight-training base. Frank moved on with our class.

Pre-Flight 1947

We took the train to the Naval Air Station at Ottumwa, IA and began Pre-Flight Training there in April 1947.

Upon arrival at Ottumwa, I had to make a decision about my future service. After World War II, the Naval Air Service had grown too large for the Naval Academy to be able to provide all the Regular Officers needed. ADM James L. Holloway, the father of a future Chief of Naval Operations, had developed a program to provide a new stream of regular officers for Naval Air. The "Holloway Plan" created "Aviation Midshipmen" whom the Navy would commission Ensigns in the Regular Navy - rather than Ensigns in the Naval Reserve - after they completed flight training. If I signed up for the program, I would gain my commission in two years regardless of when I finished flight training. As a NAVCAD, the Navy would commission me a Reserve Officer when I completed flight training. As an Aviation Midshipman, however, the Navy would commission me a Regular Officer. Being career-motivated, it was a simple choice: I changed from being a NAVCAD to being an AVMID, an Aviation Midshipman.

Pre-Flight began another screening stage in training. The intense academic, physical, and military training phase would last sixteen weeks and the Navy designed it to thin out the weak-willed and less-motivated members of the class. The minute a classmate showed any sign of weakness, the staff jumped on him as if he were a sick pigeon. If he were not strong willed, he did not last long.

Throughout the history of mankind, God has tested his creatures by visiting on them trials and tribulations in the form of floods, famines, plagues, pestilence, locusts, earthquakes, and such. In Pre-Flight School, one of God's lesser known trials and tribulations awaited the unsuspecting. God called him the Marine Drill Instructor – "the DI." Naval officers served as our academic and physical instructors, but Marine officers and Drill Instructors handled our military training. They kind of took the fun out of it and from the training I developed a strong dislike of Marines. I have always respected the Marines. They certainly knew how to train a warrior, but that did not mean that I ever had to like them.

All in all, I enjoyed Pre-Flight. I enjoyed the strict regimentation, the discipline, and the physical training. No one gets in better shape or condition than when he is a student in Pre-Flight. We studied theory of flight, aerology, engineering, Morse Code and flashing light, dead reckoning and celestial navigation, leadership, Naval organization, and much more.

After our academic classes, we fell in for Military Drill. After drilling, we ran to the gym for calisthenics, swimming, running, boxing, wrestling, or gymnastics, or we ran through the obstacle course.

We formed up by battalions for each meal every day. After supper, we had compulsory study until lights out.

While stationed in Ottumwa, heavy rains caused flooding in the city. We Pre-Flight students helped fight the flood on weekends. We filled sandbags and built dikes to hold back the flood waters. We stood guard duty, knee deep in water, to prevent looting at night. We did this for quite awhile, but the extra work did not interrupt our training. Mondays through Fridays we attended our scheduled classes.

About half way through Pre-Flight, the Navy decided to close Ottumwa and move the school to the Naval Air Station at Pensacola, FL. The flight school leaders moved quickly and flew

1947
Department of Defense

Congress established the Department of Defense in 1947.

It also established the Department of the Air Force and changed the name of the War Department to the Department of the Army.

Under the reorganization, the Secretaries of the Air Force, Army, and Navy were no longer cabinet officers, and they reported to the Secretary of Defense who became a cabinet officer.

The Marine Corps remained a separate service under the Department of the Navy.

The Coast Guard remained a separate service under the Secretary of Treasury. In time of war, the Coast Guard would serve under the Department of the Navy.

President Harry Truman appointed James Forrestal to be the first Secretary of Defense.

Joint Chiefs of Staff

Congress established the Joint Chiefs of Staff (JCS) to advise the Secretary of Defense and the President on defense matters and to control and coordinate inter-service operations.

Four officers made up the JCS:

Chairman;
Chief of Staff, U. S. Air Force;
Chief of Staff, U. S. Army; and
Chief of Naval Operations

The Commandant of the U. S. Marine Corps participated in JCS meetings that concerned the Marine Corps. He was not a voting member.

SNJ Texan

us to Pensacola. When we arrived we helped set up the new barracks. Surprisingly, we hardly missed a beat in our training. However, our life changed: Pensacola's heat and humidity shocked us all. We lived six to a room in the un-air conditioned barracks and sweltered though the last two months of our training, often not sleeping well because of the heat. I finished in August 1947. So long, DI's!

Primary

In another economy move, the Navy closed down primary flight training at our next duty station - the Naval Air Station at Corpus Christi, TX – and moved it to Corry Field in Pensacola. With our move to Texas canceled, we became the first class to start at Corry. Corry, however, was not ready for us, so the Navy put us aboard the USS *Wright*, the training command carrier, for a month of duty as deck hands.

In September, we started ground training to prepare for flight training in the North American SNJ Texan. The Texan was a big step up from the small-engine, bi-wing Stearman.

Our class started flying the first week of October 1947. Because of this, we caught up with classes that had been six months ahead of us in Pre-Flight and passed many others which had gone to Corpus Christi and were waiting in pools to start training.

We all found that the Texan was a lot of airplane to learn to fly. We had to deal with a bigger and heavier airplane with new technology: a variable pitch propeller, retractable landing gear, flaps, a BIG engine, many more instruments, and radios. The plane had a clear, two-way intercom system. We started Primary in an aircraft that Navy and Air Corps designated an advanced trainer during World War II.

I drew LT Howie Schlundt, a big German-American from the Midwest, as my first instructor. He turned out to be a great help. In the SNJ, the instructor sat in the back rather than in the front, and Schlundt's 225 pounds or so made it easy for me to get my tail down for a three-point landing.

LT Schlundt, however, had a speech impediment which sometimes hampered communications. Fortunately, we did not have to talk with one another through a GOSPORT.

Depending on the day or how much I scared him on a poor cross-wind landing, the lieutenant would have a problem with "S's," "T's," or "F's."

One sunny day in our operating area in southern Alabama, I practiced doing steep turns holding a constant altitude, flying figure eights, and using farm field boundaries to learn how to compensate for the wind. I was not having one of my better days. After another sloppy maneuver, Schlundt got angry and said: "Morin, you're all fu-fu, fu-fu, fu-fu – screwed up!"

But under his guidance, I safely soloed and moved through the precision and acrobatic stages of Primary. Since there was no one ahead of us waiting in pools, we stepped smartly from stage to stage: night flying, instruments, and radio navigation. Then, we went on to Saufley Field, a few miles away, where we entered the exciting and fun stages of formation flying, cross-country navigation, aerial gunnery, and carrier qualification (CQ). By this time, we were getting pretty salty, and we flew solo most of the time.

CQ was what Naval Aviation was all about.

We learned to fly low and slow, simulating carrier patterns, approaches, and landings. The Landing Signal Officer (LSO) stood out on the runway giving us signals with his paddles. He indicated that we were OK, HIGH, LOW, TOO FAST, OR TOO SLOW. The idea was to make the perfect approach so that the LSO would signal a "ROGER" with his paddles by holding his arms straight out. We gained a lot of status by being in the CQ stage because we had reached the top stage of our training.

Finally, the day came: after we "field" qualified, we went out to do the real thing on a carrier.

1948
NATIONAL LEADERS

PRESIDENT
HARRY TRUMAN

SECRETARY OF DEFENSE
JAMES FORRESTAL

JOINT CHIEFS OF STAFF

CHAIRMAN
*NONE

CHIEF OF STAFF, AIR FORCE
GEN HOYT VANDENBERG

CHIEF OF STAFF, ARMY
*GEN OMAR BRADLEY

CHIEF OF NAVAL OPERATIONS
ADM LOUIS DENFELD

*GEN BRADLEY BECAME THE FIRST CHAIRMAN OF THE JCS IN AUGUST 1949.

In all the years at sea that followed, the thrill of circling that first carrier waiting to land never dampened.

CARQUALS

In May and October in Pensacola, you can have the prettiest flying weather of any place in the world. On a beautiful May morning, six of us flew out to the carrier. A student led each of two "V" formations from Saufley, across the beach, and out to sea where the USS *Wright*, a light carrier, waited for us.[6] There was not a cloud in the sky and you could see forever. The blue sky above and the blue ocean below made a spectacular sight. And there ahead of us, the *Wright* knifed through the water. That sight and the moment exhilarated me. I have never forgotten them. In all the years at sea that followed, the thrill of circling that first carrier waiting to land never dampened.

The flight leader signaled us to "lower our tail hooks." We had each been in contact with the ship since we went "FEET WET," crossing the beach and heading south into the Gulf of Mexico. And then the moment came: the ship's air controller called us and said, "YOUR SIGNAL, CHARLIE." In carrier parlance, that meant "COME ABOARD." As we circled down, the carrier turned smartly into the wind. We flew up the starboard side of the ship in right echelon at about 200 feet. As each pilot peeled off from the formation, the next pilot in line started his timing to establish the proper interval of about thirty seconds before he peeled off and turned to the downwind leg.

As I came through the ninety-degree position and picked up the LSO visually, I realized that the deck was not very wide. I did not think much about the length because when your hook snares the arresting wire, you do not travel very far up the deck. As I neared the ship my approach was satisfactory, and the LSO signaled: "CUT." I closed my throttle and fell out of the sky into a three-point landing. As the hook snagged the wire, I rapidly decelerated to a stop. I then raised the hook, and the director quickly positioned me properly in the center of the deck for take-off. The Flight Deck Officer immediately signaled me to rev my engine. I nodded that everything was normal and he signaled: "GO." I applied full power and got airborne in seconds before I ran out of deck. Once airborne, I lowered the hook, picked up my interval, turned downwind, and approached the carrier to land again. To qualify, I had to make six arrested landings. On my final landing, the deck crew chocked the plane. I heard the bullhorn call out "Switch Pilots." I disconnected my parachute, radio, leads, shoulder harness, and seat belt as another student climbed on the wing to take my place. I had done it! What a great feeling!

1948 DESEGREGATION

In July 1948, President Harry Truman issued an executive order desegregating the armed services.

In Morin's Pre-Flight class there was a black aviation midshipman named Jesse Brown. He was from Mississippi.

Advance Flight Training 1948

Upon completing CQ, each of us filled out a "Dream Sheet" to select the type of training we wanted to take in our next stage, Advance Training. We had two choices available: multi-engine training for seaplanes and land patrol planes at Corpus Christi, or single-engine training for carrier flying at Jacksonville Naval Air Station, FL. Most everyone got his choice. I got my first choice to fly the F6F Hellcat fighter at Jacksonville.

BERLIN AIRLIFT 1948-49

In Europe, the Berlin Airlift gained the attention of the world when it thwarted Stalin's attempt to drive the United States, Great Britain, and France out of their occupation zones in Berlin.

By the time the airlift ended on 2 May 1949, American and British pilots had made 277,264 flights and hauled 2,343,315 tons of food, fuel, medicine, and clothing into West Berlin.

Besides gaining status as a carrier-qualified pilot, the Navy promoted me to Midshipman First Class and with my new rank I could wear brown shoes with my khakis and greens. Brown shoes were a real status symbol. Many of us bought unique, stylish boots to wear instead of so-called brown low-quarter shoes: jodhpurs, ankle high boots with a strap, or half-Wellingtons (known as slow-roll boots), fifteen-inch high boots. And we kept those beauties shined.

Flight School authorities granted us a short leave between Pensacola and Jacksonville. I went home and during this time proposed to Betty Anne Kane. She accepted my proposal knowing that our wedding had to wait until I finished training and I gained my commission. Both events were a year away.

Jacksonville personnel treated us as officers. I lived in the Junior Bachelor Officers' Quarters (BOQ), ate in the Officers' Mess, used the Officers' Club, and got a snappy salute from the

Marine sentry at the Main Gate. As First Class Middies, the Navy permitted us to have cars. Our pay, however, did not reflect the vestiges of our upward mobility. We still made only $118 per month.

JACKSONVILLE PERSONNEL TREATED US AS OFFICERS. I LIVED IN THE JUNIOR BACHELOR OFFICERS' QUARTERS (BOQ), ATE IN THE OFFICERS' MESS, USED THE OFFICERS' CLUB, AND GOT A SNAPPY SALUTE FROM THE MARINE SENTRY AT THE MAIN GATE.

Ψ

Our final stage of training required each of us to fly 100 hours in the Hellcat fighter: going from familiarization (FAM Stage) to formation flying, practicing air-to-ground weapons delivery and aerial gunnery, and mastering instrument flying and night flying. Then, we returned to NAS Pensacola for more CQ.

In those days, the Hellcat was a big airplane. I could not wait to fly this famous carrier fighter of World War II with its 2,000 horsepower engine. No more 120 knots in the Texan trainer. I was getting into something fast. On take-off, I would get fifty-four inches of manifold pressure. That was a lot of power, but once airborne I only flew around at 160-165 knots. That disappointed me. Based on my reading, I had expected to fly around at 300 knots or more.

We progressed through the hot summer of 1948. In September, I returned to Saufley for CQ in the Hellcat, the last stage of training before winning my coveted gold aviator wings. At Saufley, we became the big men on campus. We lived in the BOQ, not in barracks like the lesser lifes, and we walked around in our brown boots, pants caught jauntily on one of the half-Wellingtons.

Before we knew it, we flew back out to the ship to qualify in the Hellcats. When we completed CQ in the big fighter, we were done with the Training Command. All that was left, on our return to Jacksonville, was to attend the big ceremony scheduled for 7 October 1948 at which the Navy would designate us Naval Aviators and award us our wings. These ceremonies were supposed to big deals: bands would play marches; airplanes would fill the skies; we would be decked out in our stiffly starched choker-whites; and some beautiful celebrity such as Doris Day or Elizabeth Taylor would pin on our wings.

Where the Hellcat was disappointment Number One for me in flight school, the graduation ceremony became disappointment Number Two for me. I had to buy my wings and then had to turn them in to the office of the Commanding Officer, NAS Jacksonville. On 7 October, seven of us walked into the Office of the Base Commander in our khakis, and he handed each of us our designation certificate as a Naval Aviator and a set of wings. "Didn't he know there were supposed to be airplanes, bands, and pretty girls?" But that was it.

Naval Aviator Morin

Ψ

It had taken just under a year from my first to last hop in the Training Command. If I had been an Aviation Cadet, the Navy would have commissioned me an Ensign that day. As an Aviation Midshipman, however, I had to serve seven more months to complete the twenty-four month contract I had signed at Ottumwa before the Navy would make me an Ensign. Boy, was I had![7]

Didn't he know there were supposed to be airplanes, bands, and pretty girls? But that was it.

Sidebar Notes

1947 Truman Doctrine, etc.
See individual footnotes.

American Government Officials Spying for the Soviet Union During World War II
Haynes & Klehr, *Venona, 346-47, 352, 369.*

1947 Department of Defense
Various sources.

1948 National Leaders
Various sources.

1948 Desegregation
Various sources.

1948-1949 Berlin Airlift
Martin, *America and Its Peoples*, II, 739.

Photo Credits
Hellcats, 1943, *Wikipedia.*
NS2, Wikipedia.
SNJ2, unknown.
Jesse Brown, *Wikipedia.*
Medals, varied sources,

End Notes

[1] John Earl Haynes & Harvey Klehr, *Venona, Decoding Soviet Espionage in America* (New Haven, CT: Yale University Press, 2000), 168.
[2] *Ibid*, 9.
[3] Martin, *America and Its Peoples*, II, 736-7.
[4] *Ibid*, 738 (Kennan) and 739 (limited wars).
[5] In those days and for several more years, white people sat in the front of buses and black people sat in the back of buses in the southern states from Virginia to Texas and in Washington, DC. The civil rights movement began in 1955 when Rosa Parks, a black seamstress, refused to sit in the back of a bus in Montgomery, Alabama.
[6] Named for the Wright Brothers.
[7] Morin *Manuscript*, 3-11.

In Honor of Jesse Brown

Looking back on my training, I am reminded that during all my years of service I failed to recognize and appreciate the historical significance of certain people and events. I just did not have an eye for history until well after the fact.

In our Pre-Flight class there was an aviation midshipman named Jesse Brown. Jesse was from Mississippi. That in itself was not unusual. What was unusual (and historical) was that Jesse was the first black American to enter naval aviation training. Blacks flew and fought with the Army Air Corps in World War II, but the Navy was a pretty WASPish organization and this barrier was not pushed aside until Jesse Brown came along.

Jesse made it through training and received his wings in late 1948 after having to contend with some instructors who did not want to see him succeed. Just a couple of years later, flying a Corsair from a carrier during the Korean War, the enemy damaged his fighter, and Jesse crashed near the Chosin Reservoir. One of Jesse's squadron mates, Tom Hudner, attempted to rescue him.

President Harry Truman awarded Hudner the Medal of Honor. The citation explains what happened to Jesse.

> *For conspicuous gallantry and intrepidity at the risk of his life above and beyond the call of duty as a pilot in Fighter Squadron 32, while attempting to rescue a squadron mate whose plane struck by antiaircraft fire and trailing smoke, was forced down behind enemy lines. Quickly maneuvering to circle the downed pilot and protect him from enemy troops infesting the area, Lt. (j.g.) Hudner risked his life to save the injured flier who was trapped alive in the burning wreckage. Fully aware of the extreme danger in landing on the rough mountainous terrain and the scant hope of escape or survival in subzero temperature, he put his plane down skillfully in a deliberate wheels-up landing in the presence of enemy troops. With his bare hands, he packed the fuselage with snow to keep the flames away from the pilot and struggled to pull him free. Unsuccessful in this, he returned to his crashed aircraft and radioed other airborne planes, requesting that a helicopter be dispatched with an ax and fire extinguisher. He then remained on the spot despite the continuing danger from enemy action and, with the assistance of the rescue pilot, renewed a desperate but unavailing battle against time, cold, and flames.*

Jesse Brown died in his cockpit.

The Navy awarded Jesse Brown a Distinguished Flying Cross posthumously and in 1973 commissioned the USS *Jesse L. Brown* to honor our talented classmate.

Four

COLD WAR DUTY

China, Korea, U.S.S.R., Japan, Taiwan

Cold War 1949-1964[1]

Important Events		James Morin
	Jan 1949	**Morin joined first squadron.**
	Apr 1949	Morin promoted to Ensign.
Red Chinese took over China.	May 1949	
U. S. S. R. detonated an atomic bomb.	**Aug 1949**	
	Sep 1949	**Morin married Betty Anne Kane.**
	Jan 1951	Morin deployed to the MED on the USS *Wright.*
Korean War began.	**Jun 1950**	
	Mar 1951	James B. Morin, Jr. born.
President Dwight Eisenhower developed a "New Look" for the armed services.	1952	
Armistice signed. Korean War ended.	Jul 1953	
	Oct 1953	Sarah Elizabeth Morin born.
	Sep 1954	Morin deployed to the Pacific on the USS *Wasp.*
	Dec 1954	John Patrick Morin born.
	Apr 1955	Morin assigned to NAS Alameda, CA.
	May 1956	Morin deployed to the Pacific on the USS *Lexington.*
Hungarian Revolution erupted.	Oct 1956	
Suez Crisis developed.	Oct 1956	
	1957	Morin attended Naval Post Graduate School in Monterey, CA.
	Jun 1958	Mary Margaret Morin born.
Fidel Castro seized power in Cuba.	1959	Morin attended the University of California, Berkeley, CA.
	May 1960	Morin joined Naval AF staff in San diego.
	Aug 1960	Lydia Noreen Morin born.
John F. Kennedy elected President.	Nov 1960	
Soviets created a 3rd Berlin Crisis.	Jan 1961	
Bay of Pigs invasion.	Apr 1961	
CUBAN MISSILE CRISIS.	**Oct 1962**	
	Jan 1963	Morin deployed to the Pacific on the USS *Ranger.*
President Kennedy assassinated. Lyndon Johnson became President.	Nov 1963	
	Dec 1964	Morin deployed to the Pacific on the USS *Coral Sea.*

Cold War 1949-1964

Jim Morin entered active duty at a dangerous time. The civil war in China came to an end in May 1949. Mao Tse-dung had defeated the Nationalist forces led by Chiang Kai-shek, and Chiang with several thousand of his troops had evacuated to Taiwan where he maintained the government of the Republic of China. On the mainland, Mao established the People's Republic of China, borrowing a democratic term to describe a totalitarian Communist state. Nearby, Nationalists occupied the islands of Quemoy and Matsu which soon proved to be small battlegrounds between the two Chinese states.

MORIN ENTERED ACTIVE DUTY AT A DANGEROUS TIME.

Δ

In August 1949, American scientists "detected traces of radioactive material in the Soviet atmosphere."[2] The Soviets had developed an atomic bomb and the Cold War became a confrontation between two nuclear, superpowers. Nothing would be the same thereafter. American Soviet spies in the United States had provided critical information to the U. S. S. R., and consequently the Soviets developed a bomb "a full decade before American intelligence expected it."[3] The United States soon arrested, tried, convicted, and then executed two of those spies in 1951. The United States also arrested and tried Ivy-educated, former State Department official, Alger Hiss, for spying, but eventually gained a conviction against Hiss for perjury. Most Americans who spied for the Soviets escaped trials and few were punished.

FAR AWAY IN ASIA, THE FRENCH HAD GREAT DIFFICULTY IN REGAINING FULL CONTROL OF FRENCH INDOCHINA.

Far away in Asia, the French had great difficulty in regaining full control of French Indochina. By the end of 1946, tensions had increased between the French and the so-called nationalist group, the Viet Minh, led by Ho Chi Minh. President Charles de Gaulle had granted the northern and central states of Vietnam - Tonkin and Annam - independent status as the Republic of Vietnam under French dominion but kept Cochin China – southern Vietnam - as a separate state under the French Federation along with Laos and Cambodia. Ho Chi Minh had travelled to France in 1946 to solve the Cochin China issue, but failed to do so.

Meanwhile the Viet Minh had conducted pogroms throughout the country and slaughtered Vietnamese women who had married French men, assassinated thousands of opponents, and kidnapped tens of thousands more.

In late 1946, the returning French had "moved to take over the custom houses in Haiphong in order to reassert France's political authority in the north and to prevent military supplies from reaching the Viet Minh." A tough fight followed. The French Navy bombarded the port and killed 6,000 civilians. On 19-20 December, the Viet Minh - now openly Communist rather than Nationalist - revolted in the city. With superior fire power, the French defeated the insurgents. However, Vo Nguyen Giap "was encouraged by the fact that it took the French some seven days to rid the city of his ill-equipped forces."[4] The action began the first Vietnamese War. By 1949, the situation worsened in Vietnam because the Red Chinese began to assist the Communist-led Viet Minh.

The action began the first Vietnamese War. By 1949, the situation worsened in Vietnam because the Red Chinese began to assist the Communist-led Viet Minh.

Δ

Nearby, in Malaya, the predominately Chinese Malayan Communist Party had called for open rebellion against the British in 1948 and had begun assassinating key British leaders and pro-Chiang Chinese. The Malayan Communist Party aimed to drive the British out of Malaya, paralyze the nation's economy, establish "liberated areas, and turn Malaya into the Communist Republic of Malaya. Early on, they ambushed and killed the British High Commissioner.[5]

The British moved to undercut the Communist insurgency by establishing the Federation of Malaya while keeping the city of Singapore as a Crown Colony and by moving more and more troops into the country. By 1948, the British had 9,000 troops in country to oppose the estimated 6,000 Communist insurgents, and they brought in an experienced police leader to reorganize and improve police operations in the peninsula. The next year, Lt. General Sir Harold Briggs arrived and he developed the plan - "Operation Starvation" - that, when fully implemented by General Sir Gerald Templar, destroyed the insurgency.[6]

Δ

Shortly after the end of the war in the Philippines in 1945, the Communist Hukbalahap guerrillas had begun their insurgency against the newly established Republic of the Philippines in order to take over the government. The HUKs had been useful fighters against the Japanese and disruptive fighters against fellow guerrilla bands during the war. Early on, they conducted raids and ambushes, tried to kill government officials and police, and operated much like the Communist insurgents in Malaya. President Manuel Roxas and his replacement, Vice President Elpidio Quirino, tried to placate the HUKs and bring them back peacefully into the mainstream of Filipino life, but their efforts failed. Eventually the new government faced a dangerous insurgency. Both presidents sent in troops to sweep the HUK-controlled areas, but never succeeded in damaging the insurgents. In 1948, the HUKs committed an unpardonable act: they killed Aurora Quezon, the beloved widow of President Manuel Quezon, a Filipino Patriot leader during the Philippine Insurrection and the first President of the Commonwealth of the

Philippines.

In 1949, following the fall of China to Mao, Luis Taruc, the HUK leader, "claimed that the Philippines would have to follow the road of China." By then, key HUK leaders believed that they could take over the nation by 1951. Fortunately, they did not.[7]

Meanwhile problems festered in Korea, and on 25 June 1950, North Korean forces with Soviet tanks invaded South Korea without warning and began an ugly war that paused in 1953 with an armistice.

In July 1950, the Cold War turned a bit ugly at home. Senator Joe McCarthy, WI began a political storm that swept the country for several years when he announced:

> *While I cannot take the time to name all the men in the State Department who have been named as members of the Communist Party and members of a spy ring I have in my hand a list of 205... a list of names that were known to the Secretary of State and who nevertheless are still working and shaping the Policy of the State Department.*

Unknown to most Americans, and possibly Senator McCarthy, "The deciphered cables of the Venona Project identified 349 citizens, immigrants, and permanent residents of the United States who had had a covert relationship with Soviet intelligence agencies."[8]

The Senator eventually would look for Communists in the Army.

KOREAN WAR

25 JUN 1950
NORTH KOREAN FORCES INVADED SOUTH KOREA WITHOUT WARNING.

JUL 1950
U. S. FORCES OCCUPIED THE PUSAN PERIMETER.

15 SEP 1950
U. S. FORCES LANDED AT INCHON, DROVE INLAND, AND CUT OFF THE NORTH KOREAN FORCES.

OCT 1950
UN FORCES CROSSED THE 38TH PARALLEL INTO NORTH KOREA AND ADVANCED NORTH TO THE YALU RIVER.

25 NOV 1950
CHINESE COMMUNIST FORCES INVADED KOREA AND DROVE UN FORCES SOUTHWARD.

JUL 1951
BATTLE LINE STABILIZED NEAR THE 38TH PARALLEL.

10 JUL 1951
ARMISTICE TALKS BEGAN.

MAY-JUL 1953
COMMUNISTS LAUNCHED FINAL OFFENSIVES PRIOR TO ARMISTICE.

27 JUL 1953
ARMISTICE SIGNED.

Ψ

The Fleet

I began my active service as a Midshipman. The Navy assigned some of my classmates to squadrons of the Pacific Fleet. They went to San Diego, CA. The Navy assigned me and others to squadrons in the Atlantic Fleet. We went to Norfolk, VA.

My orders sent me to Norfolk to join Attack Squadron 45 (VA-45), which was returning from a cruise to the Mediterranean (MED] aboard the big carrier, USS *Franklin D. Roosevelt* (CVB-42), the first carrier named for a President.

Before I reported to Norfolk, I found out that the Navy had directed that my squadron and its air group (Air Group 4) relocate to Naval Air Station Jacksonville because it had decided to shut down Advanced Training at Jacksonville and move it to Corpus Christi. On the Navy's merry-go-round of bases, what goes round, comes round.

Prior to joining my first squadron in January 1949, I went home on Christmas Leave. After I got home, I gave Betty Anne Kane a diamond ring to make our engagement official. On New Year's Eve, Betty and I went to a dance at Worcester Auditorium. I wore my dress blues as a Midshipman, First Class and proudly wore my gold wings on my chest. At the dance, I met three other First Class Mids from the Naval Academy. They could not figure out how I had wings or for that matter why I was a Midshipman. These three got their appointments to the Academy in 1945 from my Congressman, the same year I sought an appointment from him. None of them stayed in the Navy. I have always felt that it was a waste to send them to the Naval Academy in the first place.

My first squadron had been flying the new AD1, the famous Skyraider. Even though it had been one of the first squadrons to fly the Skyraider, when the squadron returned home, the Atlantic Fleet scheduled its pilots to transition into the AM1, the less famous but brand new Martin Mauler. The Mauler, a big single-engine, single-pilot aircraft sported a twenty-eight cylinder Pratt & Whitney engine that developed 3,000 horsepower, a thousand more than the World War II Hellcat. When the engine worked right, the Mauler purred. Pilots found the plane to be a very good dive-bomber and very stable. However, the Mauler proved to be hard to maintain, and it did not become a good carrier aircraft.

My first Commanding Officer was CDR Leroy Swanson.[9] Swanson stressed to all of us that we were Naval Officers first and Naval Aviators second. I never forgot that. The CO, a gentleman and a great leader, became a father figure to me.

We kept very busy in the squadron. We trained intensely for deployment to the MED in January 1950 and consequently spent a lot of time at sea. During these early days, I qualified

WE KEPT VERY BUSY IN THE SQUADRON. WE TRAINED INTENSELY FOR DEPLOYMENT TO THE MED IN JANUARY 1950 AND CONSEQUENTLY SPENT A LOT OF TIME AT SEA.

to fly the Mauler off of carriers.

On 27 April 1949, I completed my twenty-four month contract, and the Navy promoted me to Ensign. The promotion brought a good increase in pay.

In September, I went home on leave to get married. Once there, I had to meet Father Curran, the old Irish priest in Betty Anne's parish, who would perform our ceremony. He was not too keen that Betty Anne planned to marry "a sailor" and "a red head to boot." Nevertheless, the good father married us on 17 September. After our honeymoon, we headed for Jacksonville. While waiting for our plane at LaGuardia, I ran into a fellow who had been in flight training with me. He had just survived a crash of a patrol plane. So the first Naval Aviator my new bride met had an arm in a sling and a head full of bandages. What an introduction to Naval Aviation.

As our cruise neared, the Navy realized that the Mauler was not going to cut it as a good fleet aircraft and directed that my squadron retrain in the AD1, Skyraider. In November, we deployed to our carrier off the coast of Jacksonville to qualify in the Skyraider. On my third landing, I came in a little high and caught a late wire. Across the flight deck on this straight-deck carrier stood barriers – raised arresting wires - that would prevent an aircraft from crashing into the aircraft parked forward on the deck. Although I had caught a wire when I landed, I had traveled far enough up the deck that my propeller had wrapped the barrier cables around it. Unexpectedly, my engine mount failed and everything forward of the cockpit fell downward. A fire started and in the few seconds that it took me to leap out of the cockpit I sustained third-degree burns on my right leg below my knee and above my sock because the heat got trapped inside the leg of my flight suit. I also sustained first and second-degree burns on my face. Fortunately, I had my goggles on and they protected my eyes so it was not too bad.

That night, my Executive Officer, LCDR Fred Kidd,[10] called Betty to tell her that I had been burned and would be flown ashore the next day. When I arrived at the Naval Hospital at

THAT NIGHT, MY EXECUTIVE OFFICER, LCDR FRED KIDD, CALLED BETTY TO TELL HER THAT I HAD BEEN BURNED AND WOULD BE FLOWN ASHORE THE NEXT DAY.

THE MED CRUISE WAS MEMORABLE BECAUSE WE DID NOT HAVE ENOUGH MONEY TO DO MUCH FLYING.

Jacksonville NAS, I look liked the invisible man – the corpsmen had wrapped my head in bandages and cut out holes for my eyes and mouth. My right shin resembled a burned pot roast. Betty, unfortunately, got a quick indoctrination into the risks associated with Naval Aviation.

I spent five weeks recovering in the hospital. That was hard on Betty. She visited me every day and then returned to an empty apartment in a strange city. I got out of the hospital just in time to get some flight time and "field qualify." Then, I headed to Norfolk during the first week of January to load aboard the USS *Midway*[11] (CVB-41) for a six-month cruise in the MED. By the time I returned from the cruise, Betty and I had been married about nine months, but we only had been together for about eight weeks. Welcome to the Regular Navy, Betty Anne Morin.

"Hey, Bambino"

We sailed for the MED in January 1950. Our Air Group 4 embarked three fighter (VF) squadrons flying F4U4 Corsairs and two attack (VA) squadrons flying AD1 Skyraiders. The Air Group also included detachments of special mission aircraft such as night fighters and early-warning, electronic-counter-measure, and photo-reconnaissance aircraft.

The MED Cruise was memorable because we did not have enough money to do much flying. However, we did visit some interesting ports and some that were not particularly interesting. Normally, we anchored a mile or two from a port, and our liberty parties traveled back and forth in large liberty barges. The Captain assigned us Air Group pilots as Boat Officers. As Boat Officers, we had to ensure the safety of the boat and its occupants and had to maintain good order and discipline. To fulfill our duty at night or during inclement weather, we would stand at the back of the boat with the Coxswain who handled the craft.

WHILE AT ANCHORAGE IN AUGUSTA BAY, SICILY, I LEARNED THE TRUE MEANING OF THE WORD "SHIPMATE."

While at anchorage in Augusta Bay, Sicily, I learned the true meaning of the word "shipmate." We all knew that sailing was a hazardous profession. History shows that. We all knew that the safety of a ship depended on the alertness, training, and sense of responsibility of the all crew members. We learned everyday to rely on others - our shipmates - for our mutual safety and the safety of our ship.

Off Augusta one day, I served as the Boat Officer on a boat that traveled into the Fleet Landing on a regularly scheduled run. After we loaded at dusk and got underway in heavy seas to return to the *Midway*, I noticed two Chief Petty Officers had taken seats in the stern sheets right below where I stood. They obviously had renewed old acquaintances at many bars in Augusta, but they sat quietly in the boat. Then, one Chief said to the other: "I'm going to be sick." His friend warned him not to throw up in the boat. We continued on toward the carrier, fighting the waves as we went. All of a sudden the Chief got sick and looked as if he would throw up in the boat. As he started to vomit, his friend immediately whipped off his uniform cap and put it under the Chief's chin. The wine consumed that day turned the white hat cover purple. When the sick Chief finished, his buddy removed the cap from under his chin, hung it over the side, washed it out, and then smartly put it back on his head. That is a SHIPMATE.

In addition to serving as Boat Officers, we also served as Shore Patrol Officers. As a twenty-three year old ensign, I looked very young. One cold night in Naples, I made the rounds of the bars with the Senior Shore Patrol Officer. Wearing my blues and the Shore Patrol brassard on my arm, I felt very official and tried to look like a tough and seasoned Shore Patrol Officer. Thinking back on it, I probably looked like Barney Fife, the deputy played by Don Knotts on the television program, "The Andy Griffith Show." As we entered one terrible place that night, one of the ladies of the night called out to the Senior Shore Patrol Officer, "Hey, Lieutenant." Then, another called out to me: "Hey, Bambino." So much for looking official.

On the *Midway*, the captain gave the Air Group officers the opportunity to participate in Officer of the Deck training.[12] I

1952
EISENHOWER'S NEW LOOK

ASSUMED NEXT WAR WOULD BE NUCLEAR.

EMPHASIZED THREAT OF MASSIVE RETALIATION.

DEVELOPED DOCTRINE OF MUTUALLY ASSURED DESTRUCTION (MAD).

EXPANDED AIR FORCES.

EXPANDED MISSILE FORCES.

CUT BACK THE ARMY.

EMPLOYED THE CIA AS A COVERT FORCE.

ENCOURAGED LOCAL TROOPS TO OPPOSE COMMUNIST INCURSIONS.

WITH THE ADVENT OF THE KOREAN WAR, THE NAVY SENT CARRIERS AND AIR GROUPS TO THE FAR EAST TO FIGHT THE WAR, AND DECIDED TO REPLACE ONE OF THE BIG-DECK CARRIERS IN THE MED WITH THE LIGHT CARRIER, USS WRIGHT (CVL-49).

took every opportunity to learn as much as possible about the shipboard and ship handling sides of our profession. The Ship's Company Officers with whom I sat with at meals always were flattered that I took an interest in what they did, and they always were quick to give me a tour of their spaces and explain their jobs. On this first cruise and on many that followed, I took the time during our months away from home to learn all that I could about the ships. Many years later, the knowledge and experience I gained on the early cruises stood me in good stead when I commanded an amphibious vessel and when I commanded a carrier.

During the MED Cruise, we learned that Secretary of Defense Louis Johnson had mandated further reductions in current defense spending. Consequently, the Navy decided to decommission our Air Group when we returned to the States.

In July 1950, the Fleet reassigned me to another fighter squadron, VF-14, to fly the F4U-5. The new Corsair looked like a fighter, felt like a fighter, and flew like a fighter. I considered it an excellent carrier aircraft. I enjoyed this assignment because I flew the new version of the Corsair, and because I served under another good skipper: CDR Bob Coats, a World War II ace who shot down five Japanese on 18 March 1945 near Kanoya, Japan. He shot down eleven more after that.[13]

With the advent of the Korean War, the Navy sent carriers and air groups to the Far East to fight the war and decided to replace one of the big-deck carriers in the MED with the light carrier, USS *Wright* (CVL-49). My squadron, VF-14, with twenty Corsairs, soon joined the *Wright* and sailed for the MED in January 1951, a year after I had left on the *Midway*.

When I left for the second cruise, Betty was with child. On 30 March 1951, she gave birth to James B. Morin, Jr. in the Naval Hospital at Jacksonville. I was not home for the birth, but that was not unusual for a Navy man.

Shore Duty

In the late summer of 1951, I completed my first tour of sea duty and headed to Pensacola to work as a flight instructor in the SNJ Texan. I found that I really learned to fly when I instructed others but found the work tedious. It wore me down. We had a large number of students going through flight training because of the Korean War, and I had to report to work at 6 a.m. in the morning, six days a week. On Sunday, I felt like a zombie; soon it was Monday again.

In those years, Pensacola was not much of a town to live in, but at least I was on shore duty. While we trained American pilots for Korea, we also trained French pilots who would be sent to Indochina (Vietnam) to fight against the Viet Minh. One day, I worked with a French officer with the Irish name of Hennessy. He spoke English with a very clipped British accent and was a member of the very famous family in France that makes cognac bearing its name. He died for nothing in Vietnam as so many Americans would do eleven to eighteen years later.

In the last part of my tour, the training command selected me to be an instructor in the unit that trained future instructors. The work turned out to be less hectic and, in fact, enjoyable. Soon my shore tour came to an end and that bogey-man, sea duty, loomed again.

Betty, however, was again with child and due to deliver just about the time we were to move on. I requested and gained a two-month extension of my shore tour. The Korean War was ending at the time, but the Pacific Fleet squadrons still needed pilots. My orders came in telling me to report to San Diego at the regular time. I immediately called the Bureau of Naval Personnel (BUPERS) to ask about my approved two-month extension. BUPERS told me that an error had been made and that they would modify my orders. Rather than just delaying my detachment date, the BUPERS ordered me to Ellison Field outside of Pensacola to go through helicopter training. That would take two months.

1954

Guatemala

EISENHOWER PERMITTED THE CIA TO OVERTHROW A LEFTIST GOVERNMENT AND REPLACE IT WITH A PRO-U. S. REGIME.

Nuclear Strike

ADMIRAL ARTHUR RADFORD, CHAIRMAN, JOINT CHIEFS OF STAFF, RECOMMENDED A NUCLEAR STRIKE TO AID THE FRENCH AT DIEN BIEN PHU.

Dien Bien Phu

APPROXIMATELY 10,000 FRENCH TROOPS SURRENDERED AT DIEN BIEN PHU. ABOUT 2,000 DIED THERE; NEARLY 8,000 VIET MINH PERISHED TAKING THE FRENCH POSITION.

I just did not want to fly helos in the Fleet. At that time, helos flew rescue and utility missions, and I wanted to stay in carrier aviation.

I just did not want to fly helos in the Fleet. At that time, helos flew rescue and utility missions, and I wanted to stay in carrier aviation. But soon I was going through helo training and enjoying it.

Sarah Elizabeth Morin was born on 21 October 1953 and at the end of November we drove cross country to San Diego. Fortunately, we did not have to deal with the desert heat during the drive. That would have been difficult for Betty and the new baby. After days of seeing nothing but boondocks and desert driving though Texas, New Mexico, and Arizona, we thought we had arrived in the Garden of Eden when we drove into San Diego and saw its green lawns and its multitude of flowers. We fell in love with California and its lifestyle.

With the Pacific Fleet

When I reported in to the Staff of the Commander, Naval Air Force Pacific Fleet at Naval Air Station North Island on Coronado Island in San Diego Harbor for duty, I was asked: "Where had I been?" The Bureau of Naval Personnel had not sent out my modification of orders to helo training. When I asked what assignment I preferred, I said: "FIGHTERS."

Fortunately, the command sent me to Fighter Squadron VF-111 at Naval Air Station Miramar, just north of San Diego, to fly the fleet's first swept wing jet fighter, the F9F-6 Cougar. The Cougar was a great carrier aircraft. It was reliable and easily maintained, but like all our aircraft at that time, it was underpowered. My new Commanding Officer was CDR Gus Knudson.[14] I was surprised to find that he was a terrible aviator. Beside my regular collateral duties[15] in the squadron I became a Landing Signal Officer (LSO) and soon found myself out on the end of a runway at Miramar day and night with other Air Group LSOs learning how to wave aircraft aboard a carrier.

In September 1954, we sailed for the Pacific as part of Air Task Group One in the USS *Wasp* (CV-18).[16] VF-52, flying the F9F2 Panther fighter, VF-151 also flying the Panther, and VA-

194, flying the AD4 Skyraider, completed the Air Group.

As the ship pulled out, my most vivid memory was that of Betty, pregnant again, standing on the pier holding the hand of three-and-a-half-year old Jim and holding eleven-month old Sarah in her arms. Boy, what we asked of our wives! The wives of our seagoing officers and men have never received the recognition and credit they so richly deserved.[17]

The Pacific Fleet was a different Navy. We always seemed to be on alert. The Korean War had just ended; the French were fighting for their existence at Dien Bien Phu in French Indochina; and Quemoy and Matsu Islands between Formosa (Taiwan) and China were a bone of contention between the Nationalist Chinese and the Red Chinese. That part of the world had been continuously in turmoil since the 1920s and remained so. We found out many years later that the French had asked President Dwight Eisenhower for a nuclear attack to relieve them of the overwhelming pressure exerted by the Viet Minh. The pressure eventually resulted in the humiliating defeat of the French. I could not understand how an ill-trained and ill-equipped peasant army could defeat a modern power. I found out ten to fifteen years later what their leadership, motivation, and bravery could do against their enemies who lacked a national will and consensus to fight an extended war.

While anchored in Hong Kong for Christmas, I received notice that John Patrick Morin, our third and last child, had been born on 27 December 1954. I returned to my enlarged family in April 1955. Once home, I found new orders waiting for me that assigned me to the USS *Shangrila*[18] as ship's company. The Air Group Commander, when he realized that all his LSOs were either getting out of the Navy or had new orders, had my orders cancelled. I had no time, however, to relax because the Navy changed my squadron's home port from Miramar to Naval Air Station Alameda in the San Francisco Bay area. After we moved north, Betty and I bought our first house in nearby Hayward.

My new Commanding Officer, CDR Glen Medick,[19] ran a good squadron. He appointed me - promoted me - to be the

1954
KHRUSHCHEV AND MAO

AMERICAN PREDATOR

"I FOUND IT PERFECTLY INCREDIBLE THAT MAO COULD DISMISS AMERICAN IMPERIALISM AS A PAPER TIGER WHEN IN FACT IT IS A DANGEROUS PREDATOR...

HE SAID TO ME, "IF WE COMPARE THE MILITARY MIGHT OF THE CAPITALIST WORLD WITH THAT OF THE SOCIALIST WORLD, YOU'LL SEE THAT WE HAVE THE ADVANTAGE. THINK OF HOW MANY DIVISIONS CHINA, THE U.S.S.R. AND THE OTHER SOCIALIST COUNTRIES COULD RAISE."

BOMB FODDER

I SAID, "COMRADE MAO TSE-TUNG, THAT THINKING IS OUT OF DATE. YOU CAN NO LONGER CALCULATE THE ALIGNMENT OF FORCES ON THE BASIS OF WHO HAS THE MOST MEN. WITH THE ATOMIC BOMB, THE MORE TROOPS ON A SIDE, THE MORE BOMB FODDER."

1,000 DIVISIONS

"LISTEN, COMRADE KHRUSHCHEV. ALL YOU HAVE TO DO IS PROVOKE THE AMERICANS TO MILITARY ACTION, AND I'LL GIVE YOU AS MANY DIVISIONS AS YOU MAY NEED TO CRUSH THEM — A HUNDRED, TWO HUNDRED, ONE THOUSAND." I TRIED TO EXPLAIN THAT ONE OR TWO MISSILES COULD TURN ALL THE DIVISIONS IN CHINA INTO DUST. BUT HE WOULDN'T LISTEN AND OBVIOUSLY REGARDED ME AS A COWARD."

Squadron Operations Officer because, although not the senior lieutenant, I had the most flight time and the most carrier landings of the pilots in the squadron. I was pleased that Medick picked me for the job because I believed that I was the best person to train the pilots to be carrier fighter pilots. In my new job, I no longer had to be the Squadron LSO. Therefore, I paid full attention to the training of pilots on the brand new Cougars, the F9F8s, which we had received directly from Grumman's factory on Long Island, and to the training of the pilots for the upcoming cruise to the Pacific on the USS *Lexington* (CV-16).[20] As it turned out, I wrote the squadron's training doctrine and the syllabus for our between-cruise training.

We sailed for the Pacific in May 1956. During the cruise, I made the 1000th arrested landing on the carrier ...

The recently re-commissioned *Lady Lex* sported steam catapults [CATS], an angled deck, and a mirror landing system. These innovations made carrier operations much safer. No more takeoffs under our own power; now a CAT hurled us off the deck. No more LSO; now we used large mirrors to fix our positions for landing. And no more barriers across the deck! We landed on the angled deck that jutted out ten degrees from the regular deck. If we did not catch a wire, we would be up on power, go around, and try again.

We sailed for the Pacific in May 1956. During the cruise, I made the 1000th arrested landing on the carrier and the captain had my name engraved on a small brass plate and placed it in the officers' mess. At the celebration that followed, I cut the cake and enjoyed hearing the compliments handed out to the pilots and crew.[21]

In the fall of 1956, the Navy selected the *Lexington* to visit Melbourne, Australia where the 1956 Olympics were to be held. Through the U. S. Embassy in Tokyo, crew members purchased tickets to events they wished to see. Prior to our sailing for Australia, the British and the French invaded the Suez Canal and created a world crisis. President Eisenhower objected, and the Russians growled. The Commander-in-Chief, Pacific Fleet (CINCPAC), cancelled our trip, ordered that we put to sea to start operations again, and directed that our cruise be extended indefinitely. I never did get to Australia for the games.

Shortly thereafter, we received another change in plans. CINCPAC directed the *Lexington* to return to the States at high speed and be home before Christmas.

1955 Summit

Eisenhower proposed an "OPEN SKIES" program; Khrushchev rejected it.

Shore Duty Again

Before leaving the *Lexington*, I received orders to the Naval Post Graduate School in Monterey, CA.

Warsaw Pact

The Soviet Union organized the WARSAW PACT as a Treaty of Friendship, Cooperation and Mutual Assistance to oppose NATO.

We could not sell our house immediately so I commuted for awhile before Betty and I found a great rental on 17 Mile Drive on the Peninsula and enjoyed Monterey very much. A year later, Betty and I drove to Kingsville, TX, where I learned to be a jet flight instructor in Panthers and Cougars. After California, and particularly Monterey, when we arrived in Kingsville we suffered a culture shock. The dusty Texas town southwest of Corpus Christi hardly had any downtown; we found no housing available. We had to contract to have a house built; in the meantime, we lived in a small Navy apartment on NAS Kingsville. We moved into our new house in May 1958, and on 7 June Mary Margaret Morin, our fourth and last child arrived.

Eight months later, the Navy ordered me to attend the University of California at Berkeley to study Business Administration. We sold our new house and headed back to California. Eighteen months later after earning a Bachelor of Science degree, the Navy ordered me to San Diego again, this time to serve on the staff of the Commander Naval Air Forces, Pacific Fleet at North Island.

In 1960, Betty and I were expecting our fifth child. On the afternoon of 16 August, Betty called me at North Island and told me she needed me to take her to the hospital. She said it was not urgent at the moment, so I cleaned up my desk and checked out on leave. I had an old 1950 Buick as a second car and while driving home, I smelt something burning. I pulled into a gas station in Mission Valley and opened the hood. Smoke billowed out. I called Betty to tell her about my problem. When I reached her, she said in an urgent voice, "Hurry." I ran back to the car, hosed down the engine compartment, and drove off. Betty was ready to leave when I got home so we got in the other car and headed for Balboa Naval Hospital near the center of the city. We arrived twenty minutes before Lydia Noreen Morin, our fifth and last child, arrived!

I found the AIRPAC Staff to be an interesting assignment. As a "detailer," I assigned enlisted personnel to the Pacific Fleet aviation squadrons and ships. And, of course, we both enjoyed San Diego and its weather again.

I found the AIRPAC Staff to be an interesting assignment. As a "detailer," I assigned enlisted personnel to the Pacific Fleet aviation squadrons and ships. And, of course, we both enjoyed San Diego and its wonderful weather again. I did have one scare while there: at Betty's insistence I had a mole removed from near my left temple and the doctors diagnosed it as malignant and gave me a 50-50 chance of living six or more months. It turned out the diagnosis was incorrect. Otherwise, the tour proved uneventful.

Sea Duty III

The career patterns of the day had pretty well shaken down so that officers would be given a command tour on their third sea tour. However, as an ex-Aviation Midshipman who had to wait to be commissioned after gaining my wings, I was too junior to be in the running for Executive Officer then Commanding Officer. I had the flight time and the carrier experience, but not the required seniority. I needed another year or so.

Working in Personnel on the AIRPAC Staff, I knew that I would have no problems getting orders to a Fighter Squadron at Miramar. But when my new orders assigned me as the Operations Officer of Carrier Group Nine at Naval Air Station Lemoore near Fresno, CA, I was crestfallen. That meant flying the A4 Skyhawk, an attack bomber! I asked my boss at AIRPAC to pull some strings for me during his visit to Washington. When he returned, he told me that the best thing for me to do was to carry out my orders. I was upset, but little did I realize how important the transfer would turn out to be for me and the impact it would have on the rest of my career.

My family stayed in the house we had bought in La Mesa when I started training because as soon as I finished my new Air Group would depart for a nine-month tour of the Pacific in the USS *Ranger* (CVA-61).[22] I attended Survival School in the mountains east of San Diego as the first phase of my training. In Survival School, we learned how to live off the land, and then we learned how to survive in a prisoner of war camp. During the POW training, the Prison Camp Staff worked us over pretty well. The staff members roughed us up and intimidated us. It was not fun. In one case, they coerced a young enlisted crewman in my class to sign a germ warfare confession. Jim Stockdale was a classmate. Years later the North Vietnamese shot him down and imprisoned him in Hanoi for seven years. Captain, later Vice Admiral, Stockdale won the Medal of Honor for his courage and fortitude in captivity.

Then, I returned to Lemoore and started flying with VA-125, the A4 training squadron. We had an interesting class of old hands and brand new Naval Aviators. I met and trained with three old friends: Gene Sizemore [23] with whom I had received my wings on 7 October 1948; Hank Urban,[24] a shipmate from the *Midway* cruise in 1950; and Paul Peck,[25] whom I had known in flight training. I flew with two nuggets – new Naval Aviators – who became casualties during the Vietnam War. The North Vietnamese shot down and imprisoned one, Bill Shankel. When he returned to the U. S., Bill graduated from medical school, and the last I knew he had rejoined the Navy and served as a doctor. The other nugget, Art Avore, died in action.

Besides a lot of ground training, we flew for about 100 hours in the A4. The training included day and night carrier qualifications. The A4 was the last "small" airplane. You strapped it to your behind and flew it! The A4 which had been designed to be a nuclear weapons delivery aircraft was not a good ship for delivering conventional weapons. We, however, did become surprisingly accurate with it in practice.

In January 1963, we flew aboard the USS *Hancock* (CVA-19)[26] for day and night carrier qualifications. We worked off San

1956
Hungarian Revolution

Following the liberalization of Soviet controls, Imre Nagy took power from a Stalinist leader in Hungary and announced that Hungary would leave the Warsaw Pact. Shortly thereafter, Soviet tanks moved into Budapest and triggered the Hungarian Revolution.

Eisenhower chose not to react.

The Red Army soon regained control. The following year the Communists executed Nagy.

1957
Space Race

The Soviet Union launched SPUTNIK into near space. The launch upstaged U. S. rocket-development efforts and created an illusion of Soviet superiority.

Eisenhower became concerned that with a similar rocket, the Soviets could launch a missile attack on the United States.

Francisco under a low overcast which made the night work seem like we were flying in the black hole of Calcutta. With the water temperature in the forties, we had to wear immersion suits (poopy suits) to protect ourselves in case we wound up in the water. The A4 cockpit was small enough; never mind having to wear a cumbersome, bulky rubber suit. The night work required a lot of training and skill. When I had to land at night, my adrenalin was pumping, but at least like many others I had my carrier experience to fall back on. I felt bad for our Air Force exchange pilot, CPT Vickery, who landed on a carrier for the first time. But the Air Force did not send second stringers on exchange tours. CPT Vickery did fine. Upon completing CQ, we all went our separate ways to our new flying jobs.

Ψ

Ranger sailed before I finished training in the A4 so I flew to Japan to join the ship in Yokosuka, a navy base south of Yokohama.

Ranger sailed before I finished training in the A4 so I flew to Japan to join the ship in Yokosuka, a navy base south of Yokohama. CAPT Bill Eason commanded my new Air Group. The group had six squadrons:

1. VF-91, a fighter squadron, flying the F8 Crusader,
2. VF-96, a fighter squadron, flying the new F4 Phantom,
3. VA-93, an attack squadron flying the A4 Skyhawk,
4. VA-94, an attack squadron also flying the Skyhawk,
5. VA-95, an attack squadron flying the Skyraider, and
6. VAH-6, a heavy attack squadron, flying the big A3 Skywarrior.

As the Air Group Operations Officer, I worked directly for the CAG, but I flew with VA-93 which Bill Gortney commanded.

As the Air Group Operations Officer, I worked directly for the CAG, but I flew with VA-93 which Bill Gortney commanded. Bill had just replaced Dutch Shultz as the CO. His Executive Officer was Ken McArthur with whom I flew many missions a few years later over North Vietnam.[27]

CAG Eason, a decorated World War II pilot, was tough on his squadron commanders.[28] One night at sea, he called a meeting of all his skippers in his stateroom and evidently gave them a tongue lashing. After the meeting, Bill Gortney stormed into my room, paced back and forth, and ranted about the CAG. He went on for ten minutes, vented himself on me, seemingly made himself feel better, and then left my room. After serving as his sounding board, I was limp as a rag.

Billy Phillips, the XO of the Crusader squadron and a colorful member of the Air Group, admitted to everyone that he was the world's greatest fighter pilot. Bill Phillips, in fact, was one heck of an airplane driver, and he flew the F8 around the ship better than anyone I have ever seen. One morning in Yokosuka, his Squadron Duty Officer came to me for help. A squadron sailor had gotten hurt while ashore and the Duty Officer did not know what to do. He did not know where his CO was, and he did not know where CDR Phillips was. Normally, one or the other officer would stay on board during the day. A few days later, seeing Billy Phillips at morning formation (quarters), I asked him: "Why didn't one of you stay aboard?" He let me know immediately, in a thousand words or fewer, that Morin, a Lieutenant Commander, did not talk to him – a Commander, in that tone of voice, particularly since Morin was an attack pilot, a second stringer, and that he was a member of the exalted fighter community. After being soundly chastised, I returned to the Air Group office knowing that the wrath of our hardnosed CAG soon would swoop down on Phillips and his CO and solve the problem. Revenge can make you feel good.[29]

Ψ

After the cruise, the Morin family moved to Lemoore, CA. Lemoore had been a quiet farming town of 2,500 people before the Navy built a master jet base there in 1960. Situated in the tremendous farming area of California, the San Joaquin Valley, Lemoore's extremely hot and dry weather in the summer differed totally from the cool coast of California. The temperature stood at about 110 degrees Fahrenheit when we drove into town for the first time. At first, we judged Lemoore not to be much of a town: it certainly was a far cry from San Diego, and the Morin

1959
CUBA

FIDEL CASTRO AND HIS FOLLOWERS SEIZED POWER IN CUBA AND MADE CUBA THE FIRST COMMUNIST STATE IN THE WESTERN HEMISPHERE.

THE UNITED STATES SEVERED DIPLOMATIC RELATIONS WITH CUBA. PRESIDENT EISENHOWER GAVE APPROVAL TO THE CIA TO PLAN FOR AN ATTACK BY ANTI-CASTRO FORCES.

1960
U2 INCIDENT

THE RUSSIANS SHOT DOWN AN AMERICAN U2 SPY PLANE OVER THE SOVIET UNION. THE INCIDENT BECAME A CAUSE CÉLÈBRE AND UNDERMINED A PROJECTED SUMMIT CONFERENCE BETWEEN PRESIDENT EISENHOWER AND SOVIET PREMIER NIKITA KHRUSHCHEV.

family was not impressed. None of us thought that when we moved three years later we would miss the place.

The Lemoore natives were a pretty clannish bunch. The economy was purely agricultural and a large percentage of the people were of Portuguese descent. The arrival of 6,000 service people and their dependents from all over the United States changed their quiet, conservative existence...

We found a nice new rental house in town and settled into an interesting neighborhood filled with Navy families. The Urban family with six children lived three doors up on one side, Bob Schwoefferman with five children lived next door, and Harry Zenner - a shipmate from the *Ranger* cruise - with six children lived nearby.[30] Other families in our neighborhood had two, three, and four children. In the evening, when the children played touch football or rode their bikes in the street, the street looked like a tenement district in New York City or Chicago. All we needed were fire hydrants squirting water.

The Lemoore natives were a pretty clannish bunch. The economy was purely agricultural and a large percentage of the people were of Portuguese descent. The arrival of 6,000 service people and their dependents from all over the United States changed their quiet, conservative existence, and it took awhile for the locals and the Navy folk to adjust to each other.

A large number of Navy people lived on base and sent their children to the public and parochial schools in town, others attended local churches, and many joined Lemoore service organizations, so the locals soon found out that we were normal people after all. Sports also helped break down barriers. Lemoore High School, which had played sports in the lowest category before the Navy arrived, profited from the influx of Navy dependents. Suddenly, with a host of new athletic boys, the school began cleaning up on nearby Hanford, Tulare, and Visalia. While we lived there, Lemoore High School gained a fullback, the son of a black enlisted man at the air station, who became known as "The Tank" as he ran over the opposing teams which for years had wiped their feet on Lemoore. And the Morin children got extra attention at the local parochial school because the Irish nuns came from the same part of Ireland as their grandmother.

Lemoore changed all of us too.

Lemoore changed all of us too. We found ourselves reading the farming section of the Fresno *Bee*, and paying attention to new subjects: alfalfa, the big money crop; the defoliation of the cotton crop; making raisins; and learning about melons that we had never heard of before.

Ψ

Back on shore, we prepared for our next cruise to the Pacific. At the time, the Navy re-designated all Carrier Air Groups as Carrier Air Wings. We, however, decided not to call the Wing Commander - "CAW," so the term "CAG" remained.

As the Wing Operations Officer, I went aboard different carriers as we conducted refresher training during day and night operations. During one of these sea periods, I witnessed the practice landing of a U2 spy plane with a tail hook.

While ashore, the wing occupied offices in a hanger at the Naval Air Station. I had an office next to the CAG. One day, CAG Eason called in my neighbor, Harry Zenner, the skipper of the Skyraider squadron, and chewed him out because of an article in the local paper about his squadron. I heard every word because the partition between my office and the CAG's office was so thin. When the CAG had finished, Zenner walked into my office and said in a loud voice; "How can you stand that a——e?" Once again, an angry skipper dumped his frustrations on me.

Early in 1964, the Navy selected me for promotion to Commander and soon after that I found out that I had screened for command. Being screened for command was big news in Naval Aviation. Failing to screen for command reduced the chance for further promotion to a minimum. Shortly thereafter, I received orders to the Silver Foxes of VA-155 as the Executive Officer. CDR Tom Russell commanded VA-155 at the time. CDR Jack Harris succeeded him.[31] After the change of command, our Wing, Carrier Wing 15, moved aboard the USS *Coral Sea* (CVA-43) [32] to train at sea for our upcoming deployment to the Pacific in December.

Ψ

1961
BERLIN

A THIRD BERLIN CRISIS DEVELOPED AFTER KHRUSHCHEV DEMANDED IN JANUARY THAT THE WESTERN POWERS "SEVER TIES TO BERLIN" AND VOWED "RESOLUTE MEASURES" IF THEY DID NOT. PRESIDENT KENNEDY MET KHRUSHCHEV IN JUNE IN VIENNA, BUT THE TWO FAILED TO RESOLVE THE GROWING CRISIS. ADVISORS TO THE PRESIDENT TOLD HIM THAT THE U. S. COULD NOT STOP THE SOVIETS WITH CONVENTIONAL FORCES AND THAT THE PENTAGON "SHOULD BE ASKED TO SUBMIT A PLAN FOR GRADUATED NUCLEAR RESPONSE."

AFTER THE CRISIS ENDED, THE U. S. REVISED ITS NUCLEAR STRIKE PLAN. KHRUSHCHEV THEN DECIDED TO INSTALL MEDIUM RANGE BALLISTIC MISSILES IN CUBA TO GAIN SOME "REAL MILITARY LEVERAGE IF HE DECIDED TO MOVE AGAIN AGAINST BERLIN.

1961
BAY OF PIGS

IN APRIL, CUBAN EXILES SUPPORTED BY THE CIA INVADED CUBA AT THE BAY OF PIGS TO TOPPLE CASTRO. THE INVASION FAILED BECAUSE PRESIDENT KENNEDY CALLED OFF THE PLANNED SECOND AIR STRIKE, CASTRO QUICKLY DEPLOYED LARGE FORCES AGAINST THE INVADERS, AND THE CUBANS DID NOT RISE UP AGAINST THE DICTATOR. FOLLOWING THE INVASION, THE U. S. DEVELOPED OPERATION MONGOOSE TO DESTABILIZE CUBA AND POSSIBLY ASSASSINATE CASTRO, AND CASTRO SOUGHT SOVIET AID.

Ed Dickson

Ed Hiebert, the Safety Officer, investigated the incident. He reported:

"After the unplug [from the tanker] the fuel entered the engine through the check intakes and caused the engine to fail. There was no reported explosion.

Rather than eject over the High Sierras in the winter, Ed turned to Nevada hoping to make it over the mountains and into the flat lands of western Nevada. He made it over the real high stuff but ran out of altitude near Monument Ski area.

Ed reported to me that he ejected at about 500 feet above the trees at approximately 200 knots. He got seat separation but no chute. The chute did not open because it was set by regulation not to deploy above 10,000 feet.

I determined later with the help of Douglas engineers and their computer that if the chute had deployed that it would not have had time to deploy fully and the streaming chute would have steepened Ed's descent angle and it is unlikely he would have survived...

The rest of the flight thought he was dead until Ed was able to shoot his little pencil flare device for all to see. The flight leader called Fallon Nevada Naval Tower who telephoned the Reno Tower who called the Mammoth Ski Lodge who dispatched a snow cat in the direction of the circling planes. It took the cat about an hour to reach Ed..."

During this training period, I learned about a remarkable incident that took place while a flight of our aircraft practiced in-flight refueling while returning to Lemoore from Fallon, NV. As the flight of A4s flew high over the snow-covered Sierras, the pilots formed up to refuel. When his turn came, LTJG Ed Dickson flew up to the tanker and inserted his fueling probe into the "basket" that trailed behind the tanker. Ed took fuel aboard and then began backing out of the basket so that the next aircraft could refuel. As he backed out, the probe pulled free properly, but a valve in the basket failed to close properly. Consequently, Ed's A4 ingested fuel through its engine intake and that triggered an immediate explosion. Ed immediately ejected from his aircraft.

Once free of the A4, Ed had the option of pulling the D-ring on his parachute harness to deploy his chute, or if he were hurt or incapacitated, he could wait for the barometric device in the parachute to deploy it around 10,000 feet. Ed did not pull his D-ring and long before he dropped to 10,000 feet he hit the side of a snow covered mountain and schussed down on his back. He landed without being injured, but then he ran into a pine tree and broke his left femur. Luckily, two skiers had seen the explosion and Ed's ejection, and they found Ed and brought him to safety. With a broken femur and dressed in his flight suit, Ed would not have survived many hours in the snow at that altitude.

Ed Hiebert concluded that if Dickson's automatic device had been set for a higher altitude and his parachute had deployed, Dickson most likely would have been killed slamming into the mountain just after the parachute opened.

According to a comrade, Ed became more and more religious after his incident. He also became a celebrity. The local newspaper wrote a story about him, complete with a photo of Ed in hospital. The national press picked up his remarkable story, and one or more of the national television stations planned to have him on programs. So many outlets wanted to highlight him that the Navy gave him the option of staying behind to be on programs or deploying with the squadron. Ed healed quickly and chose to sail with us on the cruise.[33]

Ψ

One Sunday, while the USS *Coral Sea* operated off the coast of California, our skipper, CAPT Pierre Charbonnet,[34] brought us alongside an oiler to take on fuel. I was on the wing of the bridge getting some ship handling experience at the time. Carriers normally have Protestant and Catholic Chaplains aboard, and on Sundays the Padres helo to the other ships working with the carrier to conduct religious services. Our Catholic Chaplain was Father Evan J. Greco. The wonderful priest had come into the Navy from the Franciscan order. It was not unusual to see Father Greco conducting services in a monk's hooded robe and wearing sandals. This Sunday, Father Greco had gone aboard a destroyer that was taking on fuel on the starboard side of the oiler. Wanting to return to the carrier and seeing the helo on the carrier's deck, the good Father called CAPT Charbonnet over the phone line that ran between the replenishing ships and the oiler and asked "Captain, how will I get back to the ship?" Without blinking an eye, Charbonnet replied: "Put on your sandals and walk across!"

I had joined a good squadron, staffed with pilots whom I came to admire greatly and whom I count as good friends to this day. They formed a full bell curve of quality as aviators and officers, but they jelled and formed an excellent squadron. Each pilot was a character in his own right and each had a wonderful sense of humor. Because of the spontaneous humor in the squadron, I started a "Quotable Quote" Logbook to record each humorous incident. When the Navy decommissioned VA-155 in the 1970s, I contacted the last CO and he gave me the Logbook to retain for posterity. Whenever we gather for a reunion, we sit around, read from the Logbook, relive the stories, and laugh out of control at our antics of years ago.

We sailed in early December for what we expected to be a

1962
CUBAN MISSILE CRISIS

IN SEPTEMBER, THE U. S. S. R. SIGNED AN AGREEMENT WITH CUBA TO PROVIDE ARMS, TECHNICAL ADVISORS, AND MEDIUM RANGE BALLISTIC MISSILES WITH NUCLEAR WARHEADS, AND FOREIGN MINISTER ANDREI GROMYKO ANNOUNCED IN THE UNITED NATIONS THAT ANY U. S. ATTACK ON CUBA WOULD MEAN WAR WITH THE SOVIET UNION.

IN MID-OCTOBER, THE CIA PRESENTED INFORMATION THAT SOVIET BOMBERS HAD BEEN DELIVERED TO CUBA AND SECRETARY OF DEFENSE MCNAMARA TASKED THE JOINT CHIEFS OF STAFF TO DEVELOP A PLAN TO INVADE CUBA AND REMOVE CASTRO. WHEN U-2 PHOTOGRAPHS SHOWED MISSILES IN CUBA, THE PRESIDENT AGREED TO A NAVAL BLOCKADE OF CUBA AND THE ARMED FORCES BEGAN TO DEPLOY FORCES FOR THE BLOCKADE AND THE POSSIBLE INVASION. THE JCS WANTED TO STRIKE; KENNEDY DEMURRED.

EVENTUALLY, KHRUSHCHEV AGREED TO REMOVE THE MISSILES IF KENNEDY REMOVED U. S. MISSILES FROM TURKEY. KENNEDY AGREED.

THE *QUID PRO QUO* AGREEMENT ENDED THE CRISIS AND AVERTED A POSSIBLE NUCLEAR WAR BETWEEN THE SUPER POWERS.

"ALTHOUGH CDR MORIN WAS A PRESENCE AT THE PARTIES, HE ALWAYS SEEMED TO DISAPPEAR AS THINGS WERE GETTING EXCITING."

normal WESTPAC cruise, but little did we know that we would not return until the following November. En route to Hawaii, the ship developed an engineering problem that laid us up in Pearl Harbor for four weeks. This meant that the USS *Constellation* (CVA-64), the ship we were to relieve so she could be home for Christmas, had her cruise extended. The *Connie* sent us a Christmas card that showed Santa Claus with an arrow through his heart and a caption that read: "You killed our Santa Claus." We knew we were to be gone for Christmas so we felt bad that our problem had denied the officers and men of the *Connie* their Christmas at home.

In order to maintain our flying proficiency while the ship was being repaired, the Air Wing flew ashore to Naval Air Station Barber's Point on Oahu. Because of the alleged North Vietnamese torpedo boat attacks on two of our destroyers in the Gulf of Tonkin[35] during the summer of 1964, we flew many missions at night under the control of a destroyer south of Oahu. The controller aboard the destroyer would vector us to a point where we dropped flares to light up an "attacker" so that we could attack it. This turned out to be tough flying. We flew in poor weather and at low altitudes, and we shifted constantly from visual flying to instrument flying.

One of my young Silver Foxes recalled his "comfortable" time at Barbers Point:

> *[We] flew a very relaxed schedule of navigation, tactics, and some practice bombing, strafing, and rocket shooting. It was the kind of lifestyle that the young pilots and others found to be a very harmonious fit with how they thought pilots should live. Not only was the pace very comfortable, but even the senior officers, CDR Morin in particular, were in evidence on the party scene. They allowed that we should enjoy ourselves because it wasn't likely to be quite this way once we went back out to sea. Although CDR Morin was a presence at the parties, he always seemed to disappear as things were getting exciting. Upon reflection, I believe this was practiced and conscious behavior.*[36]

Another sea story by our Navigation Officer highlighted a different twist to our pleasant stop-over.

> *During the Operation Readiness Inspection, I got to feel the wrath of Morin because of my own breach of a collateral duty. One of my collateral tasks was to function as the Squadron Navigation Officer. I worked in the Operations Department of the squadron because I was the Landing Signal Officer. Unfortunately for me, I considered the task of Navigation Officer somewhat beneath my abilities ... When called upon to furnish some specific charts during the ORI, I found myself unable to do so. In the privacy of his room on the ship CDR Morin, explained in a less than gentle manner that my job here was more than hopping in an airplane, showcasing skills, and collecting at the bar from losers. There was, in his view, a professional series of tasks to be performed and not only was his disappointment in me acute, but any recurrence of this malfeasance would result in something more than words. When I left his room I was somewhat surprised the paint was still adhering to the bulk heads. I personally decided not to test what the result of a further occurrence would be.*[37]

Finally, with the repairs completed, the *Coral Sea* sailed west in early 1965.

Ψ

By the time we sailed on, I had completed nineteen years of naval service, had been a Naval Aviator for sixteen years, an officer for fifteen years, and was serving as a Commander. I had over 3,000 hours of flying time in fighters and attack aircraft and had made over seven hundred carrier landings. Now a confirmed "attack pilot," I expected to be one for the foreseeable future. I looked forward to being the XO of VA-155 and then taking over as the CO from CDR Jack Harris. I had completed four cruises of six months or longer: two in the Mediterranean with the Atlantic Fleet aboard the USS *Midway* in 1950 and the USS *Wright* in 1951 and four in the Pacific with the Pacific Fleet. I made my first Pacific cruise in 1954 in the USS *Wasp*; the second in 1956 aboard the USS *Lexington*, and the third in 1963 aboard the USS *Ranger*. While at sea, I had sought and gained ship handling experience on the carriers, something that served

1962 Results

Three results emerged from the Cuban Missile Crisis that would affect the conduct of the war in South Vietnam.

1. JCS not Trusted.

Kennedy no longer trusted his military advisors, stating: "Watch the generals and avoid feeling that just because they were military men their opinion on military matters were worth a damn."

2. Use Force to Communicate.

McNamara believed that because of the apparent success in the crisis that the United States could communicate with opponents using force or could use gradual force as a negotiating tool.

3. McNamara became the Chief Strategist.

McNamara thought himself to be the chief Pentagon strategist.

I LEARNED THE MOST ABOUT FLYING FROM BOB COATS, MY SECOND SKIPPER, AND THE MOST ABOUT BEING AN OFFICER FROM MY SQUADRON COMMANDERS AND CAGS.

me in good stead many years later.

I had served with some World War II and Korean veterans, but unlike some of my flight school classmates such as the late Jesse Brown, I had not made the Korean War and had no combat experience. I had served under some good squadron commanders, CAGs, and carrier captains, and had learned a great deal from them. I learned the most about flying from Bob Coats, my second skipper, and the most about being an officer from my squadron commanders and CAGs. I also HAD served under some poor aviators and officers. I learned a great deal from them also: namely, what not to do. But most importantly, I had worked hard to be a good Naval officer and a good pilot.

As we headed west, I regretted that our family would not be together for Christmas. Betty Anne once again faced a trying separation, and she had to manage our home and raise our five children by herself. Fortunately, she had lots of friends and other Navy wives at Lemoore who shared her situation. At least this time no one was in diapers. Our eldest son, Jim, going on thirteen, was finishing Junior High School. Our eldest daughter, Sarah, age eleven, our second son, John, just ten, and second daughter, Mary Margaret, age eight, were in elementary school. Our youngest daughter, Lydia, was four and not in school yet.

Looking back, I could hardly believe that I had completed over half of my Navy career as we headed toward combat.[38]

Δ Under the Nuclear Umbrella

By the time Jim Morin headed for Southeast Asia, there had been four hot wars in Asia. The Communists had prevailed in China and North Vietnam, Great Britain and the Philippines had defeated their Communist insurgencies, and the United Nations had kept South Korea free but had to settle for an uneasy armistice in Korea.

In these wars, certain characteristics stood out:

1. Strong, focused leaders such as General Templar in Malaya and Secretary of Defense Ramon Magsaysay in the Philippines had brought victory.

2. Political and economic corruption as in China had undermined efforts to defeat Communist insurgents.

3. In all the countries except Korea, the local Communist Parties had organized before World War II and gained strength during the war fighting against the Japanese.

4. During the wars, the U. S. S. R. had directly aided the Red Chinese and the North Koreans.

5. And, during the wars, the Communists had fought until they gained final victory or government forces had defeated them on the ground. Only in Korea, did the Communists accept an indecisive result.[39]

Δ

Before Morin sailed west, the U. S. armed forces had become nuclear forces. The Air Force led the way with its nuclear-capable bombers in the new Strategic Air Command: initially the B29 used in World War II; then the B47 Stratofortress which had limited strategic range; and finally the B52 which could strike at unlimited distances. In addition, the Air Force had added intercontinental ballistic missiles (ICBMs) to its inventory after the Atlas missile flew in 1957.

To fight the Soviet Red Army, the Army had developed the Honest John Rocket in 1953 along with the so-called Atomic Cannon and then had replaced them with newer and longer ranged solid fueled missiles. Small yield atomic demolitions entered the Army's arsenal in the 1950s and had been deployed to Europe and Korea.

The Navy began developing sea-borne missiles and missile ships after World War II. In 1955, it deployed the sub-sonic, 500 nautical mile, nuclear-capable missile, the Regulus I on the heavy cruiser, USS *Los Angeles* (CA-135). The Regulus I could fire a forty to fifty kiloton nuclear warhead or a one to two megaton thermonuclear warhead. By 1956, the Navy had three Regulus-carrying cruisers in the Pacific Fleet and one in the Atlantic Fleet. In addition to these surface ships, the Navy had equipped ten aircraft carriers to launch Regulus I missiles.

The Navy also had begun developing missile-firing submarines after World War II. In 1947, it launched a German V-1 type rocket from the deck of a submarine. By 1953, a converted World War II submarine, USS *Tunny* (SSG-282), test fired the Regulus I in the Pacific. Three years later, another converted World War II missile submarine joined the Atlantic Fleet. By 1958, the Navy had added two more deck-firing missile submarines to the Fleet, and facing a greater Soviet threat moved the four missile submarines to the Pacific where they joined the

1963
"THE POLITICAL VIRTUOSO"

"THE WHITE HOUSE, IT IS SAID, HAS A STRETCHING, A STRENGTHENING EFFECT ON ITS OCCUPANT AND ALREADY ITS MARK IS UPON THE NEW PRESIDENT.

THERE IS IN LYNDON JOHNSON AN INNER CALM, A CERTAINTY OF PURPOSE WHICH ONLY GREAT POWER BESTOWS ON THOSE CAPABLE OF RECEIVING IT. THE SMALL VANITIES, THE OCCASIONAL PETULANCE, THE INORDINATE PREOCCUPATION WITH WHAT MEN SAY OR WRITE ABOUT HIM ARE GONE, AT LEAST FOR THE PRESENT. THERE IS ABOUT HIM NOW AN AURA OF FULFILLMENT WHICH IN ALL HIS PRIOR YEARS IN OFFICE NEVER MARKED THE MAN...

UNQUESTIONABLY, PRESIDENT LYNDON JOHNSON HAS MADE A STRONG AND WIDELY ADMIRED BEGINNING TO HIS ADMINISTRATION."

JOHN STEELE
LIFE MAGAZINE,
DECEMBER 1963

three missile cruisers.

...as CDR Morin sailed west, he faced fighting an important war under the nuclear umbrella of the two superpowers.

With the launching of the USS *Nautilus* (SSN-571) in 1954, the world's first nuclear-powered submarine, the Navy entered a new era. Nuclear-powered submarines provided many advantages: a larger platform with more space to store large missiles, greater speed, longer submersion times, greater stealth, and the ability to fire missiles while submerged. By 1961, the nuclear-powered USS *Halibut* (SSGN-58) began serving with the four other nuclear-carrying missile submarines in the Pacific. At the same time, the Navy dispatched a new nuclear submarine, USS *George Washington* (SSBN-598), to the Atlantic carrying the newly developed solid-fuel Polaris missile, an ICBM.

As the *Coral Sea* left for Southeast Asia in December 1964, the USS *Daniel Boone* (SSBN-629) began the first Polaris patrol in the Pacific. By that time, the Regulus submarines had completed forty WESTPAC patrols.[40] Also, by that time the Air Force and Navy had added nuclear-capable fighter bombers such as the F105 and the A4 to their inventories.

Δ

During these years, Korea remained divided between the Communist North and the free South, but armed clashes frequently erupted along the Demilitarized Zone. Red China remained solidly Communist and a large part of the Communist bloc. Nationalist China continued to survive on the island of Formosa.

Also, during these years, the Soviet Union had developed formidable nuclear missile and submarine forces along with nuclear-capable ground forces to match and counter the nuclear forces that the United States had developed and deployed around the world.

Δ

As CDR Morin sailed to Southeast Asia, he and the aviators on the *Coral Sea* faced fighting an important limited war under the nuclear umbrella of the two superpowers.

END NOTES

[1] Martin, *America and Its Peoples,* II, 753 (Korean War and Armistice). 753 (China and atomic bomb), 763 (New Look), 764 (Hungarian Revolution), 765 (Suez Crisis), 779 (Cuba), 788 (Bay of Pigs), 788 (Cuban Missile Crisis), 791 (JFK assassination).
[2] *Ibid,* 739.
[3] *Ibid.*
[4] Bernard C. Nalty, *The Vietnam War* (NY: Barnes & Noble, 1998).
[5] Bert H. Cooper, Jr., "MALAYA (1948-1960)," in Major J. W. Woodmansee, Jr. (ed.), *History of Revolutionary Warfare,* Volume IV, *The British Experience – "Keeping the Peace,"* (West Point, NY: Department of History, 1970), 26. High Commissioner Henry Gurney died in an ambush on 6 October 1951.
[6] *Ibid,* 27-39.
[7] Robert Ross Smith, "The PHILIPPINES (1946-1954)," in Major J. W. Woodmansee, Jr., (ed.), *Revolutionary Warfare,* Volume VI, *Three Experiences Since World War II; Greece, The Philippines, and Cuba* (West Point, NY: Department of Military Art and Engineering, 1967), 67.
[8] For Senator McCarthy's comment, see Martin, *America and Its Peoples,* II, 747. For the Venona comment, see Haynes and Klerr, *Venona,* 9.
[9] Swanson retired as a Rear Admiral.
[10] Kidd retired as a Commander.
[11] Named for the Battle of Midway which took place in June 1942. The U. S. lost the USS *Yorktown* during the battle, but Japan lost four of her six fleet carriers.
[12] Aviators on all the aircraft carriers I served on were given the same opportunity.
[13] Coats served with VF-17 (F6F Corsairs) which was embarked on the USS *Hornet.* He and his squadron commander, LCDR Marshall U. Beebe, became aces on the same day. Beebe's squadron was credited with 161 victories and twelve aces from the *Hornet.* VF-17 had a two-tour total of 313 victories. Coats retired as a Captain, USN.
[14] Knudson retired as a Commander.
[15] Each young pilot in the squadron had a collateral job where he learned how to do "officer work" in administration or personnel, in maintenance, in safety, or in some other section.
[16] This was the second *Wasp.* The first was sunk during World War II.
[17] The impact of family separations is being highlighted in 2010 with the increase of divorces and marital problems of veterans returning from Afghanistan and Iraq. Family separations have always triggered problems for spouses and their children; they will continue to cause them as service members deploy on extended assignments overseas, both in peace and war. Nothing can really prepare people for the impact of separations on their marriages and on their children.
[18] Named in honor of the Doolittle Raid. President Roosevelt announced that the B25s that flew off the *USS Hornet* and bombed Tokyo came from Shangri-la.
[19] Medick retired as a Commander.
[20] This was the second *Lexington.* It had been built during World War II. The first *Lexington,* the largest of the American carriers of World War II, sank following the Battle of the Coral Sea in May 1942.
[21] During my career, I made 963 carrier landings. That was a lot at the time. Today, by the time an aviator is appointed a squadron CO,

SIDEBAR NOTES

Korean War
Department of Military Art and Engineering, United States Military Academy, *Summaries of Selected Military Campaigns* (West Point, NY: 1956), 171-75.

1952 Eisenhower's NEW LOOK
Martin, *America and Its Peoples, II,* 763.

1954 Guatemala
Martin, *America and Its Peoples, II,* 767.

1954 Nuclear Strike
Philip Gutzman, *Vietnam, A Visual Encyclopedia* (London: PRC Publishing Ltd., 2002), 7.

1954 Dien Bien Phu
Howard R. Simpson, *Dien Bien Phu, The Epic Battle America Forgot* (Washington: Brassey's, 1994), 169.

1954 Khrushchev and Mao
"Khrushchev Remembers: Part IV, Playing for High Stakes," *Life* Magazine, December 18, 1970, 47.

1955 Summit
Martin, *America and Its Peoples,* II, 764-65.

1955 Warsaw Pact
Modern History Source Book for text at http://www.fordham.edu/halsall /mod/1955warsawpact.html., 339-41.

1956 Hungarian Revolution
Martin, *America and Its Peoples,* II, 765. Stéphane Courtois, et. al., *The Black Book of Communism, Crimes, Terror, Repression* (Cambridge, MA: Harvard University Press, 1999), 339-341.

1957 Space Race
Martin, *America and Its Peoples,* II, 766-68.

1959 Cuba
Various sources.

1960 U2 Incident
Martin, *America and Its Peoples,* II, 767-68.

1961 Berlin
Fred Kaplan, "JFK's First Strike Plan," *The Atlantic Monthly,* October 1961, found in http:// www.theatlantic.com/past/issues/2001/10/kaplan.htm (4/17/2010).

1961 Bay of Pigs
Martin, *America and Its Peoples,* II, 786-7. "Bay of Pigs," John F. Kennedy Presidential Library and Museum. http:///www.jfklibrary.org /Historical+Resources /JFK+in+History/ JFK+and=the+bay +Pigs...(4/17/2010).

Ed Dickson
Ed Hiebert Paper, subject, "Safety Investigation," no date (possession of JHB).

1962 Cuban Missile Crisis.
H. R. McMaster, *Dereliction of Duty, Lyndon Johnson, Robert McNamara, the Joint Chiefs of Staff, and the Lies that Led to Vietnam* (NY: HarperCollins Publishers, 1997), 24-32.

1962 Results
McMaster, *Dereliction of Duty,* 28 (JFK) and 62 (McNamara).

1963 "Political Virtuoso"
John L. Steele, "The Political Virtuoso Gathers the Forces to Take on the Job," *Life* Magazine, December 13, 1963, 32-3.

he normally has more than a 1,000 carrier landings.

[22] This was the second aircraft carrier named *Ranger.* The first had been sunk during the naval battles off Guadalcanal in World War II.

[23] Sizemore retired as a Rear Admiral.

[24] Urban retired as a Captain, USN.

[25] Peck retired as a Rear Admiral.

[26] Named for John Hancock, the President of the 2nd Continental Congress, a noted New England smuggler and the principal signer of the Declaration of Independence.

[27] Eason, Gortney, and McArthur retired as Captains, USN. Schultz retired as a Vice Admiral.

[28] Captain Eason who would serve as the Executive Officer on the USS *Coral Sea* while Morin was with VA-155, began his naval career in 1942 when he entered Naval pre-flight training. Commissioned an Ensign and a Naval Aviator in August 1943, Eason earned two Distinguished Flying Crosses and four Air Medals while serving in the Pacific Theater. He also earned a Navy Unit Commendation Ribbon. Eason graduated from the Armed Forces Staff College, the General Line School, and other Navy schools before becoming the CAG. After leaving the *Coral Sea,* he joined the Office of the Chief of Naval Operations in Washington. (Untitled source in possession of JHB.) Eason retired as a Captain, USN.

[29] Phillips retired as a Captain, USN.

[30] Schwoefferman and Zenner both retired as Captains, USN. Schwoefferman is deceased.

[31] Russell retired as a Rear Admiral. CDR Jack Harris died as a result of the fire on the USS *Oriskany.*

[32] Named for the Battle of the Coral Sea. This was the first battle between opposing carrier groups. The U. S. Navy lost the USS *Lexington* as a result of the battle; Japan lost a smaller carrier.

[33]Conversation with Ed Hiebert, Houston, 23 July 2010.

[34] Pierre Numa Charbonnet, Jr. retired as a Vice Admiral in 1978. He graduated from the U. S. Naval Academy in 1941. He earned a Distinguished Service Medal, a DFC, five Legions of Merit, Presidential Unit Citation, and Navy Unit Commendation Ribbon, *1993 USNA Register of Alumni,* #14494, 257. According to an untitled source in possession of JHB, he commanded Fighter Squadron 24 and Carrier Air Group 24 during World War II in the Pacific, earning two Distinguished Flying Crosses and seven Air Medals. A graduate of the Naval War College and the Advanced Management Program at Harvard, he commanded Fleet Oiler USS *Marias* (AO-57) during the Cuban Missile Crisis before assuming command of the *Coral Sea.* When relieved of command of the *Coral Sea,* he became the Chief of Staff of Carrier Division Seven.

[35] This was the famous or infamous Tonkin Gulf Incident that resulted in the Tonkin Gulf Resolution.

[36] Bill Kish, *Unpublished paper,* no date. 2-3.

[37] *Ibid.*

[38] Morin, *Manuscript,* 11-21.

[39] Characteristics of Cold Wars 1964. John H. Bradley, "Lecture on Cold War," Glasscock School for Continuing Education, Rice University, April 2010.

[40] Submarine information came primarily from Edward C. Whitman, "Regulus, America's First Sea-borne Deterrent," at http://www.navy.mil/navydata/cno/n87/usw/issue 11/regulus.html (6/26/2010).

Five

Vietnam

AIR WAR VIETNAM I

Precursors[1]

Important Events		James Morin
NVN forces captured Dien Bien Phu. Ended French suzerainty in Vietnam.	May 1954	About this time, Morin served with VF-111 at Miramar, NAS, CA.
Vietnam partitioned at 17th Parallel. Over 1,000,000 refugees - mostly Catholics - left NVN for SVN.	Aug 1954	
Southeast Asia Treaty Organization (SEATO) formed.	Sep 1954	Morin sailed in USS *Wasp* to the Pacific.
Ngo Dinh Diem established the Republic of Vietnam.	Oct 1955	
U. S. closed its Consulate in Hanoi.	Dec 1955	
	May-Dec 1956	Morin sailed in USS *Lexington* to Pacific.
U. S. Advisors assigned down to regimental level in South Vietnamese Army.	May 1959	
Viet Cong (VC) raided Bien Hoa and killed two American Advisors - the first U. S. casualties of the war.	Jul 1959	
John Kennedy inaugurated President. Robert McNamara confirmed as Secretary of Defense.	Jan 1961	
Premier Nikita Khrushchev declared that the U.S.S.R. will back all "wars of liberation" around the globe.	6 Jan 1961	
Soviets deployed IRBMs to Cuba, triggered the **Cuban Missile Crisis**.	**Oct 1961**	
U.S. Air Force pilots deployed to SVN.	Oct 1961	
First American ground troops, U. S. Army 39th Signal Battalion, arrived in SVN.	Feb 1962	
	c. Jan 1963	Morin joined USS *Ranger* for WESTPAC cruise.
SVN generals executed President Diem and his brother and took control of SVN government.	1-2 Nov 1963	
President Kennedy assassinated. VicePresident Lyndon Johnson becamePresident. McNamara remained Secretary of Defense.	22 Nov 1963	
U. S. established Military Assistance Command-Vietnam (USMACV).	Apr 1964	About this time, Morin joined VA-155.
GEN William Westmoreland assumed command of USMACV.	Jun 1964	
NVN patrol boats allegedly attacked two U. S. destroyers in the Gulf of Tonkin.	2-4 Aug 1964	
First U. S. Navy aircraft struck NVN. NVN shot down two; captured LT Everett Alvarez, USN.	5 Aug 1964	
Congress passed the Gulf of Tonkin Resolution.	**7 Aug 1964**	
	7 Dec 1964	**USS *Coral Sea* and VA-155 departed for Southeast Asia and a combat deployment.**

Cold War/Hot War 1964-1965

The large "Hot War" in Southeast Asia against North Vietnam challenged Jim Morin and the Silver Foxes.

Δ

FOLLOWING CUBA, PRESIDENT KENNEDY INCREASED HIS SUPPORT OF PRESIDENT NGO DINH DIEM IN VIETNAM.

Following Cuba, President Kennedy increased his support of President Ngo Dinh Diem in Vietnam. He had sent 3,205 military advisors to Vietnam in 1961, then had 11,500 in country in 1962, and had 16,300 in Vietnam in 1963.

In November, believing that Diem could not do the job in Vietnam, Kennedy and/or his advisors consented to the removal of Diem. The JCS opposed the action. During the coup, Vietnamese generals murdered Diem and his powerful brother. Just about three weeks later Lee Harvey Oswald, who had spent time in the Soviet Union and Cuba and had married a Russian, assassinated President Kennedy in Dallas.

The situation continued to deteriorate in South Vietnam in the months that followed.

After becoming president, Lyndon Johnson told his advisors that he was not going to be "the president who saw Southeast Asia go the way China went."[2] Therefore, he continued Kennedy's policies, but he was not devoted to the task. Basically a domestic politician, he pushed his domestic programs more vigorously than the war.

Events soon swayed him. The alleged attack by North Vietnamese patrol craft on U. S. destroyers in August 1964 led the Eighty Eighth Congress to pass the Gulf of Tonkin Resolution "To promote the maintenance of international peace and security in Southeast Asia." The Resolution gave Johnson nearly "carte blanche for the escalation of the war."[3] The Gulf of Tonkin Resolution stated:[4]

> *Resolved by the Senate and House of Representatives of the United States of America in Congress assembled, That the Congress approves and supports the determination of the President, as Commander in Chief, to take all necessary measures to repel any armed attack against the forces of the United States and to prevent further aggression...*
>
> *Whereas naval units of the Communist regime in Vietnam, in violation of the principles of the Charter of the*

WAR IN VIETNAM, IN FACT, WAS AN EXTENSION OF THE AMERICAN POLICY OF CONTAINMENT OF THE SOVIET UNION AND ITS SATELLITES.

> *United Nations and of international law, have deliberately and repeatedly attacked United Stated naval vessels lawfully present in international waters, and have thereby created a serious threat to international peace; and*
>
> *Whereas these attackers are part of deliberate and systematic campaign of aggression that the Communist regime in North Vietnam has been waging against its neighbors and the nations joined with them in the Collective defense of their freedom; and*
>
> *Whereas the United States is assisting the peoples of southeast Asia to protect their freedom and has no territorial, military or political ambitions in that area, but desires only that these people should be left in peace to work out their destinies in their own way:*
>
> *Now, therefore be it Resolved by the Senate and House of Representatives of the United States of America in Congress in Congress Assembled, that the Congress approves and supports the determination of the President, as Commander in Chief, to take all necessary measures to repel any armed attack against the forces of the United States and to prevent further aggression...*
>
> Johnson, however, failed to use his powers effectively.

Δ

Carl Clausewitz, the 19th Century German military theorist, said that "war is an extension of policy by other means." War in Vietnam, in fact, was an extension of the American policy of containment of the Soviet Union and its satellites. Clausewitz, however, could not have meant that American political leaders could use combat operations to "send messages" to an enemy, messages that they assumed the other side would interpret according to their American paradigm. But that is what the American leaders did.

Historian H. R. McMaster noted:

> *The attorneys, managers, and analysts failed to consider that Hanoi's commitment to revolutionary war made losses that seemed unconscionable to American white-collar professionals of little consequence to Ho's government. They discounted instinctively advice that questioned their assumption that Ho, when confronted with military action designed to affect his "calculation of interests," would respond as they anticipated. Lieutenant General [Andrew] Goodpaster protested to Secretary McNamara in the fall of 1964, "Sir, you are trying to program the enemy and that is one thing we must*

never try to do. We can't do his thinking for him."

McMaster added that Goodpaster's words fell on deaf ears.[5]

Moreover, Clausewitz surely did not imply that political leaders should micromanage war, but the key American leaders, President Johnson and Secretary McNamara, did just that to the detriment of the fighting forces.[6]

Δ

U. S. forces faced all sorts of restrictions in fighting the war. Ground forces had to observe "rules engagement" that were nearly comical in the Mekong Delta. For example, American troops could not use artillery against Viet Cong attackers unless the enemy fired automatic weapons. Therefore, American troops could not fire on 100 hundred Viet Cong (VC) riflemen firing two or three rounds each, but they could fire artillery at one VC crew firing a machine gun that fired fewer rounds.[7]

But nothing irritated pilots more than the President and the Secretary of Defense selecting bombing targets or Secretary McNamara requiring pilots to attack targets at the same time each day. The following highlighted the problem:

> *[Johnson] made appointments, approved promotions, reviewed troop requests, determined deployments, selected bombing targets, and restricted aircraft sorties. Night after night, wearing a dressing gown and carrying a flashlight, he would descend into the White House basement "situation room" to monitor the conduct of the conflict... often, too, he would doze by his bedside telephone, waiting to hear the outcome of a mission to rescue one of "my pilots" shot down over Haiphong or Vinh or Thai Nguyen. It was his war.*[8]

Johnson even said: "they can't bomb an outhouse without my approval."[9] And it nearly proved true.

The Rules of Engagement for pilots from 1965 to 1968, just as asinine as those for ground troops, included:

> *1. No air strikes within ten nautical miles [NM] of Hanoi and four NM of Haiphong;*
>
> *2. No air strikes within thirty NM of Hanoi and ten NM of Haiphong without White House or Pentagon approval;*
>
> *3. No penetration of the thirty NM buffer zone along the Chinese border;*
>
> *4. No strikes on SAM sites being built by Russian technicians;*

1962-1964
PRE-VIETNAM
JOINT CHIEFS OF STAFF

CHAIRMAN
GEN LYMAN LEMNITZER
GEN MAXWELL TAYLOR

CHIEF OF STAFF, AIR FORCE
GEN CURTIS LEMAY

CHIEF OF STAFF, ARMY
GEN GEORGE DECKER
GEN EARLE WHEELER

CHIEF OF NAVAL OPERATIONS (CNO)
ADM GEORGE ANDERSON

Δ

1965
VIETNAM WAR
JOINT CHIEFS OF STAFF

CHAIRMAN
GEN EARLE WHEELER

CHIEF OF STAFF, AIR FORCE
GEN CURTIS LEMAY (JANUARY)
GEN JOHN MCCONNELL

CHIEF OF STAFF, ARMY
GEN HAROLD K. JOHNSON

CHIEF OF NAVAL OPERATIONS
ADM DAVID MCDONALD

And political leaders went so far as directing pilots to attack bridges, the most difficult of all ground targets, perpendicularly, rather than attacking from more effective directions —

> *5. No retaliatory measures could be taken against the Chinese airfields on nearby Hainan Island from which Chinese MiGs often attacked naval aircraft; and*
>
> *6. No action could be taken against armed Chinese and Soviet ships moving into and out of Haiphong harbor.* [10]

On the Air Force side, a special twist was given to these rules of engagement.

> *Absolutely forbidden targets included any MiG base designated as a sanctuary, a MiG fighter that did not have its landing gear retracted, any MiG fighter not showing hostile intent (no fighter jock ever figured that one out), and any SAM site not in operation. A SAM had to be fired at a U. S. plane before the plane could fire back....* [11]

Early on, the State Department also prevented air forces from using napalm and cluster bombs (CBUs) in the north for fear of causing civilian casualties.

And political leaders went so far as directing pilots to attack bridges, the most difficult of all ground targets, perpendicularly, rather than attacking from more effective directions - longitudinally or at a slight angle - because attacking from those directions might cause civilian casualties if the bombs missed the bridges and struck the ground around them. So pilots assumed greater risks to themselves when they attacked bridges in order to prevent civilian casualties. The enemy quickly took advantage of the restriction. Subsequently, they placed their anti-aircraft units up and down river from bridges and forced American pilots to fly a gauntlet of fire to try to destroy a bridge. The American political leaders also required pilots to attack only military trucks at safe distances from inhabited areas.[12]

But these types of bombing restrictions were not new. During the first months of World War II, Britain's Royal Air Force bomber pilots labored under the following absurd rules of engagement: "The Government's 'Revised Instruction' to the Air Ministry in June 1940 insisted that the attack must be made with reasonable care to avoid undue loss of civil life in the vicinity of the target. Their order emphatically stated that if they could not identify their target they were to come home with their bombs." Consequently, some crews failed to bomb and in other cases, "Crew after crew crashed attempting the difficult landing with an explosive load aboard." Heavy casualties resulted. In one case, air officials disciplined a crew for bombing a steel factory because they could have killed civilians.

Bombing became a political as well as a military matter. President Franklin Roosevelt, at the beginning of the war in Europe, appealed "to the belligerents to renounce the bombing of civilian targets." Britain accepted his proposal. Later, Roosevelt ignored his own appeal as did Britain when Prime Minister Winston Churchill permitted Royal Air Force Bomber Command to begin its war-long campaign of area bombing. [13]

Attack pilots such as Morin flew with these restrictions and in the course of the war the Navy lost 257 A4s, most of them lost to ground fire.[14] With so many military restrictions, the American leaders, regrettably, surrendered any chance that the air forces (and the ground forces) could have gained a devastating victory over the North Vietnamese. The fact that Morin and his pilots fought so well and flew so many missions is a testament to their sense of duty and loyalty to each other.

During their tour in 1965, VA-155 pilots and their comrades participated in several major air operations.

- On 7 February, FLAMING DART I sent the pilots against enemy barracks and facilities at Dong Hoi, North Vietnam.
- Three days later, FLAMING DART II sent them to attack targets in the panhandle of North Vietnam.
- In April, STEEL TIGER operations deployed the Navy pilots against targets in Laos.
- In August, OPERATION IRON HAND directed them to destroy SAM sites in the north.
- In September, the Navy pilots began attacking targets in the ROUTE PACK VI area near Hanoi and Haiphong.[15]

Navy pilots flew under the command of Navy Task Force 77.

While Navy pilots flew these missions, U. S. Air Force pilots concentrated on attacking targets in South Vietnam and then began OPERATION ROLLING THUNDER on 2 March 1965 when they attacked targets in North Vietnam. The first Air Force strike by thirty F105s from Thailand and other assorted aircraft from South Vietnam proved costly. The North Vietnamese shot down seven aircraft that day. The Air Force rescued three F105 pilots.[16] Soon Navy pilots joined Air Force pilots in OPERATION ROLLING THUNDER.

As the war continued, American pilots faced ever more effective anti-aircraft (AAA) and missile fire over all targets. While the Soviet SAM missiles proved dangerous and Soviet-supplied MiG fighters pounced occasionally, ground fire from automatic weapons and anti-aircraft guns caused most of the air losses. The restricted rules of engagement enhanced the effectiveness of all the gun and missile fires.

Jim Morin would fight in this manufactured environment.

1965
ROLLING THUNDER

SECRETARY MCNAMARA DEFINED THE PURPOSE OF THE OPERATION:

"US OBJECTIVES ARE NOT TO DESTROY OR OVERTHROW THE COMMUNIST GOVERNMENT IN NORTH VIETNAM. THEY ARE LIMITED TO THE DESTRUCTION OF THE INSURRECTION AND AGGRESSION DIRECTED BY NORTH VIETNAM AGAINST THE POLITICAL INSTITUTIONS OF SOUTH VIETNAM."

Δ

BREAKOUT & COMMAND OF AIR OPERATIONS

ROUTE PACKAGE 1

CG, USMACV
SAIGON

ROUTE PACKAGES II, III, IV, AND VIB

COMMANDER, TASK FORCE 77
YANKEE STATION

ROUTE PACKAGE V AND VIA

CG, USAF
THAILAND

B52 OPERATIONS

CG, STRATEGIC AIR COMMAND (SAC)
OMAHA, NE

1965
FLAMING DART I

On 7 February 1965, Task Force 77 ordered a three carrier strike – named FLAMING DART I - against two targets in North Vietnam.

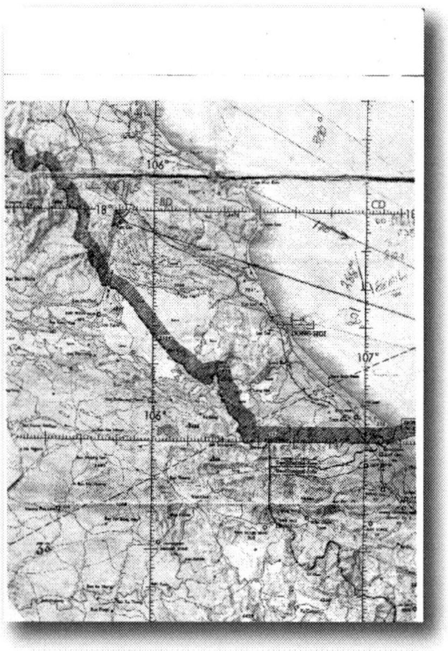

Dong Hoi
Flight Map of LTJG J. B. Mattly

Ψ
Yankee Station

South East Asia was heating up so CINCPAC[17] ordered the USS *Coral Sea* to that part of the world. President Lyndon Johnson was looking for reasons to get the United States more involved in the growing conflict to stiffen the backbone of the South Vietnamese and somehow punish the North Vietnamese for supplying and backing the Viet Cong in the South. He got an excuse when the Viet Cong killed and wounded some U. S. Army Military Advisors during an attack on Pleiku in the Central Highlands of South Vietnam.[18]

On 7 February 1965, Task Force 77 ordered a three carrier strike – named FLAMING DART I - against two targets in North Vietnam. The Strike Force Commander assigned specific targets to the attack squadrons from Air Wing 9 in USS *Ranger*, Air Wing 15 in USS *Coral Sea*, and Air Wing 21 in USS *Hancock*. We commenced planning our tactics and air wing coordination immediately. Then, following weather and intelligence briefings in the Intelligence Center, the strike force pilots returned to their respective Ready Rooms to finish the final briefings and to wait for the command from the ship: "Pilots man your planes."

My squadron, VA-155, had an unusually high percentage of pilots who were Catholics. While we waited to be called away, Father Evan J. Greco walked in and announced: "OK guys, anyone wanting battlefield absolution, let's go to the back of the Ready Room." The Catholics going on the mission got up and joined the padre. The Protestants and those of other or no persuasions turned around to watch Father Greco do his thing. Shortly thereafter, we were ordered to man up.[19]

The weather at YANKEE STATION, the area in the South China Sea where Task Force 77 steamed, was clear and hot. The weather in the target area, however, was predicted to be poor. The weather in South East Asia is affected by the monsoon. From about mid-April to mid-October the weather over

NVN can be clear to the moon while in SVN there is a lot of thunderstorm activity. From November to April, the reverse is true: NVN is usually under a low overcast while the South is relatively clear.

On this day, the strike forces from each carrier launched, rendezvoused, and headed into the targets. The *Coral Sea* and *Hancock* attack squadrons "were to conduct a retaliatory air strike"[20] against the barracks area and the port facilities at Dong Hoi, a small city about twenty miles north of the border between SVN and NVN. To avoid radar detection, we ran in low. I was so low that I was afraid I was going to put my wingman, LTJG Jerry Chasko, who was stepped down on me, in the water.

As we approached the coast, we could see that we would have to go in under a low ragged overcast which was at about 800 to 1,200 feet. Visibility was poor. That complicated our navigation to the target. Twisting and turning and trying to get oriented and find the target, made flying tough for a gaggle of aircraft close to the ground. Flying under the clouds and with lousy visibility, our coordinated tactics all went to hell.

By sheer luck, I came up on an athletic field that I recognized from the briefing and found the barracks that I had been assigned to hit dead ahead of me. I climbed to just under the clouds and dropped my bombs. Being so slow, I was amazed that our own fragments did not hit us. With my bombs away, I immediately turned to an easterly heading to get back out to sea. Flying out, the situation appeared chaotic. I saw A4s everywhere. Some were attacking enemy PT boats in the river. Enemy 37mm gun emplacements from the port and the boats in the river threaded the sky with their red tracers.

Jim Mattly, one of our young pilots, with other aircraft and FLAK[21] to worry about, commented "I hope I don't hit some damn bird."

A later report accurately stated that, "The weather conditions forced the attacking aircraft to remain in the area of concentrated anti-aircraft and small arms fire for an extended period of time to ensure the success of the strike."[22]

Dong Hoi

"The flight seemed pretty much like a normal hop except for the number of aircraft involved. The North Vietnam coast was what I expected it to be like. As we approached the target area, the first two groups of aircraft said that they had missed the target and were aborting their runs. For this reason our flight had to make two 360 degree turns to allow for the other flights to reenter. It was during one of these turns that my section leader said he thought he was hit. I moved up and looked his aircraft over, but saw no visible damage.

We proceeded to the target itself. We were flying 4 aircraft abreast... On signal from the leader we all started pickling our bombs. About the time my section leader pickled his first bomb, his port wing just outboard of the wheel well caught fire. I notified him of this and he continued to drop the rest of his bombs before turning to sea and away from the target area.

The last I saw of the aircraft was as I followed it into the clouds. When I came out on top, it was not there. I found out later that the pilot had ejected into the sea. This was seen by the rest of the flight.

I was about to make the necessary emergency transmissions concerning the ejection when I heard the division leader doing so. After learning this I proceeded to rendezvous with the tanker aircraft as my fuel was getting low. ...

The inflight refueling and the remainder of the hop were uneventful."

LTJG J. B. Mattly

Ed Dickson

Posthumously awarded the first of 120 Navy Crosses awarded to U. S. Navy personnel during the Vietnam War.

Postscript

Ed Hiebert recalled that the North Vietnamese released a photograph of Ed Dickson's body and that there appeared to be a gunshot wound above Ed's right eye.

If it were a gunshot wound, an enemy soldier must have shot Dickson while he descended in his parachute or after he landed.

On the radio, we heard something about a Silver Fox getting hit, but until we regrouped and counted noses we could not be sure. As it turned out, LT Ed Dickson, who had survived the ejection over the Sierras, had been shot down and killed in action. CDR Pete Mongilardi, Commanding Attack Squadron 153, reported that he "heard Lieutenant Dickson call that he had been hit but was pressing on with the attack." He added after he completed his attack that "I heard Commander Jack Harris tell Lieutenant Dickson that he was on fire and to prepare to eject." Mongilardi wrote: "We all said a silent prayer for his safe recovery."[23] Harris confirmed the action:

> *On our run in, Lieutenant DICKSON'S aircraft was hit by anti-aircraft fire. His aircraft caught fire as we started hitting the targets. Knowing he was on fire, Lieutenant DICKSON stayed in formation until he had dropped his bombs, then headed for sea. His aircraft was burning fiercely when I told him to eject. We saw him eject but lost sight of him in the clouds before we could see his chute open.*[24]

A few days later, CDR Harris informed the squadron that "We have received information via Chinese news broadcasts that the North Vietnamese have recovered Lieutenant DICKSON's body."[25]

Ed became the first pilot killed in action in 1965 – the only one while I served with the squadron. He was to be followed by many others, mostly attack pilots, in the months and years ahead. The Navy awarded Ed a posthumous Navy Cross.[26]

U. S. Air Force pilots, launching from Danang in South Vietnam, were to strike into NVN right after us. Because of the weather, they aborted and attacked the following day. Shortly thereafter, the *Stars and Stripes* newspaper carried an article about the attack by the Air Force. Photos showed a general right after the completion of the mission pinning Air Medals on the pilots and Air Force Commendation Medals on the enlisted men who loaded the ordnance. That was a far cry from the Navy's parsimonious attitude about awards which plagued us all through 1965. But that is its own story.

As a side note, I immediately tried to get LTJG Chasko awarded an Air Medal for the Dong Hoi strike. The command did not approve the award, but the command awarded the Armed Forces Expeditionary Medal - not a Navy Commendation Medal - to all the members of the squadron.[27]

Months later, after evaluating the circumstances surrounding the loss of Ed Dickson and after having flown many more attack missions, I wrote the following to the Commander of the U. S. Naval Aviation Safety Center:

> *During peace time operations a damaged aircraft which is on fire is normally abandoned immediately. In combat this action is not practical over enemy territory.*
>
> *We have observed that this fact is not always considered by new combat pilots. Peace time training instills peace time pilot reactions, which are not always applicable in combat. This squadron lost a pilot ... possibly because he ejected from his burning aircraft before it stopped flying. We had not weighed the advantage of reaching friendly areas, risking the possibility of an explosion, against falling into enemy hands. It was felt at this time that a burning aircraft would explode. Since this time, CVW-15 and other air wings have flown burning aircraft hundreds of miles... To my knowledge not one of these aircraft have exploded.*
>
> *The fires have either extinguished themselves or depleted the flammable material. It has been observed that burning fuel from a ruptured wing, although a frightening sight, does not burn in the wing but [burns] one to two feet aft of it. As soon as the fuel is exhausted the fire ceases with little fire damage. Hydraulic fires normally are short lived due to the small amount of fluid available. Residual fires do not burn well above 15,000 feet.*

Morin concluded: "The A4 seems to be able to take an extremely large amount of battle damage and still make it home" and "Having experienced the foregoing this squadron has adopted the philosophy of staying with the aircraft until well into friendly territory or until the aircraft is physically uncontrollable. We feel that the odds of survival are in the pilot's favor."[28]

Ψ

APRIL 1965 CHAIN OF COMMAND

COMMANDER, VA-155 SQUADRON
USS *CORAL SEA*
CDR JACK HARRIS

COMMANDER, CARRIER WING 15
USS *CORAL SEA*
CDR H. P. GLINDEMAN

CAPTAIN, USS *CORAL SEA*
CPT GEORGE CASSELL
YANKEE STATION

COMMANDER, CARRIER DIVISION ONE (TASK GROUP 77.5)
RADM EDWARD OUTLAW
YANKEE STATION

Ψ

COMMANDER , 7TH FLEET
VADM PAUL P. BLACKBURN
JAPAN

COMMANDER, PACIFIC FLEET
VADM ROY L. JOHNSON
HONOLULU

COMMANDER-IN CHIEF, PACIFIC (CINCPAC)
ADM ULYSSES S. GRANT SHARP
HONOLULU

CG, 7TH AIR FORCE
THAILAND

B-52 OPERATIONS

CG, STRATEGIC AIR COMMAND (SAC)
OMAHA, NE

1965
FLAMING DART II

AFTER THE 7 FEBRUARY STRIKE, THE VIET CONG ATTACKED A BARRACKS IN QUI NHON IN SVN AND KILLED AND WOUNDED MORE AMERICANS.[29] PRESIDENT JOHNSON ORDERED ANOTHER ATTACK ON NVN FOR 11 FEBRUARY, FLAMING DART 2.

Air-to-ground attack is the name of the game in carrier aviation. The fighters protect the strike groups and the ships of the force, but it is the attack people who have to put the ordnance on target. And as mentioned, the attack pilots suffered the most casualties.

Until the Atlantic Fleet carriers started showing up, nearly all the losses - killed, missing, or captured - came from one little farming town in the San Joaquin Valley. When the Base Captain's sedan showed up at a school or in a neighborhood, it meant that another Dad had been shot down. The children and wives would see it and hold their breaths. The townspeople of Lemoore, loyal Americans, realized now the seriousness of this foolish war, long before any other place in the U. S. did. The relationship between the town and the Navy became one of cooperation, admiration, and respect. The small size of a town like Lemoore magnified the impact of the losses. As the A4 squadrons from NAS Cecil Field in Jacksonville, FL and the A6 squadrons from NAS Oceanna in Virginia Beach, VA became involved, those communities also got to feel the pain of the losses.

To try to help the families at home, I wrote the following in the February edition of *The Silver Fox News*:

> *Many of us have learned through experience that some of our shipmates put information in their letters that they shouldn't. This includes classified information, wild rumors, out and out distortions or lies, and tales of wild liberty on the part of some shipmates. Wives at home are anxious for their husbands out here; those letters are most important to them. But wives talk, they expand the rumors or lies, and they love to hear gossipy tid-bits about so and so's carousing on the beach.*
>
> *If you're not sure if some information is classified or not, ask your shop Petty Officer, or any of the CPOs, or your division officer. If your story is a rumor or you're fabricating a story to glamorize it or because you've got a grudge – don't do it! And, finally what one of us does on the beach is our own business and not for public consumption back at Lemoore. I've seen many cases where someone writes home about someone else, when actually the writer is one of the world's biggest beach rats and he's trying to justify his own actions. Don't do it."*[29]

After the 7 February strike, the Viet Cong attacked a barracks in Qui Nhon in SVN and killed and wounded more Americans.[29] President Johnson ordered another attack on NVN for 11 February, FLAMING DART 2. The three carriers planned the strike on a supply depot and barracks at Chanh Hoa. After the mission briefing, we all went to our Ready Rooms for the final brief and waited to be called away. Once again, Father Greco walked in and motioned us to the back of the Ready Room. Our non-Catholic shipmates, feeling the loss of Ed Dickson just four days earlier, said, "Wait a minute, we want a piece of the action" and came back with us for absolution and a pre-mission prayer.

I cannot remember a great deal about this attack. The weather was not a factor; I cannot remember any anti-aircraft fire. But I do remember being impressed with the accuracy of the Skyraider squadron (VA-165) led by CDR Ken Knoizen.[31] We knocked down a bunch of flimsy buildings and returned to our ships. We lost one F8 Crusader from *Coral Sea* on the mission. Those buildings were not worth the cost of the F8. The pilot, Bob Shumaker,[32] ejected safely. The North Vietnamese captured him and he became the first pilot POW in 1965. Bob spent eight years in North Vietnamese prison camps.

We flew no missions for ten or so days. Then, we began flying into Laos to run the roads on which the supplies flowed into SVN from NVN. We were not supposed to be flying over Laos, a neutral country, but we did. There were a lot of silly rules we would encounter in this war.

President Johnson wanted to punish the North Vietnamese some more; so in early March 1965, we began the ROLLING THUNDER missions. We flew a series of retaliatory strikes against selected targets in NVN to convince the North Vietnamese leaders that backing and supplying the Viet Cong in the South were foolhardy things to do. As the number of strikes slowly increased, we leaked off the North Vietnamese and provided them with the back-bone "stiffner" that we were trying to provide to the South Vietnamese. The North Vietnamese became a brave and tenacious people.

1965 Laos

TWO SILVER FOXES WERE ACCUSED OF MAKING A NAVIGATION ERROR AND BOMBING AND KILLING SOME FRIENDLY LAOTIANS.

LAOS WAS TECHNICALLY NEUTRAL AND BECAUSE OF THE POLITICAL CONDITIONS AT THE TIME, THE COMMAND GROUNDED BOTH MEN AND THEN CONVENED AN ARTICLE 32 INVESTIGATION TO DETERMINE WHETHER TO PREFER COURT- MARTIAL CHARGES AGAINST THE TWO.

THE OTHER SILVER FOXES, CDR MORIN, AND CAPTAIN CASSELL SUPPORTED THE TWO PILOTS. APPARENTLY, THE TASK FORCE 77 COMMANDER, THE CAG, AND THE SQUADRON CO DID NOT. ONE OF THE PILOTS STATED, "CDR MORIN WENT SO FAR AS TO SUGGEST THAT THIS WHOLE THING STUNK," BECAUSE THE PILOTS "WERE DOING THEIR JOB AS BEST [THEY] COULD," AND "WHY COULDN'T THE PEOPLE AT THE TOP SHOW SOME LOYALTY."

THE BOARD OF INQUIRY EXONERATED THE TWO PILOTS.

WHEN ONE OF THE PILOTS COMPLAINED TO CDR MORIN ABOUT THE CONDUCT OF THE TASK FORCE COMMANDER IN THE AFFAIR, MORIN "RESPONDED THAT SINCE I HAD BEEN LOAFING FOR TWO WEEKS I WOULD GET TO FLY EXTRA MISSIONS TO CATCH UP BECAUSE WE HAD A SQUADRON POLICY THAT EVERYONE SHOULD HAVE THE SAME OPPORTUNITY FOR HEROIC ACTION."

1962-1963
Problems at the Top

President Kennedy distrusted the JCS he inherited from President Eisenhower. He considered the Chiefs "more an impediment than an asset." He particularly disliked GEN LeMay, the Air Force Chief, because LeMay insisted on using bombing to solve military problems.

Kennedy pushed out GEN Lemnitzer, the JCS Chairman, because he did not like his advice to use "full force" before intervening anywhere, and he pushed out GEN Decker, the Army Chief of Staff, because Decker, a World War II Pacific Veteran, told McNamara "we cannot win a conventional war in Southeast Asia." McNamara got ADM Anderson, the CNO, replaced because Anderson told him during the Cuban Missile Crisis that "the Navy had been conducting blockades since the days of John Paul Jones and suggested that McNamara return to his office and let the Navy run the operation." McNamara told Anderson that the operation was "not a blockade but a means of communication between Kennedy and Khrushchev." Then, McNamara and Kennedy ran the blockade.

Kennedy brought retired Army Chief of Staff Maxwell Taylor back as his Military Advisor and then made him the Chairman of the JCS. The JCS soon suspected that Taylor did not accurately present their views to Kennedy. Moreover, Taylor clashed with LeMay, whom he considered "politically naive," LeMay, in turn, hated Taylor.

No one could figure out the rhyme or reason for the targets we had to strike. The admirals on the carriers and the Fleet Commander did not designate them. Sadly, they had no input. The President and his political staff half way around the world in Washington selected the targets, and President Johnson and Secretary McNamara micromanaged the air war. Their micromanagement dismayed us because we knew that it was a costly and ineffective way to fight an air war and it added to our many frustrations as we continued flying into North Vietnam.

Saigon

In early 1965, Task Force 77 sent a senior aviator from one of the carriers on Yankee Station to serve with the Air Division Staff, U. S. Air Force at Tan Son Nhut airfield in Saigon. Officers rotated in for a month at a time. My turn came in mid-March. If the Air Force wanted the Navy to fly sorties against targets in SVN, the Task Force 77 Rep would make the necessary arrangements. My counterpart with the Air Division Staff, an Air Force colonel, felt that SVN was an Air Force air war and he did not need any help from the Navy. Begrudgingly, he would give us targets when he could not supply enough Air Force sorties, or if the targets were too far away.

At this time, the Viet Cong offered a bounty for the capture of U. S. military officers so I never felt safe once I was off the Air Base. I kept a pistol in my pocket and a sawed-off carbine in my brief case. I had seen plenty of John Wayne movies, so I was ready.

Meanwhile, my squadron mates were flying missions up North, and I wanted to get back with them. While I was in Saigon, the North Vietnamese shot down my CO, Jack Harris, but a submarine rescued him. Jack, later, died in the fire on the USS *Oriskany*.

When it was time for me to leave Saigon, the *Coral Sea* sent its mail plane (COD, Carrier on Board Delivery) to pick me up. It arrived around noon and I was at Base Operations with all my gear ready to go. The pilot said we had a 6:00 p.m. recovery time at the ship so there was no rush. Then, he told me that he and the copilot wanted to go into town to buy some bush hats and porcelain elephants. I knew we would have to leave around 3 p.m. to make our recovery time and when they had not even shown up by that time, I got antzy.

They arrived at 4 p.m., and both had consumed too many beers. I had never flown this twin-engine prop aircraft but I told the young copilot to go in back and sleep, and that I would fly copilot. The pilot, in his stupor, started the engines. We completed the checklist and I told him that I would take it off and fly it, and to just back me up. We got airborne and started the long run to the ship. The pilot, mercifully, fell asleep.

Near 8 p.m., just after sundown, I came up on the ship and checked in. Arriving so late, however, the ship was not ready for us. We were told that they would take us aboard in about thirty minutes by which time it would be dark. I was not sure this guy could hack a day landing, never mind a night landing, so I told the ship: “Negative, thirty minutes. I have a pilot emergency and must come aboard immediately.” I got the expected answer, “ROGER, WAIT!” In a minute, the ship called and said, “YOUR SIGNAL, CHARLIE.” The pilot, by now he was awake, said, “OK, I got it.” He lowered the wheels, flaps, and hook and he brought that thing aboard as if it were nothing. He could fly better under the influence than I could cold sober. I was so appalled by his total lack of professionalism, however, that I went right to the bridge and reported to CAPT George Cassell[33] and explained why I had insisted on landing immediately.

April-May 1965

Secretary McNamara’s program of applying increasing pressure on the North Vietnamese to make them cry “uncle” completely flopped. As I said, all it did was make the enemy mad. As the days wore on, we increased our number of sorties and we attacked targets further up into NVN, into what we

1964-1965 LBJ and the JCS

President Lyndon Johnson inherited all the personal distrust and antagonisms between the members of the JCS , between the military leaders and Secretary McNamara and his whiz kids, and between the military leaders and many other Kennedy officials who remained in the Cabinet and in key advisory positions.

Like Kennedy, Johnson distrusted the JCS, and “brought with him to the presidency a low opinion of the Nation’s top military men and a long history of taking positions on military issues to enhance his political fortunes.”

President Johnson’s biggest personal flaw was that he wanted advisors to tell him what he wanted to hear, not the truth.

His biggest political weakness was that he was more interested in promoting his programs for his “Great Society” than he was in prosecuting the war.

1965
SURFACE TO AIR MISSILES (SAMS)

BETTY WROTE AND ASKED ME WHY WERE WE LETTING THEM BUILD THEM. SHE COULD NOT UNDERSTAND WHY WE DID NOT HIT THEM BEFORE THEY WERE OPERATIONAL. NEITHER COULD WE. BECAUSE WE DID NOT DESTROY THEM, WE PAID THE PIPER IN THE MONTHS TO COME.

called "Indian Country." Initially we had good success, and caught trucks and trains out in daylight. We were so successful, in fact, that soon we could not find a thing on the roads in daylight: no trucks, no people, not even an animal.

Besides major strikes on specific targets, we flew road reconnaissance missions around the clock. One carrier would fly midnight to noon; the other noon to midnight. We also flew strikes deep into Laos, day and night, trying to interdict the Ho Chi Minh Trail and other roads that the North Vietnamese used to move troops, equipment, and supplies into South Vietnam. We all noted the futility of the operations: we would crater the holy hell out of the trail only to find out the next day that the North Vietnamese had filled in all the craters during the night. In 1972, a high ranking enemy general commented on the bombing of the trail.

> *Gen Dong Sy Nguyen said that the American bombings even at their peak were effective (meaning that they actually hit provision stores) only 0.19 percent of the time. In other words, of every one thousand bombs dropped, only two hit their targets. This was because bombs dropped from an altitude of 3,000-5,000 meters and at a speed of 400 kilometers per hour were trying to hit a target that was never more than 5-6 meters wide.*[34]

But the Ho Chi Minh Trail was critical to the North Vietnamese. They feared an attack on the trail by ground forces, but they did not fear air attacks, even by B52s.

Ψ

We were specifically forbidden to attack anything that was not an approved target. Secretary McNamara would dispatch a list each week through channels to the carrier commanders telling them what targets to attack and when to hit them.

Several new targets appeared in 1965 which we could not hit: the new SAM sites the Russians began constructing in NVN. *Time* Magazine in May showed photos of the ground-to-air missile sites being built around Hanoi. Betty wrote and asked me why were we letting them build them. She could not understand why we did not hit them before they were operational. Neither could we. Because we did not destroy them, we paid the piper in the months to come.

1965

SURFACE TO AIR MISSILES (SAMs)

A frustrated General William Westmoreland commented on the construction of SAM sites in his memoirs.

> *In 1965, we observed the construction of the first surface-to-air [SAM] sites in North Vietnam, and the military sought permission to attack them before they were completed to save American casualties.*
>
> *Assistant Secretary of Defense for International Affairs John McNaughton ridiculed the idea. "You don't think the North Vietnamese are going to use them!" he scoffed to Seventh Air Force Commanding General Joseph H. Moore. "Putting them in is just a ploy by the Russians to appease Hanoi."*
>
> *It was a matter of signals, said the clever civilian theorist in Washington. "We won't bomb the SAM sites, which signals to North Vietnam not to use them."*[35]

ASSISTANT SECRETARY OF DEFENSE FOR INTERNATIONAL AFFAIRS JOHN MCNAUGHTON RIDICULED THE IDEA. "YOU DON'T THINK THE NORTH VIETNAMESE ARE GOING TO USE THEM!"

It was a good thing that we attack pilots did not know about MacNaughton's comments.

Ψ

As we moved farther up into North Vietnam, the anti-aircraft fire intensified.

At this time, we flew a lot of missions and this meant we worked the carrier's crew and the squadrons' crews to a frazzle moving and loading ordnance. We used up ammo and fuel so fast that the carrier had to replenish every fourth or fifth day. Task Force 77 got so sophisticated at underway replenishing that while the *Coral Sea* would be refueling from an oiler or resupplying from an ammo ship, we would launch and recover aircraft without missing a step.

Harry Thomas, the XO of the Blue Tail Flies of VA-135, our sister squadron, became the Commanding Officer of the Blue Tail Flies after CDR Pete Mongilardi moved up to be the CAG.

Harry had been an Aviation Midshipman with me in training. Harry and the new XO, Dave Leue,[36] another ex-Aviation Midshipman, proved to be innovative tacticians and very brave attack pilots.

TOP TO BOTTOM

- MATTLY
- FIDELIBUS
- BLOSSER
- CDR JACK HARRIS

During this period, the pilots of our squadrons continued to knock down narrow bridges in our target areas even though they found the bridges hard to hit and they faced heavy North Vietnamese anti-aircraft defenses. It was not easy.

VA-153 was a "Bullpup" squadron. The Bullpup was a radio controlled guided missile that, when fired, would rocket out ahead of the aircraft. By observing a flare in the tail of the Bullpup, a pilot could steer the missile to a target with pinpoint accuracy. The North Vietnamese soon figured out that the aircraft firing the missile stayed in its dive to follow and guide the missile. Therefore, they would concentrate all their fire on the aircraft to break the pilot's concentration.

Once our guys noted this, we changed our tactics: we had the firing aircraft launch its missile and then maneuver to draw away the enemy's fire while a trailing A4 would take over steering the missile. We constantly had to modify all our tactics because the North Vietnamese would counter them. Next, we would counter their counter; then, they would counter the counter-counter. And so it went in the deadly cat-and-mouse game in the air.

We flew a lot of missions in May. I flew thirty-six according to my log book and, since we worked at keeping the missions even, I am sure every Silver Fox had about the same number.

During the month, we lost pilots from the carriers, and we learned important lessons on how to survive as we flew our missions:

1. Never duel with a FLAK sight,

2. Do not make a second dive on a target - do it right the first time, and

3. Keep jinking the aircraft when coming off a target in order to spoil the aim of enemy gunners.

Finally, and most importantly, in case you were hit and had to eject, get out over the water as quickly as possible where the Task Force always deployed several destroyers to rescue downed aviators.

CO

When our normal WESTPAC cruise ended, we headed for Yokosuka, Japan for a week in port before heading for the States. We Silver Foxes had survived after losing just Ed Dickson, and we went on the usual shopping spree in Japan to buy things for our wives and children. The ship off-loaded everything that the relieving carrier could use when it came through Japan. On the next-to-last day in port, the command notified all hands that plans had changed and we were going back on the line. The decision had been made to go from a three-carrier war to a four-carrier war, and the simplest way to do that quickly was to turn *Coral Sea* back around. We reloaded all our supplies and then sailed for the Philippines. We knew that there would be a lot of heartbroken families when the news reached them at Lemoore. Of course, we were not too thrilled ourselves.

This decision fouled up a lot of plans. I was to relieve Jack Harris as CO once we returned to Lemoore. Jack had orders to a Carrier Division Staff so we decided to hold the Change of Command Ceremony at Cubi Point in the Philippines before we were to go back on the line in June.

When we sailed to fight again, I commanded VA-155.

Down to fifteen pilots, we faced tough going until we got some replacements. Getting the replacements proved difficult. The Commander scheduled to come into the squadron behind me as XO died in a training flight in California. The Navy ordered an old friend from Memphis days, John Duck,[37] to become my XO, but he had a ways to go yet in the A4 training squadron at Lemoore. When John got to the CQ stage, his aging eyes let him down and he could not night qualify. The Navy ultimately canceled his orders to VA-155.

JUNE 1965 REMAINING VA-155 PILOTS

1. GEORGE BLOSSER
2. GERRY CHASKO
3. BILL FIDELIBUS
4. DICK FLOREANI
5. GEORGE HART
6. ED HIEBERT
7. KEN HOLTHAUS
8. JIM HUGHES
9. ALBIE HYDE
10. BILL KISH
11. ERNIE MARES
12. JIM MATTLY
13. JIM MORIN
14. WENDY RIVERS
15. JAY STARLING

Sidebar Notes

1962-1964 Pre-Vietnam Joint Chiefs of Staff, and 1965 Vietnam War Joint Chiefs of Staff
Varied biographical sources.

1965 Operation Rolling Thunder
Gutzman, *Vietnam, 341.*

Breakout & Command of Air Operations
Nichols and Tillman, *Yankee Station,* 23 (Map).

Dong Hoi
LTJG James B. Mattly, *Narrative statement concerning action against enemy forces during a strike over DONG HOI, North Vietnam, 7 February 1965,* no date.

Ed Dickson
For Navy Cross, see Ronald E. Fischer, "The Navy Cross," *Journal of the Orders and Medals Society of America,* Vol. 45, No. 2, March 1994. (Source of table statistics) and *Coral Sea Website,* http://www.usscoralsea.org/pages/pow/mia.php For photograph information, Ed Hiebert, *Conversation,* Houston, 23 July 2010.

April 1965 Chain of Command
Sea Breeze, January 1966 and various sources.

1965 Laos
Bill Kish, *Letter to John Bradley,* no date. *Conversations with RADM James B. Morin,* 20 October 2010, Tallahassee, FL.

End Notes

[1] Gutzman, *Vietnam,* 7-11

[2] H. R. McMaster, *Dereliction of Duty, Lyndon Johnson, Robert McNamara, the Joint Chiefs of Staff, and the Lies that Led to Vietnam* (NY: Harper Perennial, 1997), 48.

[3] *Ibid,* 108.

[4] *"Transcript of Gulf of Tonkin Resolution (1964),"* www.ourdocuments.gov (7/9/2010).

[5] McMaster, *Dereliction of Duty,* 163.

[6] Joe Patrick, in "Testing the Rules of Engagement," Email message dated 27 November 1999, 1, wrote: "The war in Vietnam was a strange war, indeed. It was a conflict that should not have been lost. But the men who ran that war were politicians and bureaucrats, not military professionals. Men like Secretary of Defense Robert Strange McNamara and President Lyndon Baines Johnson, along with Department of Defense bureaucrats, civilian and military called the shots."

[7] Personal knowledge. JHB.

[8] Martin, *America and Its Peoples,* II, 793

[9] John B. Nichols and Barrett Tillman, *On Yankee Station, The Naval Air War over Vietnam* (Annapolis, MD: Naval Institute Press, 1987), 16.

[10] *Ibid,* 17 (Map) and 17-21. See Chapter 2 for a useful discussion of Rules of Engagement.

[11] Joe Patrick, "Testing the Rules of Engagement."

[12] Nichols and Tillman, *Yankee Station,* 25-6 (napalm and CBUs), 26-7 (bridges), 28 (trucks).

[13] Max Hastings, *Bomber Command,* (NY: The Dial Press/James Wade, 1979), 53 (FDR), 68 (disciple), 89 (restrictions), 90 (failure to bomb), 135 (area bombing).

[14] Nichols and Tillman, *Yankee Station,* 165. The Navy and Marine Corps lost 711 aircraft in flight actions.

[15] *Ibid,* 152-154.

[16] John Harrington, *Email to LTC William B. McGee,* USAFR-Ret., 1 Aug 2010. "Our take off time was 14:04 and our TOT [time over target] was 14:48. Robbie Risner was leading the force of, I believe, 30 F-105s. Other units from South Vietnam flying F-100s, B-57s, and South Vietnamese Air Force A-1s also participated. Seven fighters were lost that day, our 1st day - Three F-105s, all from the 67th, [Spagnola, Baird, and Pannis, all rescued], two F-100s, 1 B-57, and 1 VNAF A-1."

[17] CINCPAC. Commander in Chief, Pacific Area Command. CINCPAC was a unified commander who controlled Army, Navy, and Air Force units. He reported to the Joint Chiefs of Staff in Washington. A naval officer, however, he did not as CINCPAC report to the Chief of Naval Operations. The Commanding General, Military Assistance Command, Vietnam (MACV) reported to the CINCPAC.

[18] Gutzman, *Vietnam,* 316. "The capital of Pleiku province is located in the Central Highlands of II Corps Tactical Zone [Vietnamese Region II]. Located at the junction of north-south National Route 14 and east-west National Route 19, it was an important regional market for the Montagnards. On February 7, 1965 the Viet Cong guerrillas attacked the US helicopter base, Camp Holloway, and Pleiku's airfield, killing nine and wounding 128 Americans."

[19] Chaplain (CDR) George B. Riley served as the Protestant chaplain.

[20] *Letter from Commanding Officer, Attack Squadron ONE HUNDRED*

FIFTY-FIVE to Commander in Chief, U. S. Pacific Fleet, 16 February 1965. Proposed citation for the award of an Air Medal to LTJG Gerald Joseph Chasko.
[21] FLAK was originally derived from the German term, FLAKCANNON.
[22] *Ibid.*
[23] "A Navy Carrier Pilot's Personal Story," *Sea Breeze*, January 1966, 15.
[24] *THE SILVER FOX VA 155 NEWS*, Issue #5, February 1965, 1.
[25] *Ibid.*
[26] Notes from Gerald J. Chasko, 8 July 2004. (Possession of JHB).
[27] See footnote 19 (Chasko) and footnote 17 (medal).
[28] *Letter, Commanding Officer, Attack Squadron 155, to Commanding Officer, U. S. Naval Aviation Safety Center, Subject,* "Combat Mishap Report," no date, 2-3.
[29] *THE SILVER FOX VA 155 NEWS*, Issue #5, February 1965, 2.
[30] Gutzman, *Vietnam*, 327. "Qui Nhon was one of the first US air bases and a major supply base and transportation terminal in II Corps where the east-west National Route 19 intersected with north-south National Route 1 on the seacoast. It was one of five coastal surveillance centers supporting Operation MARKET TIME, the naval interdiction operation along the South Vietnamese coast. A major logistics installation was also located there."
[31] Arthur Kenneth Knoizen retired as a Rear Admiral in 1982. He graduated from the U. S. Naval Academy in 1949. He earned a Distinguished Service Medal, Legion of Merit, Distinguished Flying Cross, Navy Commendation Medal, Navy Unit Commendation Ribbon, and Joint Staff Commendation Medal. *1993 USNA Register of Alumni*, #01006, 300.
[32] Shumaker retired as a Captain, USN.
[33] Cassell retired as a Rear Admiral.
[34] Bui Tin, *From Enemy to Friend, A north Vietnamese perspective on the war* (Annapolis, MD: Naval Institute Press, 2002), 74-75.
[35] Quoted by Patrick in "Testing the Rules of Engagement," 3. See also Nichols and Tillman, *Yankee Station*, 55.
[36] Leue retired as a Captain, USN.
[37] Duck retired as a Captain, USN.

Map Note
Vietnam, 107
Department of History, U. S. Military Academy, *Revolutionary Warfare*, Volume V, (West Point, NY:1970).

1962-1963 Problems at the Top
McMaster, *Dereliction*, 22 (Anderson, Decker, Lemnitzer), 30 (Anderson), 43 (Taylor, LeMay), 50 (LBJ), 52 (LBJ), 61 (LBJ).

1964-1965 LBJ and the JCS
McMaster, *Dereliction,* 52 (low opinion), 61(wanted to hear).

Six

USS Coral Sea[1]

AIR WAR VIETNAM II

Vietnam 1965[2]

Important Events		James Morin
Buddhists disrupted all major cities in SVN.	12-27 Jan 1965	
	23 Jan 1965	USS *Coral Sea* joined U. S. Seventh Fleet and arrived on Yankee Station.
Viet Cong attacked Pleiku.	7 Feb 1965	
Operation FLAMING DART I began.	**7 Feb 1965**	**Silver Foxes conducted first combat strike against Dong Hoi.** **NVN shot down and killed LT Ed Dickson.**
U. S. dependents ordered to leave SVN.	8 Feb 1965	
South Korean troops arrived in SVN.	26 Feb 1965	
Operation ROLLING THUNDER began.	2 Mar 1965	Silver Foxes committed to ROLLING THUNDER operations.
U. S. Marines landed at Danang.	8 Mar 1965	
Operation MARKET TIME - the interdiction of enemy shipping - began.	11 Mar 1965	
	29 Mar 1965	Silver Fox 1, CDR Jack Harris, ejected and was rescued by submarine.
U. S. Army 173rd Airborne Brigade arrived in SVN.	3-12 May 1965	
U. S. Intelligence confirmed that the Soviets had begun building AAA positions around Hanoi and Haiphong.	22 May 1965	
	May 1965	USS *Coral Sea* departed Yankee Station en route to Japan and the U. S.
	28 May 1965	Gordon Byron Hiebert born.
	Early Jun 1965	U. S. *Coral Sea* and VA-155 returned to Yankee Station.
U. S. began Operation ARC LIGHT. B52s bombed VC positions near Saigon.	17 Jun 1965	
	25 Jun 1965	NVN killed CDR Peter Mongilardi, CAG.
173rd Airborne Brigade & ARVN units conducted first joint ground operations.	28-30 Jun 1965	
U. S. Marines conducted Operation STARLIGHT, first all-U. S. ground operation.	18-24 Jul 1965	
	12 Aug 1965	NVN severely damaged LT Bill Fidelibus' A4. Fidelibus ejected & was rescued.
	13 Aug 1965	NVN shot down and killed CDR Harry Thomas of Blue Tailed Flies.
	10 Sep 1965	NVN shot down & captured LCDR Wendy Rivers.
	Sep 1965	LCDR Paul Moore joined Silver Foxes.
	14 Oct 1965	Silver Foxes conducted last air strike on Dong Hoi.
	17 Oct 1965	USS *Coral Sea* departed Philippines for US.
	1 Nov 1965	USS *Coral Sea* arrived San Francisco. Ended 331 day deployment.
1st Air Cavalry Division conducted Ia Drang Valley Campaign against NVA forces on Cambodian border.	14-17 Nov 1965	

Cold War/Hot War Mid-1965

By June 1965, the war had changed. Against the advice of General Maxwell Taylor, the Ambassador in South Vietnam, President Johnson deployed combat troops to South Vietnam in 1965. When the *Coral Sea* returned to Yankee Station, Marine units had landed in Danang in March and the 173rd Airborne Brigade began landing in June.[3] Offensive ground operations soon began and expanded the war.

The first Army-ARVN operation hit thin air; but the first all-American operation, a spoiling attack launched by Marine units - Operation STARLIGHT - against a NVA regiment near Chu Lai, devastated an enemy regiment.[4]

In June, Air Force B52s from Guam bombed Viet Cong areas near Saigon, beginning the first of many sporadic raids by the heavy strategic bombers against targets in South Vietnam and then in North Vietnam.

About the time that the *Coral Sea* left Yankee Station for the United States, the newly organized 1st Air Cavalry Division, formerly the 11th Air Assault Division,[5] arrived in country. In November, the cavalrymen landed by helicopter near Chu Peh Mountain on the Cambodian Border and fought a vicious battle there against larger NVA forces. After a three-day fight for survival, one battalion moved to a planned extraction site and triggered a brutal NVA ambush. The American battalion barely survived. By the time the Battle of the Ia Drang Valley ended in November, both sides had suffered heavy casualtics, and thc *Coral Sea* was home.

JUNE 1965
AIR WING 15

CAG
CDR PETER MONGILARDI, JR.

CO, VF-151
CDR JOSEPH M. SCHNEIDERS

CO, VF-154
CDR WILLIAM N. DONNELLY

CO, VA-155
CDR JAMES B. MORIN

CO, VA-153
CDR HARRY E. THOMAS

CO, VA-165
CDR ARTHUR K. KNOIZEN

CO, VAH-2
CDR ROBERT M. DILORENZI

OFFICER IN CHARGE, VFP-63 DETACHMENT
CDR LEE C. MILES

OFFICER IN CHARGE, VAW 11 DETACHMENT
LCDR CHARLES L. MCGATHY

***KIA**

Ψ

Back on Line

We set sail in the early part of June from the Philippines to begin a lengthy line period. *Coral Sea,* from this time on, fought as the midnight-to-noon carrier a great deal of the time. We flew a lot of missions and, with fifteen pilots, we got tired. En route to Yankee Station, our new Carrier Division Commander, RADM Ralph Cousins[6] visited each Ready Room to be briefed by each squadron commander and to meet the pilots. When I had finished my briefing and answered his questions, he asked me if I

A4 EN ROUTE TO A TARGET

THE YOUNG GUYS WANTED US OUT THERE IN FRONT, AND THE CAGS, COS, AND XOS TOOK THE HARD MISSIONS. AND THE SENIOR PEOPLE PAID A HIGH PRICE AS SHOWN BY THE LARGE NUMBER OF COMMANDERS WHO DIED IN ACTION, SUFFERED WOUNDS, AND BECAME POWS.

had any questions. I told him that the major question was: "When are we going home?" We were, at that point, the Air Wing with the most combat experience. The pace of the war was demanding increased efforts by the pilots, the squadron crews, and the ship's company. I told him a lot was being asked of us.

We would meet the challenge but my people had to have a goal. We had to know how much longer we would be out there. Our families, with the constant worry for our survival, had to have a date also. He answered, "I don't know." And we were not to know for quite awhile which made it tough on all of us. He said, "Everyone back in the States thinks your morale is high. You might as well think so too." To which I replied, "Aye, aye, Sir." He was really a great Carrier Division Commander, concerned for all of us. I returned one night around 3 a.m. from a recce [reconnaissance] mission and went down to the War Room to brief the Staff Duty Officer. There was ADM Cousins and he asked, "How are you feeling, Jim, and how are the young guys doing?"

Ψ

As a CO, I had the leadership task of keeping everyone's spirit up by stressing the importance of what we were doing, and letting them know I was honestly concerned for their health and welfare and for their safe return home. As far as the pilots were concerned, they saw the foolishness of the way the war was being handled by Washington but, to their credit, they did their jobs.

We senior aviators had to lead by example. The young guys wanted us out there in front, and the CAGs, COs, and XOs took the hard missions. And the senior people paid a high price as shown by the large number of Commanders who died in action, suffered wounds, and became POWs. In World War II, enemy gunners aimed at the lead aircraft, consequently the young guys in back took the hits. Now, twenty years later, with improved accuracy, enemy gunners could hit the strike leader. In addition, the volume and density of the anti-aircraft fire we faced increased the chances that the enemy would hit the leaders.

JUNE 1965
SILVER FOX LEADERS

CO
CDR JIM MORIN

XO
NONE

ACTING XO & OPERATIONS OFFICER
LCDR WENDY RIVERS

ADMINISTRATIVE OFFICER
LCDR ERNIE MARES

SAFETY OFFICER
LT ED HIEBERT

MAINTENANCE OFFICER
LT BILL FIDELIBUS

The Silver Foxes

I was blessed with a squadron of great officers and men. By the time I took command, the pilots had flown at least fifty combat missions. I could trust my skilled veterans with any mission. They never disappointed me. They did everything I asked of them, always supported each other, made me look good, and got few rewards for their courage and perseverance.

We got to the point where the Silver Foxes seldom got out of their flight suits. They flew at least one mission every day, many times two, and spared [backed-up] another. With briefing times thrown in, they spent many hours under stress. And some of the pilots performed other duties. Wendy Rivers served as my Operations Officer as well as my XO. Those duties added hours to his days. Ernie Mares, as the Administrative Officer, put in extra hours of work when not flying. So did Bill Fidelibus and George Blosser who oversaw our maintenance.

With fifteen pilots, I could not afford to have anyone grounded. There was no slack. If a pilot fell behind in sorties, I would double him up until he caught up to the average again. I wanted my Silver Foxes to earn all those gold stars for their Air Medals! When Wendy Rivers fell behind because of his heavy dental work, he scrambled to make up his missions. Bill Fidelibus, who got banged up when he ejected, knew it was grim to fall behind in missions, so as soon as he could, sore neck and all, he flew again.

My sailors also worked their butts off. When the ship was flying twelve hours a day, the squadron sailors worked sixteen to eighteen hours a day, and they performed physically demanding work. My plane captain, AN Jenkins, a young, uneducated, but extremely intelligent black sailor from Watts, CA, who had a very quick mind and a terrific sense of humor but could not read, typified so many of our hard-working crewmen.

Deck Crews

"IT'S DIFFICULT TO EXPLAIN THE HECTIC ACTIVITY AND COORDINATION THAT TAKES PLACE ON A CARRIER DURING WARTIME CONDITIONS.

LONG BEFORE THE PLANES WHICH ARE GOING TO BE USED ON A MISSION HAVE RETURNED FROM THEIR PREVIOUS HOP, A NEW GROUP OF PILOTS [IS] BUSY PLANNING, AND BRIEFING. WHEN THE PLANES LAND THEY ARE FIRST MET BY THE EVER PRESENT PLANE CAPTAINS WITH THEIR HEAVY BAG OF CHAINS READY TO LASH THE PLANE DOWN TO THE SWAYING DECK. OUR MAINTENANCE CHIEF, *CHIEF SHELLABERGE* IS ON THE SCENE TO TALK TO THE PILOT BEFORE HE EVEN UNSTRAPS TO FIND OUT WHAT MECHANICAL OR ELECTRICAL WORK HAS TO BE ACCOMPLISHED, IF ANY, BEFORE THE PLANE IS LAUNCHED IN ABOUT TWO HOURS.

BEFORE THE PILOT CAN SECURE THE ENGINE SOME OF OUR ORDNANCEMEN *AUTREY* AO2, *BRADLEY* AO2, *MILES* AO3, *STANFILL* AO3, AND SOME OF OUR "IMPRESSED" ORDNANCE HELPERS *WHITE* PN2, *BOREN* AK3 AND *PATTISON* ATN3, JUST TO MENTION A FEW, CAN BE SEEN WHEELING BOMB LADEN TRUCKS BENEATH THE WINGS PREPARING TO START YET ANOTHER LOAD."

THE FLIGHT DECK FOR THE NEXT SEVERAL MINUTES BETWEEN LAUNCHES IS AN AMAZING SCENE OF ACTION, WITH HUNDREDS OF MEN HURRYING ABOUT THEIR JOBS."

One hot day when we both were tired and probably had a ways to go before we saw a cold beer again, Jenkins strapped me into my cockpit and said to me, "Skipper, there ain't no segregation problem in this squadron. You treat everybody like niggers!" That went into the Quotable Quote Logbook. Unfortunately, Jenkins constantly got into trouble, and the Navy barred him from re-enlisting. I learned later that Jenkins died while still a young man.

Tired and overworked, the only time we aviators and our support crews knew what day of the week it was, was when we heard Church Services being called away on Sunday.

The Grind Continues

The June-July line period lasted over five weeks. On 25 June, the very first day on the line, the North Vietnamese shot down and killed our CAG, Pete Mongilardi, as he flew an armed route recce mission.[7] That was not a great way to start our extended cruise. The senior squadron CO acted as CAG until a replacement could be sent out to us. Our new CAG was to be Wes MacDonald,[8] a former A4 skipper who had been involved in the Gulf of Tonkin incident in 1964.

I flew thirty-eight more missions of all kinds during this period: road recce, missions into Laos, and designated, heavily defended targets in North Vietnam. Because we flew A4Es with a more fuel-efficient engine than our sister squadron with A4Cs, we flew recce routes way up in Northern Laos. At the time, we had another silly rule imposed on us: WE COULD NOT FLY OVER NVN TO GET INTO LAOS. Why? Were we afraid we would be put on report for violating their airspace after we had been putting bombs all over their country? Did someone think that the North Vietnamese did not know that U. S. pilots were trying to bomb their trucks coming down through Laos? From the ship, we had to fly just south of the Demilitarized Zone (DMZ) into Laos before turning north.

I led my usual division with Jim Hughes on my wing,

George Hart as section leader and Jerry Chasko as number 4. Flying from Yankee Station, over the DMZ into Laos, then up into Laos was a long trip. Once into Laos, the sky was full of building cumulus clouds so that we could not see the ground. I used DANANG TACAN as long as I could; then used time and headings the rest of the way. We had to maneuver around some towering cumulus en route to the area. When I figured it was time, we picked our way down through the clouds and came out below the clouds right over the town of Sam Neua, the starting point of the road recce. It was pure luck but my Marine had no doubt that I was the greatest navigator since Magellan. Running the roads in Laos was not easy. What our maps showed as roads were really narrow dirt roads, more like trails, that would disappear in the jungle. We even did this sort of thing at night, dropping flares to try to orient ourselves. Laos's scenery is beautiful, lush green with Karst Mountains jutting up. I never saw a thing there other than, at the end of a road segment, a gunner hidden by the trees who tried to reach us with his mount. We were out of his range so, without any serious threat, we set up a routine training racetrack pattern and saturated the area from which the tracers had come. His buddies were probably mad at him for shooting at us.

Running the roads in the daylight in the North was a waste of fuel and ordnance. It did keep the North Vietnamese from using their roads to move supplies or to repair their bridges but we could have accomplished that with a fraction of the effort. We were generating sorties for the sake of sorties, bean counting. The allegation has been made that we and the Air Force were in a sortie competition. But, as we lightly said, it was experience you could not buy (except at the cost of your life). North Vietnam became as familiar as anyplace we had ever flown over in the United States. At night, the light from the stars reflecting off a river was all that was needed to know where you were.

Ψ

DECK CREWS

THE FUELERS WEARING PURPLE SHIRTS ARE FUELING PLANES. THE ORDNANCEMEN WITH THEIR RED SHIRTS ARE LOADING BOMBS. THE TAXI DIRECTORS WITH YELLOW SHIRTS ARE BUSILY ENGAGED IN REPORTING AIRCRAFT FOR THE UPCOMING LAUNCH. OUR GREEN SHIRT MAINTENANCE PERSONNEL ARE DOING LAST MINUTE REPAIRS TO THE AIRCRAFT. *REESE* AM3 IN HIS SNOW WHITE UNIFORM WILL BE FOUND SERVICING THE AIRCRAFT WITH LIQUID OXYGEN. AND THE EVER PRESENT PLANE CAPTAINS - *GUENTHER* ADJ3, *JENKINS* AN, *OSTENSON* AN, *SANTIAGO* AN, *SMITH ADJ3, TYRRELL ATN3, WHITEHEAD* AO3, *CASEY* AMS3, AND *ZUNIGA* AN ALONG WITH THE OTHER PC'S - WILL BE FOUND CAREFULLY CHECKING OVER THEIR AIRCRAFT FOR ANY BATTLE DAMAGE, AND ENSURING [EACH AIRCRAFT] IS IN 4.0 CONDITION FOR THE NEXT LAUNCH.

WHEN THE PILOTS COME TOPSIDE TO MAN PLANES MOST OF THE HECTIC ACTIVITY HAS SUBSIDED. *CHIEF WADDELL* AND *THOMAS* ABH2 ALONG WITH SOME OF THE TROUBLE SHOOTERS - *ENOS* ATN3, *GERHART* AME3, *WINNETTE ADJ3, AND VUKONICH* AE3 - ARE THEN BUSILY ENGAGED IN SCURRYING FROM PLANE TO PLANE MAKING SURE ALL IS IN ORDER.

FINALLY WHEN THE PLANE IS HOOKED UP TO THE CATAPULT THEY GIVE STILL ANOTHER CHECK BEFORE IT IS SUCCESSFULLY SHOT OFF THE DECK, TESTIFYING TO THE HARD WORK AND EFFORT OF MANY MEN."

SILVER FOX VA-155 NEWS, 2-3
20 MAY 1965

When we flew, the doctors wired us for sound. They measured and recorded our blood pressure, heart rate and, I think, perspiration.

About this time, the Navy's Senior Flight Surgeon and a team of doctors arrived to check on our physical condition and our mental frame of mind. The Flight Surgeon interviewed me, and asked if I had seen any manifestation of combat fatigue among my people. I explained to him about target selection, rules of engagement, bean counting, going against heavily defended targets time and again that were not worth one of my guys and his aircraft, and how all these things frustrated us and made us angry. I told him that if being leaked off all the time characterized combat fatigue, then we all had it.

Coral Sea served as the night carrier while the doctors were aboard. When we flew, the doctors wired us for sound. They measured and recorded our blood pressure, heart rate and, I think, perspiration. The reason I say "think" regarding perspiration is that in that hot, humid climate, strapped into an A4, wearing a harness/survival vest that felt like a blanket, we sweated profusely. I am surprised we did not short everything out. The resultant measurements showed we all had a certain physiological reaction to flying a combat mission. However, once we crossed the beach outbound and set up for the night holding pattern, our approaches and landings, the measurements showed that our minds and bodies really geared up for the most challenging part of our profession: coming aboard at night. The instruments did not record fear, but rather showed that all our faculties registered at 100%.

Letting off steam at the wedding

The Wedding

As was stated earlier, our cruise extension fouled up a lot of plans. Bill Kish planned to marry Nancy Kneeland, a former Rose Bowl Queen, when we returned from our normal cruise. When we were extended, they decided to go forward with their plans. Nancy flew to the Philippines so that the wedding would take place between line periods. Father Greco conducted the ceremony at the Subic Bay Naval Station Chapel. We dressed in our dress whites or civvies and I filled in for Nancy's father and gave the bride away.

It was a happy wedding after which we repaired to the club for the reception. The guys consumed large quantities of champagne and let off lots of steam. Cake ended up all over many faces and not in mouths. Soon, people got tossed into the pool. These college kid antics sound foolish now but, at that time, all the young Foxes had been through trying times and had worked under a lot of pressure, and they saw no end in sight. As Wendy Rivers mused about the cruise extension, "All you have to do is bear it, the grin is optional." From the Philippines, we got a week in Hong Kong which was always one of my favorite ports.

The SAMs

En route back to Yankee Station, we got the word of the first SAM kills. In one twenty-four hour period, 12-13 August, the North Vietnamese bagged a lot of airplanes. We had learned to stay high enough on a road recce to keep the small stuff from hitting us, particularly the 37mm shells which could knock an airplane out of the air. Above 5,000 feet, however, we worried about the 57mm guns because of their rapid rate of fire and the size of the shells. The enemy grouped these weapons around their valuable positions. As we approached and rolled in on these targets, we always saw enemy guns blinking away at us. I had many orange tracers and shells go right by me in my attacks. If the North Vietnamese had had proximity-fused 37mm and 57mm shells that would have sensed a shell's closeness to a target aircraft and exploded, they would have hit a bunch of us, including me. For high altitudes, the North Vietnamese used 85mm guns, but the rate of fire was slow and they had a hard time hitting a fast-flying jet. However, they often fired a carpet of 85s in hopes that we would fly into the saturated area. But we knew nothing about the SA-2 missile. Until we did and until we could formulate appropriate tactics to counter it, we were told to stay low. Staying low put us back down where the small stuff could get us.

1965
SAM Sites

"Surface-to-air missiles appeared early in the North Vietnamese defense net, and remained for the ensuing eight years.

Photorecon planes [one, a *Coral Sea* Crusader] confirmed the presence of SA-2s southeast of Hanoi in April 1965, and the Soviets announced this fact to the world in May.

Despite this hard intelligence, no action was taken against SAM sites until several had been completed. We lost our first plane to the new threat on 24 July 1965 when an air force F-4C was bagged."

Ψ

"By July a ring of SAM sites surrounded Hanoi and part of Haiphong."

To our surprise, we came upon a large number of North Vietnamese, mostly farmers, walking down the road or pushing their bicycles. As we approached them, the surprised North Vietnamese dove into the ditches, and those up ahead of us who were armed started shooting at us.

The SAMs that got our planes did not come from the hard missile sites up around Hanoi but from mobile sites. They would fire at us, lie low, and then reposition during the night. Our only hope was to spot some trails left by their vehicles. That, however, proved hard to do when zipping along down low.

Ψ

One afternoon while conducting a search mission, I decided to take my division over into Laos, proceed north to pick up my route, pop over the mountains, and then run the road eastwards. I had Al Hyde, Jerry Chasko, and Virgil Jackson[1] with me. Jackson, a volunteer temporary replacement from Lemoore, saw the nitty-gritty for the first time.

To our surprise, we came upon a large number of North Vietnamese, mostly farmers, walking down the road or pushing their bicycles. As we approached them, the surprised North Vietnamese dove into the ditches, and those up ahead of us who were armed started shooting at us. Virgil Jackson, seeing all the tracers, commented, "It sure is hot down there!" Jerry Chasko, his section leader who had been on a lot of missions, thought Virgil was referring to the outside temperature, simply replied "ROG."[11] Jerry's casualness under fire impressed Virgil.

As if in slow motion, I saw a man armed with a rifle up ahead of me squatting by the road in a typical oriental squat position. When I passed him, I heard a bang, similar to the sound I would hear if I had fired off one of my own 20mm rounds, and I felt something hit the plane. I told Albie that I thought I had been hit. Then, as soon as I dared, I started climbing and headed for the beach. Al joined on me and made a visual check all around and under me. He reported seeing nothing.

It was now getting dusk, I decided that it was useless to go back and finish running the road, so I told Albie we would dump our rockets on Hon Me Island off the coast. The island was crisscrossed with "W" trenches. Al and I split up and fired from different directions. We felt good when we saw our 2.75 inch rockets hit the trenches. Continuing on, we landed on the ship after dark. After I had taxied forward, I told my Maintenance Chief that I thought I had been hit. About an hour later, he told me that the crew had found nothing. The next day, however, a crewman discovered a thirty-caliber bullet hole in the fuselage. The bullet had hit my ejection seat about a foot below my butt. The bullet impacting the seat had been what I felt when I thought that I had been hit. It was sobering to think that a rifleman could knock a plane down by hitting the pilot or some vital component of the fuel control.

On a similar mission, around mid-August, the North Vietnamese shot down and killed Harry Thomas, the CO of the Blue Tail Flies when he ran across one of their heavily defended areas. Dave Leue, the XO, acted as the CO until Ken McArthur, my shipmate from *Ranger* days, flew out from the States and took command.

APR-23

The radar in the A4 was not very good. By today's attack aircraft standards, it was pathetic. There was one feature of our electronics package that we did not know a great deal about, the APR-23. The radar receiver picked up an enemy radar emission and depicted it as a strobe of light on our small radar scope. The strobe would depict a relative bearing off the aircraft's nose to the enemy radar, but it could not determine the distance to the North Vietnamese radar that was targetting us.

The enemy's SA-2 missile employed a radar, code-named FANSONG, to find targets. The radar would search for an aircraft, and when it found one, the radar antenna stopped rotating and locked on the target. FANSONG then provided bearing, distance, and altitude information to the missile firing site. When the target came in range of the missile, the radar pulses doubled and provided very accurate information to the fire control computer. The missile would be launched and guided by

Navy Aviator Missing in Action Identified

"The Department of Defense... announced today that the remains of a U. S. serviceman, missing in action from the Vietnam War, have been identified...

He is Navy Commander Peter Mongilardi, Jr., of Haledon, New Jersey. He will be buried on April 11, 2007, at Arlington National Cemetery near Washington, D.C.

On June 25, 1965, Mongilardi departed the USS Coral Sea in his A-4C Skyhawk on an armed reconnaissance mission over North Vietnam. His flight encountered bad weather and enemy fire over Thanh Hoa Province, causing the wingman to lose visual and radio contact with Mongilardi. Contact was never reestablished and the aircraft failed to return to the carrier...

His wingman was Paul Reyes. A second 2-aircraft section led by Commander David Leue was briefed at the same time. Cmdr. Leue's wingman was forced to return to the carrier...so David Leue joined Peter Mongilardi's section for the remainder of the mission. This was the first day of flight operations after the ship returned from Japan to Yankee Station.

As the pilots searched out targets of opportunity CDR Leue spotted a power plant through the clouds... Mongilardi ordered him not to strike the plant because it was a denied target... Leue immediately broke of his attack run, then rejoined the flight. As he did so Mongilardi radioed "I'm rolling in on a little bridge," then as he pulled off target, "Flak!" Leue heard the flight leader's aircraft hit by enemy fire as he made his last transmission. The others next heard him take a couple of deep breaths while his mike was still keyed."

radio control to its target.

He listened to us on our frequency as we teased the FANSONG. Later, during our debriefing aboard the ship, Bill told us that he had nothing but the greatest respect for thee attack pilots, crazy as we were.

The SAMs had to have the radar's information and one day Dave Leue and I went hunting for a radar site. We each had a wingman and flew up to the Red River Delta, south of Haiphong. We orbited off the coast, listening. Then, I had the sections take turns being the "bait" for the suspected radar. One section would run in low and fast until a little ways inland, then pop up to a medium altitude to see if it could get the FANSONG to lock on. The other section's pilots, if the enemy radar locked on the "bait" and they, in turn, got a strobe light on their own radars, would head towards the enemy radar site. As they attacked, the "bait" dived for the deck to keep from getting zapped. The attackers always ran in on the last known bearing because they expected that the enemy radar would shut down. Luckily, we flew in an area where there were no gun emplacements. I do not remember how many times we went in and out that day. We burned our fuel fast at that low altitude, and we saw nothing that we could identify as a radar site. I do remember passing a grass hooch a few times out on the flat delta. I figured it was some poor fisherman's house and I did not want to hurt civilians. But looking back on it, that could have been where the FANSONG was. We saw little else out there and we should have smoked it. While we tried to find a target, Bill Donnelly,[12] the F8 squadron CO, orbited off the coast as combat air patrol (CAP) to protect us in case MiGs attacked us. He listened to us on our frequency as we teased the FANSONG. Later, during our debriefing aboard the ship, Bill told us that he had nothing but the greatest respect for the attack pilots, crazy as we were.

The Thanh Hoa Bridge

Indo-China had been a French colony. In the big cities, French influence was evident in the boulevards, buildings, parks etc. Across the countryside, they built a road and rail system and many bridges. As we started knocking down the bridges to stop the flow of supplies, we found one bridge that would not go down, the bridge at Thanh Hoa. Thanh Hoa was a town well south of Hanoi. The bridge there, built by the North Vietnamese themselves, was narrow and hard to hit, and it stood in one of the most heavily defended areas in NVN. I hate to think how many times we went against that bridge and the price we paid going after it. We hit the bridge girders and the bombs ricocheted off and exploded in the river. We changed our bomb fusing to no avail. Bomb damage photography showed we had bent the girders, but the bridge stood. It became a symbol to the North Vietnamese of their resoluteness. The enemy surrounded the bridge with 57mm gun sites which blinked away at us as we dived down to pick up speed before entering our final dives. We always fanned in around the target, the idea being to get in and out as quickly as possible, jinked wildly off target, and headed back out to sea. Before we attacked, our FLAK suppressers attacked just ahead of us.

One day, I led a strike group and Dave Leue led a division of FLAK suppressers. Dave's guys carried 1,000 pound bombs fused to go off above the ground. I followed Dave downhill and he cracked his bombs flush on a site. The guns quit firing at that particular site long enough for us to get in and out. But we did not have a sufficient number of aircraft to go against the large number of sites situated all around the countryside. After countless bombing attacks on the bridge, the area around it resembled a moonscape.

In 1967, we started to use smart bombs. Finally a new 2,000 pound bomb called Walleye damaged the bridge. Each Walleye had a TV camera in its nose. The attacking pilot would identify his target on TV, and tell the bomb, "that's your target." After the pilot dropped the free-falling bomb so it could reach its target, the bomb guided itself to the designated aim point. The bridge, however, had many lives after VA-155 attacked it.

Years later, on 13 May 1972, fourteen Air Force strike aircraft attacked the Thanh Hoa bridge with nine 3,000-pound Laser Guided Bombs (LGBs), fifteen 2,000-pound LGBs, and forty-eight 500- pound conventional bombs. By the end of the

1965
A NEW THREAT: MIGS AT THANH HOA BRIDGE

ON 4 APRIL 1965 AS AIR FORCE F-105S ATTACKED THE BRIDGE, "FOUR MIGS CAME OUT OF THE CLOUDS WITH CANNON BLAZING. MIGS HAD BEEN SEEN ON PREVIOUS MISSIONS, BUT THIS WAS THE FIRST MIG ATTACK OF THE WAR...AND IT WAS OVER ALMOST AS RAPIDLY AS IT BEGAN.

THE ENEMY AIRCRAFT HAD USED A DIVING, HIGH SPEED PASS, COMING IN BEHIND THE BOMB-LADEN THUDS TO NAIL THE FLIGHT LEADER AND HIS WINGMAN. THERE WAS NO CHANCE TO GET EVEN, AS THEY CONTINUED STRAIGHT AHEAD AND OUT OF THE AREA AT MAXIMUM SPEED. BOTH F-105 PILOTS WERE LOST.

THE MIGS HAD COME TO PROTECT THE PRESTIGIOUS BRIDGE AND ... ADDED A NEW DIMENSION TO WHAT WAS NOW A RAPIDLY EXPANDING WAR."

1967
WALLEYE

ON 12 MARCH 1967, THREE A4S FROM VA-212 ATTACKED THE THANH HOA BRIDGE WITH WALLEYES AND "ALL THREE WEAPONS IMPACTED WITHIN FIVE FEET OF EACH OTHER ON THE DESIGNATED AIM POINT; BUT THE BRIDGE STILL STOOD."

CITATION

FOR EXCEPTIONALLY MERITORIOUS SERVICE DURING THE PERIOD 7 FEBRUARY TO 18 OCTOBER 1965 WHILE PARTICIPATING IN COMBAT OPERATIONS IN SOUTHEAST ASIA IN SUPPORT OF UNITED STATES NATIONAL POLICY. CARRYING OUT OVER 10,000 COMBAT SORTIES DURING A SINGLE CRUISE, THE USS CORAL SEA AND HER EMBARKED AIR WING EXECUTED A SERIES OF DEVASTATING AIR STRIKES AGAINST MILITARY AND LOGISTIC TARGETS IN NORTH VIETNAM, CONTRIBUTING IMMEASURABLY TO THE EFFORTS OF THE UNITED STATES IN RESTORING PEACE AND POLITICAL FREEDOM IN THE AREA. THE OUTSTANDING PROFESSIONAL AND TECHNICAL COMPETENCE, ESPRIT DE CORPS, EFFECTIVE TEAMWORK, AND VALOR DISPLAYED BY THE OFFICERS AND MEN OF CORAL SEA AND EMBARKED CARRIER AIR WING FIFTEEN WERE IN KEEPING WITH THE HIGHEST TRADITIONS OF THE UNITED STATES NAVAL SERVICE.

ALL PERSONNEL ATTACHED TO THE USS CORAL SEA, OR TO HER EMBARKED AIR WING, AND SERVING ON BOARD THIS VESSEL DURING THE PERIOD DESIGNATED ABOVE, OR ANY PART THEREOF, ARE HEREBY AUTHORIZED TO WEAR THE NAVY UNIT COMMENDATION RIBBON.

SECRETARY OF THE NAVY PAUL NITZE

attack, photography revealed that the:

> *western span of the bridge had been knocked completely off its 40 foot thick concrete abutment and the bridge superstructure was so critically disfigured and twisted that rail traffic would come to a standstill for at least several months.*

The Navy flew eleven more missions against the Thanh Hoa bridge to hinder any repairs and the Air Force flew two similar missions before President Nixon stopped all bombing of North Vietnam in October 1972.[12]

More Gold Stars on Our Air Medals

In the Awards System, the Navy gave a gold star for each subsequent award of a medal after the initial award, and issued a silver star to be worn on a ribbon in lieu of a fifth subsequent award. Commanders awarded the Air Medal for heroism or aerial achievement and in Vietnam they awarded it for missions flown. We called it a "strike-flight air medal." In World War II, five missions got you an Air Medal. In Vietnam, we had a point system that would get an attack pilot a strike-flight award for ten missions. But no one anticipated the number of missions that people would fly. The Navy – as did the Army and Air Force — finally went to a numerical system to indicate the number of strike-flight awards, and kept the gold and silver stars for additional awards for heroism or achievement.

We were pleased however that Paul Nitze, the Secretary of the Navy cited the *Coral Sea* (CVA-43) and Carrier Air Wing Fifteen (CVW-15) for its service.

The missions continued, but now on the major strikes further north we had to be aware of the missile threat. We needed to get more intelligence on the missile's strengths and weaknesses. The intell people worked to get more information as we continued to fly our missions.

Our new Wing Commander (CAG), Wes McDonald, arrived. Having been an A4E skipper, he flew most of his missions with the SILVER FOXES while his Wing Operations Officer, CDR John Harris[13], flew with the Blue Tail Flies. Because we were so short of pilots, we were glad to have the CAG fly with us. A day after he arrived, McDonald led sixteen A4s from both squadrons to strike a target south of Hanoi. En route, we passed south of the city of Vinh below Hanoi, and we saw what appeared to be three landing ships in the river near Vinh. As we flew towards our target, clouds began to hide the ground and by the time we got to our target, the clouds had totally obscured the area.

CAG McDONALD ABORTED THE MISSION AND DECIDED TO GO AFTER THE THREE LANDING CRAFT IN THE RIVER. HE GAVE US A FRAGMENTARY BRIEFING IN THE AIR, BUT WE DID NOT MAKE WELL-COORDINATED STRIKES ON THE SEPARATED SHIPS.

CAG McDonald aborted the mission and decided to go after the three landing craft in the river. He gave us a fragmentary briefing in the air, but we did not make well-coordinated strikes on the separated ships. The FLAK gunners in and around Vinh tried to distract us, but they did not stop us. We attacked, came close, but unfortunately did not hit the very narrow ships. However, I am sure we scared the hell out of the crews.

As we flew back to the *Coral Sea*, I began to monitor reports from the guys that they had crossed the beach to safety. Normally at the time, each pilot would call and announce: "SILVER FOX 502, feet wet." I would check off each guy's name on my knee pad because it would take awhile for them to reform and head for the ship.

After a few minutes, I had accounted for everyone except for LTJG Jim Mattly so I called him and asked him where he was. He told me he was still circling the target, getting ready to roll in. In my mind, I could see this single A4 making an attack with every enemy gunner shooting at him. I told him I would come back and FLAK suppress for him to give him some help. By then, Al Hyde was back on my wing so I told him: "Albie, we're going back to strafe the FLAK sites on the south side of the river to make them keep their heads down."

WAS IT ST. AUGUSTINE WHO SAID THE GREATEST THING WAS TO LAY DOWN YOUR LIFE FOR YOUR FRIEND? EASY FOR HIM TO SAY: HE WAS NOT AN A4 DRIVER ON AN EXTENDED CRUISE.

We headed back inland and started burping out our puny 20mm rounds, breaking one of the commandments about never getting down there among them with our cannons. We had completed one pass, east to west, and were about to come back the other way, when I called Jim to ask him where he was and what he was doing. When he told me he was ten miles out to sea already, I could have strangled him. I still give him a ration about this whenever I see him. Was it St. Augustine who said the greatest thing was to lay down your life for your friend? Easy for him to say: he was not an A4 driver on an extended cruise.

The fighter types were not getting their air medals as quickly as we were because if they did not escort us into the target and waited off shore flying a barrier combat air patrol (BARCAP) they needed to complete twenty missions to earn a strike flight award. But it was not the medals that concerned them. The VF pilots wanted to get more involved to help out. We knew or sensed that North Vietnamese MiG fighters would be airborne when we flew north but, at that time, they stayed away from us. Sort of gentlemanly, wasn't it? With no air-to-air mission at hand, Jake Ward,[14] the F4 squadron CO, talked CAG McDonald into letting the F4 guys get into the air-to-ground business even though they had never bombed with the F4. With a decent sight and some pilot training, the F4 made a good bombing platform. But it was all new to the pilots in VF-151, Jake's squadron.

As luck would have it, and rightly so, Jake Ward flew his first air-to-mud mission against the Thanh Hoa Bridge. He was the senior pilot in the air, the mission leader, but I was the senior attack pilot, the strike leader.

As luck would have it, and rightly so, Jake Ward flew his first air-to-mud mission against the Thanh Hoa Bridge. He was the senior pilot in the air, the mission leader, but I was the senior attack pilot, the strike leader. Jake briefed the mission then I briefed the attack. He led a flight of four F4s that were to FLAK SUPPRESS for us. I explained the approach to the target, how we would start motoring down from altitude to build up some energy, then on my signal Jake was to attack the FLAK sites I wanted hit. I stressed to him the need to get steep, to get the speed they wanted, and to pickle off their bombs at a specific altitude. The F4 pilots - working with the Weapons Training Officers of both A4 squadrons - had somehow figured out a way to use their SIDEWINDER missile sights to aim their bombs.

We sneaked up on the bridge, built up our speed, and watched all the flashes from the 57mm sites.

We sneaked up on the bridge, built up our speed, and watched all the flashes from the 57mm sites. I was surprised to spot a site one to two miles west of the bridge blasting away. I told Jake to go, and the Phantoms accelerated quickly away from us. I was shocked to see them dissipate their altitude so quickly, causing them to end up in a piddly twenty-degree dive (we would be somewhere from forty-five to sixty degrees). The F4s got to their pickle altitude too soon and dropped all their bombs far short of their intended targets. The FLAK site well west of the bridge was obliterated. Luckily we all got in and out safely with no suppression on the guns around the bridge.

WE LEARNED THAT THE MISSILE DID NOT HAVE A GOOD LOW ALTITUDE CAPABILITY, AND WE LEARNED THAT, AS LONG AS WE COULD SEE THE MISSILE, WE COULD EVADE IT.

SAM Encounters

On the ALFA Strikes near Haiphong and Hanoi, we met our first missiles. After initial contacts, we started to receive more intelligence about the SA-2. We learned that the missile did not have a good low altitude capability, and we learned that, as long as we could see the missile, we could evade it. To aid us avoid the missiles, Electronic Counter Measure (ECM) aircraft accompanied each the strike group. They would:

1. Alert us to the FANSONG radar's emission,

2. Give us coordinates from where we could expect a launch,

3. Tell us when the FANSONG went to its fire control mode, and

4. Tell us about the SAM's guidance signal at launch.

At this time, we had not received our own ECM equipment for the A4, called SHOEHORN, and had to rely on the ECM aircraft with our strike group to provide us with this critical missile firing information.

The counter, counter-the-counter, and counter the-counter-counter game started anew with the missiles. The North Vietnamese learned that when they went to their fire control mode (called Rattlesnake Alpha or Rattlesnake Bravo) our ECM aircraft would notify us (verify) and we would start making three dimensional changes in altitude, heading, and speed to con-

ON ONE MISSION, THE ENEMY LAUNCHED THIRTEEN MISSILES AT OUR STRIKE GROUP. I SAW NINE OF THEM AND ALWAYS HAVE WONDERED WHERE THE OTHER FOUR WENT.

found the enemy's fire control solution. When the ECM aircraft notified us of a missile launch, we would look to the appropriate sector and start evading.

Later, the North Vietnamese began launching missiles without using their fire control mode. They also took new actions. First, they fired on a "shoot on search" mode. Next, they fired missiles from a site other than the one associated with the radar we confronted. Finally, they fired more than one missile at a time. So we engaged in a deadly "cat and mouse" game over North Vietnam.

On one mission, the enemy launched thirteen missiles at our strike group. I saw nine of them and always have wondered where the other four went.

Air Wing 16 from the *Oriskany* worked with us at this time. The pilots were an aggressive and brave bunch of guys.

Many of my old AVMID friends served in the attack squadrons on that carrier. They included: Ron Caldwell, Paul Engel, and Henry Jenkins. Caldwell previously had served as our Wing Operations Officer.[15] Wynn Foster, soon to be severely wounded, also served with Air Wing 16. One day we followed behind Air Wing 16 on a two-wing strike. Sadly, that day the North Vietnamese shot down CAG Jim Stockdale.

Jim Stockdale became an American hero.[16] When he returned to the United States after eight years as a POW, the President awarded him a Medal of Honor for his leadership in the Hanoi Hilton and other POW sites. Air Force COL Robbie Risner wrote the following about Stockdale.[17]

> *My next-door neighbor was Navy Commander Jim Stockdale, who had been there for awhile, too. How I longed to see that kindly face again, weathered by years of sun and salt-water spray, with crow's feet around the eyes from squinting through the windscreen of his jet fighter. He had a shock of grey hair falling down over one eye, which gave him a remarkable likeness to the writer Carl Sandburg. He loved philosophy and intrigue and was an outstanding leader...*

Humor

After many harrowing missions, my guys still maintained a sense of humor. We should have sent some their stories to *Readers' Digest* for publication in the section, "Humor in Uniform."

Several stories stand out.

Ken Holthaus asked out loud one day: "Why are we fighting here? We don't need the rice do we?"

Dick Floreani returned to the Ready Room after eating a meal and muttering to himself that he was the unluckiest guy on the ship. He explained that, "I fell down the Executive Officer's ladder (the stairs by the XO's cabin) and didn't even hurt myself."

On another day, a tired and lonesome Silver Fox, sitting in his Ready Room chair and feeling sorry for himself, asked out loud facetiously, "How am I going to stand not being shot at?" He continued, "How am I going to stand having a cold beer? He finished his musings with: "How am I going to stand going home each night and how am I going to stand sleeping with my wife every night?" From the back of the Ready Room, an unknown comrade answered: "I don't know – I've never met her."

On different day, our Ready Room clock had stopped and the Squadron Duty Officer was told to get it fixed. He thumbed through the ship's phone directory and put through a call. When someone answered the phone, the Duty Officer said, "This is Ready Room 5, we have a clock here that doesn't work." The person on the other end of the line replied, "This is the Photo Lab, you want a picture of it?"

By this time, I was tired and so were my pilots. One day, I made a dumb statement during a mission to Vinh and amused them. We approached the city from the south, most likely "sneaking up" on the Thanh Hoa Bridge to attack it one more time. Up ahead, the enemy gunners near Vinh filled the sky with black bursts from their 85mm guns, saturating a rectangular area about two thousand feet below us. One of my guys called me and reported, "FLAK at twelve o'clock skipper."[18] Thinking about other things - obviously not a smart thing to do

July 1965 Flights

Day	Hyde	Mattly
1		2
2	1	
3	1	
4		
5	1	
6		1
7		1
8	1	1
9		1
10	1	1
11		1
12		1
13	1	
14	1	
15	2	
16	1	1
17		1
18		1
19	1	1
20	1	1
21	2	1
22	2	1
23	2	1
24		1
25		
26		
27		
28		
29		
30		
31		
Missions	18	18
Hours	37.5	38.8

flying into an attack — I replied: "ROG, I'll go under it." "Under it!!" I can not believe that I said that.

My guys continue to remind me of my ridiculous comment on any occasion that they can. At least, I lightened up their day and added to our squadron's lore.

Parrode and Marsh flew right on the deck and looked in the doors and windows of the farmhouses for any sign of Wendy. While they were searching, a pesky 4.7mm quad-mounted machine gun fired at them, giving them Problems.

Wendy Rivers

It was the last strike of the last day on line in September 1965. If fact, we should have been done at noon, but for some reason, the commander ordered us to strike the rail marshalling yards or some other inane target in Vinh. The Commander of VA-153, Ken McArthur, was to lead the strike. Because of operational policy, Wendy Rivers led the Silver Foxes because two squadron commanders could not go on the same mission. That day, I led four planes up the coast on an IRON HAND mission to attack any missile sites that came on the air so that we could support the McArthur-Rivers strike.

The enemy hit Wendy's aircraft during his dive and caused it to catch fire. Wendy ejected. As soon as I heard this, I turned south toward Vinh and notified the Strike Leader that as the senior aviator airborne, I was taking over the Search and Rescue (SAR) effort. I learned that Wendy had ejected successfully and that he had landed southeast of Vinh, on the south side of the river. Within minutes, I spotted Wendy lying on the ground with his parachute streamed out behind him. After I reported to the Strike Leader that I had visual contact, I called in the two Skyraider pilots, Harry Parrode, the Squadron XO, and his wingman, LTJG Dan Marsh,[19] who were flying off the coast to support any SAR mission. Next I told the A3 tanker to stay on station for us.

The Skyraider pilots and I immediately hopped over the low hills and made a 360 degree turn to find Wendy again. Wendy, however, had not called on his emergency radio, and when we completed our turn, I could not see him. I figured that in those few minutes he had rolled up his parachute and was in hiding.

As we circled, the North Vietnamese anti-aircraft batteries across the river near Vinh began lobbing shells at us. They exploded above us at about 1,500 feet. The gunners did not care that the shrapnel rained down on the farming area below us. Fortunately, none of the shell fragments hit our planes.

Parrode and Marsh flew right on the deck and looked in the doors and windows of the farmhouses for any sign of Wendy. While they were searching, a pesky 14.7mm quad-mounted machine gun fired at them, giving them problems. Harry Parrode called me, and said that he had to get rid of that gun site before he could continue his search. He and Marsh broke off and quickly zapped the gun and then returned to searching for Wendy.

Shortly thereafter, we received word that two Air Force Jolly Green Giant rescue helos had been dispatched from Danang to help, and that a Grumman SA16 seaplane also was coming in. That was good news, but we kept searching because the three aircraft would not arrive for some time. Meanwhile, the North Vietnamese continued to shoot at us. There was no concentrated fire, but they did hit both Skyraiders. Both began losing oil, so I told Parrode and Marsh to head for Danang or to the ship if they could make it.

As we flew around low, we consumed fuel pretty fast, so I told Dave Leue, the XO of the Blue Tails, who was part of my four-plane flight, to head for the tanker with his wingman to refuel. When they returned, my wingman, Jay Starling, and I left to take on fuel.

The seaplane and the two helos and their two Skyraider escorts checked in with me when they arrived off the coast. Helo rescue pilots were a brave bunch of guys but they were not foolhardy. Under the Air Force rescue policy, the pilots told me that they could not go inland for a rescue unless the downed pilot was in sight. By this time, we had been there for hours and had exposed ourselves to a lot of risk, so I was not too pleased to hear about their policy. I swore some, and told them that I thought that Wendy would show himself when he heard helos. Busting their rules, the Jolly Greens flew in hugging the hills

HANOI HILTON

"I WAS IN THE HOA LO PRISON – THE HANOI HILTON... ONLY A FEW MILES FROM DOWNTOWN HANOI. MANY OF THE HIGH-RANKING [NVN] OFFICIALS HAD BEEN THERE FOR TEN TO TWELVE YEARS THEMSELVES, IMPRISONED BY THE FRENCH AND JAPANESE.

I LOOKED AROUND MY CELL; SUSPENDED FROM THE CEILING WAS AN OMINOUS-LOOKING HOOK. THE FLOOR WAS DIRTY AND THERE WERE SOME BLOODY BANDAGES IN ONE CORNER, A DESK, AND A CLOTH CURTAIN IN THE MIDDLE OF THE ROOM."

CRIMINAL

"THE RABBIT'S APPROACH WAS TOTALLY DIFFERENT: IF I DID NOT TALK, I WOULD NEVER SEE MY FAMILY AGAIN. THEN HE TOLD ME I WAS NOT A PRISONER OF WAR BUT A CRIMINAL... HE WENT ON TO TELL ME THAT I HAD VIOLATED SOVEREIGN AIR SPACE OF THE DEMOCRATIC REPUBLIC OF VIETNAM. I WAS CONSIDERED IN THE SAME WAY AS THOSE WHO USED TO PREY ON THEIR COASTS. *YOU ARE THE SAME AS A PIRATE. YOU ARE GOING TO HAVE TO TALK.*"

TORTURE

"ONE DAY, WITHOUT WARNING, WE HEARD SOMEONE SCREAM FROM THE MEAT-HOOK ROOM AT THE OTHER END OF THE COURTYARD, ABOUT SEVENTY-FIVE TO A HUNDRED FEET FROM US. WE COULD HEAR HIM GROAN AND HOLLER EACH TIME THE PRESSURE WAS APPLIED."

DARK ROOM

"THEN ONE EVENING..., AFTER ALMOST TEN MONTHS OF DARKNESS AND SOLITARY CONFINEMENT, THE BUG CAME IN. HE SAID, *FOLD UP. YOU ARE GOING TO MOVE.*"...

THE NEXT MORNING I COULD HARDLY WAIT FOR THE SUN TO RISE... I HAD MADE IT. THE NIGHT HAD PASSED AND THE SUN HAD RISEN."

R. RISNER

Our efforts were for naught. We found out after Wendy was released from POW camp that in the few minutes that it took us to circle back over him, North Vietnamese farmers had grabbed him and his parachute and hid him in a nearby farmhouse until we left.

with the two Skyraiders weaving over them. Enemy gunners opened fire and hit one Skyraider shortly after it came inland. Badly hit, the pilot immediately headed back to sea where the seaplane orbited and ditched his plane. The seaplane picked up the pilot and his crewman. Meanwhile, the helos' noise did not flush Wendy, and after taking several hits from small arms, the Jolly Greens beat a hasty retreat to the safety of the sea. Then, with the remaining Skyraider as an escort, they headed for Danang.

Topside, the North Vietnamese hit Dave Leue's wingman's A4 in one wing and shortly thereafter hit my wingman's plane it its wing. Both aircraft began leaking fuel because the A4 Skyhawk carried most of its fuel its wings. I sent both pilots back to rendezvous with the tanker. It now was late in the afternoon. Soon the tanker pilot, who had been talking with the *Coral Sea* all this time, passed on an order from the ship for me to return to the carrier. I called the seaplane pilot as we headed out and told him that he could head for home too as soon as Dave Leue and I were feet wet.

Our efforts were for naught. We found out after Wendy was released from POW camp that in the few minutes that it took us to circle back over him, North Vietnamese farmers had grabbed him and his parachute and hid him in a nearby farmhouse until we left.

The initial After Action Report submitted to the Fleet Commander described the mission and reported that Silver Fox One, another *Coral Sea* squadron commander, had been shot down. Until the follow-up report stated that Wendy had been lost, many people thought that the North Vietnamese had shot me down.

After landing back aboard, I went to the bridge to brief the Captain and the CAG. I told them that I was going to recommend everyone for an award. We had very little success during the entire cruise getting the Awards Committee at Pacific Fleet Headquarters in Honolulu to approve any of our recommendations. Air Medals and Navy Commendation Medals had come through, but only a few DFCs had been approved. This rescue attempt rated a Silver Star or even the Navy Cross, but we had such little success in the past that so we went for the DFC. Thankfully, the Navy awarded everyone a DFC.

Our complaints about the award system at a later date helped loosen up the system and they benefited the Air Wings that followed us. Those Air Wing commanders, however, learned from us that they needed to jazz up their recommendations to get awards for their pilots and crews. Unfortunately, the Navy failed to recognize many of the heroic deeds performed by the pilots of Air Wing 15. [20]

The Last Line Period

After we lost Wendy, the ship headed for the Philippines for a little rest. We needed it. While in port at Subic Bay, the maintenance crews installed Electric Countermeasure (ECM) equipment - called SHOEHORN - on our aircraft. By today's standards, the ECM device was unsophisticated, but it was a start. The technical people did not give us much information about the equipment and what it would do, however, they explained to us what the warning lights would tell us. The lack of critical information reminded me of the lack of information we had received about the enemy's SA-2 missile. I wonder how many people we lost because of these oversights and other ones.

At Subic, our first replacement, LCDR Paul Moore, joined our squadron. He became a highly respected combat aviator on our next cruise and subsequent ones.[21] Then, CAG McDonald told CDR John Harris, his Wing Operations Officer, to start flying with our squadron because we had lost Wendy. We learned at this time that our next line period would be our final one and that we would be home in November. This line period, however, was not going to be any easier.

After a week's rest, we headed back to Yankee Station. Our new strike missions sent us into Indian Country. One "BIG GO" directed us bomb a bridge between Hanoi and China. CAG McDonald led the raid. Both of the VA squadron skippers flew the mission. We coasted in north of Haiphong on the deck, flying

Meaningful Awards

Nearly every combat veteran can identify comrades who deserved high awards and did not get them and can identify others – generally officers – who did not deserve their high awards.

A significant award action took place in the 9th Infantry Division in 1967. One platoon had been out all night fighting and had done an excellent job. As soon as it returned, the Brigade Commander, Colonel William Fulton, decided that the entire platoon deserved Bronze Stars. He ordered his staff to get the medals immediately and to write up the needed citations later. Around noon, Fulton presented the BSMs to each soldier. Fulton's action provided the soldiers with his immediate appreciation of their performance and contrasts to the situation the Navy air commanders faced on Yankee Station.

In 1965, there were two and three-star admirals on station who should have had the authority to award Commendation Medals, Air Medals, Bronze Stars, and DFCs, and even Silver Stars. Sending recommendations for awards back to staff people in Honolulu guaranteed that the awards would be temporized and delayed.

By the 1970s, the award of the Air Medal had spiraled out of control in SVN because of extensive helicopter operations. Many Army helicopter pilots had earned twenty or more Air Medals. Consequently, the Army replaced oak leaf clusters with numerals because many recipients could not fit a sufficient number of clusters on their ribbons.

Many Navy fliers earned similar numbers of awards.

John Bradley

over hilly country and below some puff cumulous clouds in fair weather. As we flew in, I heard all sorts of missile warnings. Then, I saw something explode on the side of a hill to my left. I thought that someone had flown into the hill, but it turned out that a missile had struck the ground. The North Vietnamese apparently had fired at us hoping to hit somebody.

As we flew north, we could not see our target from our altitude. We flew farther north, but for the moment missed our designated target. We then saw a huge, lucrative target of opportunity: the Thai Nguyen Steel Mill, with big factory buildings and a large railroad marshalling yard. But, we could not attack it because Mr. McNamara had not included the mill on his list of approved targets.

We knew we probably had stirred up a hornet's nest and we still had to run the gauntlet outbound. Going up, each leader was on his own. One of my pilots spotted a railroad bridge south of the steel mill, so I told the Foxes that we would set up and go against the bridge. After our attack, we flew out at low level in twos and fours, and we stayed low even when we got over water until we got beyond the range of the missiles. We knew the MiGs were up that day, but they stayed at high altitude and did not bother us.

One of my pilots spotted a railroad bridge south of the steel mill, so I told the Foxes that we would set up and go against the bridge. After our attack, we flew out at low level in twos and fours, and we stayed low even when we got over water until we got beyond the range of the missiles.

Ψ

Probably one of the most remarkable feats of flying occurred during this line period when Paul Moore flew a reconnaissance mission with Jack Harris up in the delta. The weather over the ocean was clear, but a low overcast hung over the Red River Delta. Soon after coasting in, the enemy hit Paul's A4. One or more bullets hit the fuel line leading to Paul's engine and made his throttle useless. The hit caused his engine to stabilize at an RPM that was barely sufficient to keep the aircraft flying at around 180 knots. Paul immediately dumped his ordnance and external fuel tank to lighten the aircraft and headed back out to sea. At the time, I was flying a recon mission about

Probably one of the most remarkable feats of flying occurred during this line period when Paul Moore flew a reconnaissance mission with Jack Harris up in the delta.

twenty miles south of Paul, but aborted it and headed up the coast toward him. En route, I notified the A3 tanker that Paul would need him, and then I spotted Paul coming out from under the overcast. His aircraft had lost so much fuel by then that he could not gain any altitude. After I spotted the tanker, I directed the tanker pilot to where he could join up with Paul while Paul headed for the nearest rescue destroyer.

The problem now was – how was Paul going to refuel in flight when he had no throttle response and while his remaining fuel supply continued to drop rapidly? The A3 pilot positioned his big aircraft ahead of Paul and asked Paul for his airspeed. The tanker pilot planned to decelerate below Paul's airspeed and allow Paul to catch up to him. This was not a comfortable speed for the heavy tanker. As Paul lined up his fueling probe with the trailing basket hanging under the A3 tanker, he saw that he was closing faster than he wanted. He yelled "Speed up, speed up" to the tanker pilot. As his probe entered the basket, Paul flew under the A3 to avoid colliding. Immediately, his probe bent up at a severe angle and then popped out of the basket. In front, the A3 pilot could not see behind him and so had no idea of Paul's relative closure rate.

They set up again. Paul called out his airspeed more often. This time it worked. When contact occurred, Paul kept telling the tanker pilot what to do with his speed so that Paul would not overrun or back out of the basket. The A3 started transferring fuel as we came up on the first rescue destroyer. I asked Paul what he wanted to do. He was taking fuel so he wanted to push on to the second rescue destroyer about thirty miles farther south. Like a baby at a mother's breast, Paul sat there taking nourishment, enjoying the security of being near friends and a source of fuel. Paul's aircraft slowly got heavier and heavier. Before he told the A3 to stop transferring fuel, Paul started to slow down. He had insufficient power to keep flying. Things got critical. Paul popped out of the basket and started losing altitude. He could not lighten up the A4 so he could stay airborne because the fuel was not leaking out fast enough. Consequently,

Combat SAR

One of the unique aspects of the air war was the Search and Rescue (SAR) operations conducted by the Navy and the Air Force.

During the war the Navy rescued about one sixth of the pilots who went down in North Vietnam or adjacent waters.

Early Navy and Air Force efforts were poor. The Navy "possessed no organization dedicated to combat SAR." The Air Force had not maintained "a wartime Search-and-Rescue effort beyond 1958."

Fortunately, the Navy eventually organized a Helicopter Support Squadron to rescue downed flyers, and the Air Force used squadrons from their Aerospace Rescue and Recovery Service to rescue downed crews with their "Jolly Green Giants."

The North Vietnamese captured nearly three-quarters of the downed airmen within about thirty minutes after they landed on the ground as they did with Wendy Rivers.

Some helicopter rescues, however, took a long time – perhaps a full day or even two or three days.

A typical Air Force rescue task force included one or two "Super Jolly Green Giants" and four to eight propeller driven attack aircraft.

Ψ

Rescued Silver Foxes

CDR Jack Harris
LT Bill Fidelibus
LCDR Paul Moore

Paul had to eject at a low altitude. Once in the water, he had faced new problems: the parachute filled with water and the parachute suspension lines had entangled him. Fortunately, he got free of the suspension lines before the parachute pulled him under. The destroyer soon came up and rescued Paul. I got the A3 driver a DFC to recognize his outstanding airmanship. Paul? I got him another A4 so he would not fall behind in missions.

I got the A3 driver a DFC to recognize his outstanding airmanship. Paul? I got him another A4 so he would not fall behind in missions.

Ψ

As the years have gone by, I often have looked back and evaluated how we flew our night road reconnaissance missions. Flying them took a great deal of skill and superb airmanship, even though we just took them as a matter of course at the time. The command normally assigned a section of A4s, two planes, to reconnoiter a single route package. This would keep two or more sections separated while they flew their individual route packages in the dark with no lights on. We had enough to worry about flying the road reconnaissance missions, but at night we had an additional worry: a mid-air collision. To help prevent that, we stayed on the same radio frequency so that when someone dropped a flare or encountered FLAK, we could look over to the appropriate route package and see it.

The command normally assigned a section of A4s, two planes, to reconnoiter a single route package. This would keep two or more sections separated while they flew their individual route packages in the dark with no lights on.

I could have strangled Jim Hughes one night when he flew with me. I heard the leader of another section of A4s south of me call, "Flare away." I looked back and saw it light off and thought no more of it. Now a flare hanging in its parachute looks a lot like a missile coming right at you. Jim did not hear the "flare away" call. When he spotted the bobbing light in the sky, he radioed me, "Look out, Skipper, here comes a missile!!!" He broke for the ground, going into a high "G" turn. I could not see a thing other than the flare and started to get worried. The best Jim could do, pulling so many Gs, was to key his mike and grunt a few times.

Needless to say we got separated at this point. Therefore, I told Jim to head back to the ship by himself. After I landed aboard, I remained mad and planned to give Jim hell for scaring the bejabbers out of me. After I got to the Ready Room to watch the rest of the recovery, I discovered that Jim chose this night to forget all he had learned about night landings. He made four approaches but missed the arresting cables each time. We called these misses "bolters." Soon Jim was the only pilot still trying to get aboard and he was getting very low on fuel. He worried me, and I said, "Please Lord, get him aboard and I won't be angry with him." The next approach, the Lord made a good one and Jim landed safely. I kept my promise.

HE WORRIED ME, AND I SAID, "PLEASE LORD, GET HIM ABOARD AND I WON'T BE ANGRY WITH HIM." THE NEXT APPROACH, THE LORD MADE A GOOD ONE AND JIM LANDED SAFELY. I KEPT MY PROMISE.

On a night recce mission, two A4s would rendezvous off the ship's TACAN and head for the "coast-in" point. The leader carried flares and rocket pods; the wingman carried rocket pods and/or bombs. Approaching the coast-in point, the leader would signal to the wingman to separate while descending to 5,000 or 6,000 feet. The wingman then would turn 360 degrees and put himself five to six miles behind his leader. With all lights off, we flew at a specified air speed. Air-to-air TACAN told us how far apart we were. We flew a very specific route: we followed a prescribed heading between designated points for a planned time; then we turned to a new course. The leader would call POINT ALFA, POINT BRAVO, etc. as he passed the pre-designated points. These calls gave the wingman an idea where his leader was, and the wingman would "time down" to the turn point and then turn to the new heading. If the leader dropped a flare at a pet river crossing or road intersection, the wingman could then line himself up on his leader again.

I liked to light up certain points where bridges were down to see if we could catch trucks crossing on any newly-constructed pontoon bridge. I knew that the North Vietnamese defended these points, so as soon as I released my flare I reduced power and turned away. When the flare lit off, the enemy anti-aircraft gunners at the site normally started blasting away trying to hit me. Al Hyde, who usually flew with me, then would attack the FLAK site. This maneuver made the North Vietnamese very angry.

I LIKED TO LIGHT UP CERTAIN POINTS WHERE BRIDGES WERE DOWN TO SEE IF WE COULD CATCH TRUCKS CROSSING ON ANY NEWLY-CONSTRUCTED PONTOON BRIDGE.

After I dropped my flare, I turned south. Almost immediately, the enemy 37mm sites opened fire. Bill called a warning to me. I looked in my mirrors and saw all the tracers arcing up at me. I went to full power and dove for the deck. The tracers reminded me of water from a garden hose following me.

We did not catch many trucks at night, but every now and then we did.

One night as we waited to launch, the recce pilots already out radioed back that they had found a lot of trucks and that they were working them over. The air commander changed our mission: he directed us to attack the trucks. When we got to the area, I saw fires and saw Skyraiders and A4s attacking the column of vehicles. I also found that we had a sky full of airplanes milling around with their lights out. I immediately became an air controller. I called each section leader and assigned him altitudes and areas around the target, and then I cleared each section leader to attack from a specific heading and then told him to return to an assigned holding area. We smoked a lot of trucks that night.

Another night, near the end of the line period, Bill Kish flew with me on a route package up in the Delta. That night we had a clear night and a full moon. At 3 o'clock in the morning after I had just started my route, I saw headlights north of me. I told Bill that I had spotted a truck and would light him up for him to attack. After I dropped my flare, I turned south. Almost immediately, the enemy 37mm sites opened fire. Bill called a warning to me. I looked in my mirrors and saw all the tracers arcing up at me. I went to full power and dove for the deck. The tracers reminded me of water from a garden hose following me. I kept turning away from the tracers, looking back over my shoulder rather than looking ahead of me. I wanted to head back out to the coast but found myself running southwest and inland to avoid the fire. On this heading, I flew over a very heavily defended bridge and the gunners there immediately opened fire on me. Finally, I got out of the gunners' range and headed east for the safety of the sea. After I landed aboard, my Maintenance Chief found that my engine had been badly damaged by a hit in the compressor section. I surprised him when I told him that I did not know I had been hit and that the engine had not operated in an unusual manner.

About this time, journalists wrote up an old friend of mine

from flight training, Harry Southworth, in *Time* Magazine. Harry commanded one of the A4 squadrons on the USS *Independence.*[22] According to the story, he and his guys had caught a mobile missile site out in the open on one of their recent recce missions. We got a lot of satisfaction when we read how their attack had caused missiles to spin on the ground like pinwheels and explode.

The Final Strike

The last day of the last line period arrived. Each of us thought: "We've made it this far. If I can get this next one, I'm going home." The Admiral, the Task Force Commander, and his staff transferred from the *Coral Sea* to the *Oriskany*. It seemed that, in recognition of all that we had endured, the Staff assigned us a "piece of cake" target with no missile threat and little other opposition – Dong Hoi, the place where it had all started for us in February 1965.

CAG McDonald briefed the mission late in the evening for a launch at first light. Everyone was in high spirits wondering how he would aim if he dropped his bombs from straight and level at 20,000 feet or what sight setting he would have to use to pickle his bombs at 10,000 feet in a forty-five degree dive. Since we had to get up at 4 a.m., people started going to bed after the CAG completed his briefing.

Unfortunately, the mission changed.

Oriskany's A4 pilots flying up around Hanoi had found a concentration of trucks trying to hide in the foliage around a small Karst mountain. The weather was not good up in the Delta so the Admiral decided to send the *Coral Sea* against the trucks at first light. Subsequently, CAG McDonald and I heloed over to the *Oriskany* to be briefed. We returned very late to the *Coral Sea*, woke the pilots up, and briefed them for the new mission in Indian Country.

Having mentally adapted to an easy strike at Dong Hoi, gearing up for a run into Hanoi on the last mission did not set

Electronic Countermeasures (ECM)

"Electronic warfare came of age during the late 1960s. Events in Southeast Asia and the Middle East proved that modern air combat required [ECM] and even counter countermeasures [ECCM) for both offense and defense....

The unique wrinkle over North Vietnam was the surface to air missile. Because SAMs directed strike warfare tactics, defeating or confusing missile scanning and guidance became paramount...

The U. S. Navy entered the war with one aircraft devoted to ECM: [the propeller-driven Douglas EA-1 Skyraider]...

The "Electric Spads"... flew with a pilot, .. a naval flight officer (NFO), and two enlisted men as ECM operators.

Under any but ideal circumstances, optimum jamming range was twenty to twenty-five miles, though nearly twice that distance was possible. Consequently ECM coverage of strikes was usually conducted three to four miles offshore after 1965. A *Midway* EA-1F was clobbered by AAA in April of that year, and the valuable ECM birds were kept out of harm's way as much as possible after that.

With the standard two ECM aircraft airborne, and each carrying two jammers, up to four [enemy] radars could be neutralized simultaneously."

well with my people. Since it was so late, I went to my room to lie on my bunk and did not bother to get out of my flight suit. Before I fell to sleep, Bill Kish knocked on my door and walked in. "Skipper," he said, "this change is B.S. We have been out here longer than anyone else and they pull this on us. I will go tomorrow but that is the last time."

Bill did not mean it, but he was distressed so I tried to console and reason with him. "Bill," I said, "look at it this way. Here I am a Commander, a CO with seventeen years as a Naval Aviator. I have a wife and five kids at home. If anyone gets bagged tomorrow, wouldn't it be better if it were you rather than me?" Well, that hardly consoled Bill. He turned and left my room.

I dozed a little before 4 a.m. came around. Then, I went down to get some breakfast and returned to the Ready Room. There was Bill, sitting in his chair where he probably had been since he left my stateroom. His complexion matched the color of the Ready Room bulkheads – a weak green.

The OPS people conducted the final briefing and we waited to be called away. (Where was Father Greco when we needed him?) The *Oriskany* pilots reported bad weather over the beach so Operations delayed our launch until we could send a weather reconnaissance flight up to the target. The XO of the F4 Squadron, CDR John Chambers,[23] got the mission.

All the Ready Room boxes were interconnected. Soon people were calling Ready Room 1:

"Free beer for a year if the target's obscured."

The bidding increased:

"A case of Chivas if you scrub the mission for weather!"

We waited patiently after John Chambers launched. In about thirty minutes, Chambers radioed back that a low overcast covered the entire Delta area and that the strike group would be unable to get in. With comments of "Aw shucks" and "Shucks," the Silver Foxes manned up to go against Dong Hoi.

After the attack, while we flew back to the ship, one F8 Crusader pilot had to eject. A rescue helo picked him up and returned him to the *Coral Sea*. Once safely aboard, we were done,

Die Hop

"On the night before our last mission prior to heading back to the States, the JCS (Joint Chiefs of Staff) target planners came up with a real beauty. It was some truck park in a very inaccessible place and had "DIE HOP" written all over it.

We were all fatigued at this point and I went down to [CDR Morin's] room and blew up about what an unfair deal this was and why didn't some else get this, after all we had been out here for what seemed an eternity. We had more than 100 trips in the North, blah blah blah.

He looked at me with that cockeyed Boston grin and said: "Look I have a wife and 5 kids and you are a young man. Wouldn't it be better if you died than me?"

I said, "Skipper I never looked at that way and I apologize for being so narcistic."

Bill Kish

at least for 1965.

On the long trip to the States, CDR John Harris received orders to VA-155 as my XO, to relieve me as CO in June or July 1966. CAG McDonald flew off the ship to Honolulu as we passed Hawaii in order to register complaints to the fleet commander and his staff about the unfairness of the awards system.

When we finally got home and I could sit and relax, I looked out our rec room window to see a bunch of neighborhood children, noses pressed against the window, looking in to see a real Dad.

"These were hard years on the families"

As we neared California, I led the fly-off of the A4s to Lemoore. There were only a few of us because we had left most of our planes in the Philippines as replacements. I experienced a lump-in-the-throat moment as I saw California ahead of us after so many months away. So did the others.

Because we had been extended, Lemoore had not seen any returning squadrons in quite awhile. As we taxied in, we saw a crowd of people and heard bands playing. Even the good nuns from town came out for the excitement. After eleven months and the constant worry, I had a very emotional reunion with Betty and the children.

When we finally got home and I could sit and relax, I looked out our rec room window to see a bunch of neighborhood children, noses pressed against the window, looking in to see a real Dad. These were hard years on the families.

"There comes my daddy."

Weary, Happy LNAS Pilots Return After 11 Months in Vietnam

By Bob Good
Fresno Bee staff writer

LEMOORE NAVAL AIR STATION – A group of weary but happy attack bomber pilots touched down yesterday afternoon, following 11 months of duty in the flak filled skies over North Viet Nam.

On hand to greet them were their equally happy families most of whom smiled their welcomes through a veil of tears. Other greeters included the colorful Porterville High School band, Capt. Edward Lawrence of COMFAIR, Alameda and Capt. Howard Boydstun, base commander.

"There Comes Daddy"

Shouts of "There comes my daddy" were echoed and re-echoed as the Viet vets dipped low over the runways of this master jet station prior to landing.

Among the anxious returnees was Lt. Cmdr. Ernest Mares, operations officer of VA 155, who was greeted by his pretty young wife, two sons and his tiny daughter, who was just 8 weeks old when he last saw her.

The Viet Nam war was a "fantastic pace of operations," declared Cmdr. James Morin, skipper of VA 155...

"She's just beautiful," he declared. But the baby cried lustily when confronted by the strange man she will soon come to know as father.

Fantastic Pace

The Viet Nam war was a "fantastic pace of operations," declared Cmdr. James Morin, skipper of VA 155, who said that his pilots averaged 115 sorties during their tour of duty.

End Notes

[1] U. S. Navy photo.
[2] Gutzman, *Vietnam,* 12.
[3] *Ibid.*
[4] *Ibid,* 371-72 (Operation STARLIGHT).
[5] The Army organized 11th Air Assault Division (Test) at Ft. Benning, GA to test new air mobile and air assault doctrine. Named for the 11th Airborne Division of World War II lineage, the Army re-designated it as the 1st Cavalry Division when it decided to make the division a permanent one. The Army transferred the colors from the 1st Cavalry Division in Korea to the mobilized division, and then transferred the colors of the 2nd Infantry Division, stationed at Ft. Benning, to Korea and re-designated the 1st Cavalry Division as the 2nd Infantry Division. The 2nd Infantry Division remains in Korea in 2010.
[6] Ralph Wayne Cousins retired as an Admiral in May 1975. He earned a Navy Cross, Distinguished Service Medal, and Legion of Merit during his career. *1993 USNA Register of Alumni,* #12549, 242.
[7] "A Navy Carrier Pilot's Personal Story," *Sea Breeze,* January 1966, 15.
[8] McDonald reached the rank of Admiral. He would help Morin get additional awards for his pilots.
[9] Jackson retired as a Captain, USN.
[10] ROG or ROJ, short for ROGER, code word for "understood."
[11] Donnelly retired as a Captain, USN.
[12] Waddell and Wood (eds), *Air War Vietnam,* 39-40 (MiG attack), 85-6 (destruction), and 87 (additional missions).
[13] Deceased.
[14] Ward retired as a Captain, USN after commanding the USS *Constellation* and dealing with a race riot.
[15] Engel later became a Rear Admiral. Jenkins got shot down and captured in 1965. Caldwell and Foster retired as Captains, USN.
[16] James B. Stockdale graduated from the U. S. Naval Academy in 1947. He retired as a Vice Admiral in September 1979. Stockdale earned a Medal of Honor, three Distinguished Service Medals, four Silver Stars, two Distinguished Flying Crosses, two Bronze Stars, and two Purple Hearts. *1993 USNA Register of Alumni,* #18729, 285.
[17] Robinson Risner, *The Passing of the Night, My Seven Years As a Prisoner of War of the North Vietnamese* (Old Saybrook, CT: Konecky & Konecky, 1973), 100.
[18] Twelve o'clock meant straight ahead.
[19] Parrode retired as a Captain, USN; Marsh was a Rear Admiral when Morin wrote his manuscript.
[20] The heavy bomber pilots of World War II must have shaken their heads in disbelief at the number of decorations awarded during the war.
[21] Moore retired as a Captain, USN.
[22] Southworth retired as a Captain, USN.
[23] Deceased at the time of this writing.

Sidebar Notes

June 1965 Air Wing 15
Carrier West Supplement, USS *Coral Sea,* 1964-1965.

June 1965 Silver Fox Leaders
Varied sources.

Deck Crews
SILVER FOX VA-155 NEWS, 2-3, 20 May 1965.

1965 SAM Sites
Nichols and Tillman, *Yankee Station,* 54-5 and Nalty, *Vietnam War,* 158.

Navy Aviator Missing in Action Identified
"Peter Mongilardi, Jr., Commander, United States Navy," http://arlington-cemetery.net/peter-mongilardijr.htm.

1965 New Threat: MiGs at Thanh Hoa Bridge
Waddell and Wood (eds), *Air War Vietnam,* 39 (first MiG attack).

1967 Walleye
Waddell and Wood (eds), *Air War Vietnam,* 57-9.

Citation
Secretary of the Navy Announcement, no date.

July 1965 Flights
Flight Logs of Albie Hyde and James Mattly. (Copies in possession of JHB).

Hanoi Hilton

Risner, *The Passing of the Night,* 19 (Hanoi Hilton), 23 (criminal), 107 (torture), and 180-81 (dark room). The North Vietnamese comments that the American pilots were criminals and not POWs was similar to the Japanese comments to American POWs in 1942 that they were not POWs and would be treated as "captives." Lt. Col. Wm. E. Dyess, *The Dyess Story* (NY: G P Putnam and Sons, 1944), 99.

Fortunately, the NVN did not emulate the Japanese in China who executed three of the eight American pilots whom they captured after the Doolittle pilots had bombed Tokyo in 1942. The Japanese in China also tried the Americans as "war criminals." Gavan Daws, *Prisoners of the Japanese* (NY: Quill, William Morrow, 1994), 277.

Meaningful Awards

Prepared by JHB. Personal knowledge of COL Fulton's actions.

Combat SAR

Nichols and Tillman, *Yankee Station,* 119-121. See also Waddell and Wood (eds), *Air War-Vietnam,* fn 5, 134.

Electronic Countermeasures [ECM]

Nichols and Tillman, *Yankee Station,* 87-9.

Die Hop

Kish, *Letter to Bradley,* no date.

Seven

AIR WAR VIETNAM III

Vietnam 1966[1]

Important Events		James Morin
NVN increased deployment of AAA guns during Christmas bombing halt.	Jan 1966	CDR Morin began weapons training with VA-155 at Fallon, NV.
Operation MASHER/WHITE WING.* VC attacked Pleiku. 1st Cavalry Division conducted the first "search and destroy" operation in the war against the VC and NVA regiments in II Corps area.	24 Jan-6 Mar 1966	
*"*President Johnson insisted that the name of the operation be changed because MASHER was too crude a title for a nation-building operation and would have an adverse effect on public opinion.*"		
U. S. resumed Operation ROLLING THUNDER.**	31 Jan-1 Feb 1966	
"Secretary McNamara refused to unleash total air power...forbidding attacks on Hanoi and designating buffer zones to prevent strikes from violating the Chinese border or endangering ships unloading in Haiphong."		
NVA troops overran U. S. Special Forces camp in the Ashau Valley.	9-11 Mar 1966	
U. S. B52s bombed NVN for the first time.	12 Apr 1966	
Two A6 Intruders from the USS *Kitty Hawk* bombed the Uongbi power plant near Haiphong and knocked out one-third of NVN's electric power production.	18 Apr 1966	
SVN expelled six American pacifists who attempted to organize anti-war protests in Saigon, rather than execute them.	21 Apr 1966	
	May 1966	**Morin deployed with VA-155 on USS *Constellation* to Southeast Asia.**
	27 Jun 1966	NVN shot down and killed Silver Fox LCDR Gene Smith.
	30 Jun 1966	Morin turned over command of VA-155 to CDR Charles Peters.
	1 Jul 1966	NVN shot down and killed Silver Fox 1, CDR Charles Peters.
	Jul 1966	**Morin arrived in the U. S. and ended his last combat tour.**
U. S. Navy began Operation SEA DRAGON against NVN shipping below 17 degrees and 30 minutes of latitude.	Oct 1966	
U. S. troop strength reached 385,000 in South Vietnam.	Dec 1966	

Cold War/Hot War 1966

BY 1966, THE WAR BECAME REAL TO THOSE CLOSE TO IT WHO WAITED AT HOME OR LOST HUSBANDS, BROTHERS, SONS, OR FATHERS IN THE EXPANDING GROUND AND AIR WARS...

By 1966, the war became real to those close to it who waited at home or lost husbands, brothers, sons, or fathers in the expanding ground and air wars.

Δ

Unlike the losses of individual pilots the losses of ground troops normally numbered in the tens or hundreds after tough fights, and unlike the personal visit of the Base Captain at Lemoore to a new widow, the early notifications to new widows or next of kin elsewhere proved impersonal.

Back at Fort Benning and Columbus, GA, where many of the wives or the next of kin remained after the 1st Cavalry Division deployed, anonymous taxi drivers delivered the first telegrams notifying them that their relatives in the 1st and 2nd Battalions of the 7th Cavalry[2] had been killed in action. Often the shocked next of kin - especially the young wives of enlisted soldiers - did not know what to do until the local command assigned an officer to help each surviving wife, parent, or other family member.

Soon the Army began sending an officer and a chaplain to notify the next of kin of a combat death. Like the Base Captain's sedan at Lemoore, the sight of two officers walking toward a house or apartment terrified worried family members who realized immediately that the officers brought bad news. At best, they could report a soldier missing in action or wounded. To comfort and help the survivors more, Mrs. Hal Moore, wife of the Commanding Officer of the 1st Battalion, started visiting the new widows with another battalion wife.

The reports of deaths travelled quickly through the civilian and military communities at troop posts such as Ft. Benning and highlighted the change in the war. Moreover, soldiers, wives, and young children on the posts often saw the hearses and honor guards and heard the firing parties and the gun salutes from day-to-day as more and more families buried their dead in the post cemeteries. Soon, some young children after learning about war casualties said to their mothers after their fathers left for Vietnam, "My daddy is not coming home." Others, terrified that they would lose their fathers, denied that their fathers even served in Vietnam.

SOON, SOME YOUNG CHILDREN AFTER LEARNING ABOUT WAR CASUALTIES SAID TO THEIR MOTHERS AFTER THEIR FATHERS LEFT FOR VIETNAM, "MY DADDY IS NOT COMING HOME."

One wife recalled those days.

> *The houses were pregnant with families and with men who were all training and training for something new. Then suddenly the whole division left. The men donned*

ENEMY ANTI-AIRCRAFT GUNFIRE (AAA) DOWNED MOST OF THE AMERICAN AIRCRAFT DURING THE WAR AS SHOWN BELOW.

Losses to AAA Fire

Service	**Percent AAA Losses** to **Total Losses** All Causes (known & unknown)	**Percent AAA Losses** to **Known Losses**
USN	58%	77%
USAF	73%	87%
USMC	64%	97%

*Army aircraft losses were mostly helicopter losses.

their khaki armor, were dubbed the Air Cavalry, mounted their helicopters, and left on their crusade to South Vietnam. But first they massed the whole division of helicopters in a farewell fly-by over the housing area as we all stood by our kitchen doors ... and watched them go. They were followed by moving vans salting out the houses and carrying off their goods and children and wives to wherever they would be dumped. Some went to the substandard housing complex which sat... between us and the cemetery.

It seemed that we were not back in our houses very long before the funereal guns began sounding and their insistent pounding beat on our days. The taxis began delivering their midnight death telegrams about then until finally the horrified authorities began to intercept them and the remaining men began pulling duty to deliver them by daylight.

About then we had to stop watching the early news because our four year old began to understand what we did not want him to ever have to, but most certainly not at four. Almost all the children had gone and all the daddies had gone off to *war and none had returned except to the cemetery whose gates we had to pass whenever we left the housing area and got caught in the traffic jam of cars with young women and small children going in and out to gravesites.*[3]

Δ

As the ground war intensified, the air war changed. In January 1966, Americans sighted newer MiG 21 fighters away from the Hanoi area. The first sighting occurred ninety miles east of Thanh Hoa; the second twenty-five miles southeast of Thanh Hoa. These showed that the North Vietnamese Air Force planned to extend its operations over the Tonkin Gulf.

In addition to the MiG sightings, the North Vietnamese increased their deployment of anti-aircraft guns, much of it during the bombing pause from 24 December 1965 to 31 January 1966.[4]

"IT BECAME OBVIOUS BY EARLY 1966 THAT THE NORTH VIETNAMESE WERE BEING LIBERALLY SUPPLIED WITH VAST AMOUNTS OF AAA."

It became obvious by early 1966 that the North Vietnamese were being liberally supplied with vast amounts of AAA. Before the December standdown, only one mission in twelve - 8 percent - had been engaged by North Vietnamese anti-aircraft fire. However, by late February 1966, AAA defenses had increased to the point where one mission in four was being engaged...

A part of this rapidly expanding AAA and MIG threat was a more sophisticated radar-controlled environment which was capable of providing the range, altitude, speed, and azimuth of US aircraft to enemy gunners or pilots.[5]

Air Force and Marine Corps pilots suffered higher losses to ground fire because they flew mostly over South Vietnam and flew many close-support missions for ground troops.[6]

The Navy should have expected such results because the Communist AAA units in Korea had shot down 540 Navy and Marine aircraft.[7]

All the services should have expected to face heavy anti-aircraft fire because Soviet Army doctrine strongly emphasized AAA defense, the Soviet Army had many more AAA units than the U. S. Army, and the Soviets supplied and supported the North Vietnamese during the air war.[8]

On 1 February, the Air Force and the Navy re-started Operation ROLLING THUNDER after a thirty-seven day pause.[9]

Ψ Turn Around

We had only four and one-half months to conduct training before we were scheduled to sail again to Vietnam in the USS *Constellation* (CVA-64). We had much to do so I told my new XO, CDR John Harris, to take two weeks leave while I worked. Then, I planned to take two weeks around Christmas when he came back. When I returned, I would take VA-155 to Fallon, NV for weapons training around the first of January 1966.

Just after settling into Lcmoore, Navy Air Force Pacific Fleet Headquarters (NAVAIRPAC)[10] called the Wing Commander and the Squadron Commanders to San Diego to brief the Admiral and his staff about the war, the tactics we employed, what we needed to do to improve training, and what kind of new or additional equipment we needed. I also visited the Douglas Aircraft Company in the Los Angeles area and told the folks there about the A4 in combat.

The Navy decided to put some new blood into our A4 squadrons. The command notified me that five young aviators from Cecil Field in Jacksonville, FL would join my squadron, and told me to transfer four of my young veterans to the Cecil Field squadrons.

I had a difficult time picking out the four officers to be sent

AVIATORS

"THE AVIATORS HEADED FOR COMBAT IN THE MID-1960S SHARED COMMON CHARACTERISTICS. ALL WERE WELL EDUCATED, SOME WITH ADVANCED DEGREES, AND MOST WERE STABLE FAMILY MEN WITH CAREER GOALS. SOME COULD BE DESCRIBED AS CEREBRAL, AND THERE WAS A SPRINKLING OF GENUINE PHILOSOPHERS. IN CONTRAST, AND SOMEHOW COMPLETELY MELDED, WAS THE FIGHTER JOCK MENTALITY. BUT MAKE NO MISTAKE ABOUT IT, MANY CARRIER-BASED ATTACK PILOTS HAD THE VF [FIGHTER PILOT] ATTITUDE JUST AS MUCH AS THE PHANTOM AND CRUSADER DRIVERS.

HARD CORE AVIATORS WOULD RATHER DIE THAN LOOK BAD. THEY WOULD SHUN THE ADAGE, *DEATH BEFORE DISHONOR*, AS PRETENTIOUS, BUT THAT'S EXACTLY WHAT MANY BELIEVED.

THERE WERE ALSO SOME OFFICERS, INCLUDING A FEW OUTSTANDING STICK-AND-RUDDER MEN, WHO FELT FEW OF THESE MOTIVATIONS. THEY WERE SIMPLY CAREERISTS WHO WANTED THEIR TICKETS PUNCHED ALONG THE WAY. AND ONE OF THE STOPS ON THE ROAD TO REAR ADMIRAL WAS A COMBAT TOUR.

OTHERS, FEWER YET, NEVER MADE A WESTPAC DEPLOYMENT. SOMETIMES WIVES DUG IN THEIR HEELS, AND, IN EFFECT SAID, *ME OR THE WAR. TAKE YOUR PICK*."

to Cecil Field because those I did not choose would deploy to Vietnam again with me in four and one-half months. I finally selected George Blosser, Jerry Chasko, Bill Kish, and Jim Mattly. The east coasters who replaced them were good pilots. A couple of them tried to impress us with how good they were, but we were not impressionable after our combat tour. They were joining the Pacific Fleet, the real Navy.

At this time, Ken Holthaus reached the end of his obligated service and left the Navy. When I first joined VA-155, I figured one of us was not going to last long and I knew it was not going to be Morin who would go. But after eighteen months with the "King," I hated to see him leave. I knew that I had made him change his ways, and, in his own inimitable way, he must have changed me. We had a farewell beer-bust for him at the Lemoore Club and at the end of the party Ken had his say. He could have thumbed his nose at us, called us a bunch of dumb bells for putting up with the malarkey, but the past cruise had made its mark on him too. Ken's farewell remarks ended up in the *Quotable Quote Logbook*: "I really had a great time in the Navy the last five years. I wouldn't trade a minute of it. I liked eighty percent of it, but it was the other twenty percent that really offended me."

More Foxes left us at Lemoore. Some left for new assignments; some left at the end of their obligated service. Dick Floreani, Ed Hiebert, Ernie Mares, Chris Morss, and Jay Starling departed. So did George Hart. Captain Ed Huff, another good Marine, replaced Hart.[11] Hardworking Jim Tillis left too.

More Foxes left us at Lemoore. Some left for new assignments; some left at the end of their obligated service. Dick Floreani, Ed Hiebert, Ernie Mares, Chris Morss, and Jay Starling departed. So did George Hart.... hardworking Jim Tillis left too.

The A4 training squadron sent me four nuggets: LTJG Stu Fitrell, LTJG Neil Holben, LTJG Glenn McGeorge, and ENS Butterfield. In addition, the squadron sent me a new Operations Officer, Tony Karpaitas.[12]

The A4 training squadron sent me four nuggets: LTJG Stu Fitrell, LTJG Neil Holben, LTJG Glenn McGeorge, and ENS Butterfield...

Gene Smith, from the Mormon Smiths, joined and I appointed him my Administrative Officer. Bill Cook[13] and W. P. Smith[14] also transferred into the squadron. Paul Moore became the Maintenance Officer, relieving Bill Fidelibus who had orders to a new assignment.

Ψ

Already Died

Peters, a giant of a man of American Indian heritage, filled up the cockpit of an A4. When he closed the canopy, his shoulders touched both sides of the cockpit.

I finally got my chance to go on leave. Around 4:00 p.m. on my first day off, a few days before Christmas, the Squadron Duty Officer called and told me that the XO was not feeling well and was going to the Dispensary. Early the next morning, I got another call that the XO was in critical condition. Betty and I went out to the base. Shortly thereafter, John Harris died of meningitis. Now that was not fair after what he and his family had been through the previous months. His tragic death sure ruined the Christmas season for everyone.

Once again, I did not have an Executive Officer as we approached the start of training. I knew that a Commander in the A4 training squadron, Chuck Peters, had been the XO of a squadron. He had broken an ankle in a crash at sea when his squadron was leaving for Vietnam. Sent home to recuperate, he still walked with crutches. After we lost John Harris, Peters pleaded to be assigned to VA-155. I got him assigned as my XO. Even though he could not fly for four weeks, Chuck deployed to Fallon with us.

Peters, a giant of a man of American Indian heritage, filled up the cockpit of an A4. When he closed the canopy, his shoulders touched both sides of the cockpit. He told me that he was anxious to get into combat because he told me that he had already "died."

Chuck's story was incredible. Late one night during preparation for a cruise, he was flying an A4 tanker and was circling the ship waiting for the last jet to land. After the last jet landed, he got an immediate "CHARLIE." To save time, he flew a tear drop approach, descending outbound to half the altitude, and then turned inbound to the ship. He had to dump fuel in the external tank before landing, but the tank dumped slowly – agonizingly so – while he had the aircraft in a nose-down attitude. As Chuck flew inbound to the ship and approached the level-off attitude, he became so engrossed in securing the fuel dump valve that he flew smack into the water at 240 knots. When the

BESIDES DEALING WITH IMPORTANT TRAINING, I FOUND I HAD TO DEAL WITH PAINT: "SOMEONE" ON HIGH DECIDED TO PAINT OUR AIRCRAFT WITH CAMOUFLAGE PAINT TO SEE IF IT WOULD REDUCE THE CHANCE OF BEING HIT.

aircraft hit the water, the impact bent the nose upward, breaking Chuck's lower leg. The phosphorescence of the disturbed water cast an eerie glow as the aircraft sank. Still on oxygen, he realized that he was still alive. He immediately decided to open the canopy and swim to the surface. Unable to free himself from his seat, Peters unzipped his survival vest and harness in order to get out of the sinking aircraft. He got to the surface, but he did not have his flotation and survival gear. Fortunately, an external fuel tank broke free from the A4 and surfaced nearby. Chuck immediately wrapped himself around the tank to keep himself afloat. Someone said that a large fish bit him on the behind while he held on to the fuel tank, but he never confirmed that. Luckily, a destroyer found Chuck a few hours later and pulled him out of the cold Pacific.

Ψ

During the weeks we trained at Fallon, we concentrated on firing rockets, dive bombing, and strafing. We flew race track patterns, and with no one shooting at us, we found that we were extremely accurate. I found that I could put my bombs on average within eighty-one feet of a target. Before the war, the fleet average accuracy was 125 feet. Not very good when trying to hit a fourteen-foot-wide bridge while getting shot at. These numbers, of course, were practice range numbers.[15]

We returned to Lemoore and started to practice our night carrier landings because soon we expected to go aboard the *Connie* for refresher landings and work ups. In the four and one-half months between cruises, we spent half of our time away from Lemoore.

BESIDES DEALING WITH IMPORTANT TRAINING, I FOUND I HAD TO DEAL WITH PAINT...

Besides dealing with important training, I found I had to deal with paint: "someone" on high decided to paint our aircraft with camouflage paint to see if it would reduce the chance of being hit. The command decided to paint the airplanes in my squadron and two more that were joining the Air Wing: VA-65, the A6 Squadron from Oceanna at Virginia Beach, commanded by Bill Small;[16] and RVAH-5, a heavy reconnaissance squadron, commanded by Charlie Smith[17] I argued against the camouflage paint, telling NAVAIRPAC that it would make us more visible from the ground. To prove my points, I had pictures taken of an A4 from the Blue Tail Flies and one of my newly painted aircraft flying together. From the ground, it was hard to see the Blue Tail's A4, but my A4 stood out like a sore thumb. From above, the camouflage paint worked, but our threat was from below, not from above.

By the time we flew out to the *Constellation*, our replacement pilots had integrated themselves into the squadron very well, and each aviator had sharpened his skills for the coming combat cruise.

Ψ

The time to sail west came up very fast. In early May 1966, we left California. After our Operational Readiness Inspection in Hawaii, we deployed to Vietnam. A civilian author, Mr. Frank Harvey, who was writing a book about the Navy and Air Force participation in South East Asia came aboard in Hawaii. CAG McDonald assigned him to me and told him to go wherever Morin went. Harvey's book, *Air War, Vietnam* contains a section about his visit with us on the *Connie.* There are quotes attributed to me in his book where I was cited as saying that I was "utterly fearless" and that "I just bend my bird around" to evade a SAM. Poetic license, I guess, because those were not my words.

We went through Japan where the outgoing wing on the USS *Independence* briefed us. Pretty soon, we were back at it, flying strikes against the Thanh Hoa Bridge.[18] The Bridge had remained a problem since 1965. Air Force and Navy aircraft attacked it time and time again without real success.

> *Attacking the target with small strike forces of two to four aircraft, the Navy hit the bridge some 24 times with a total of 65 aircraft between June 1965 and the end of May1966. The ordnance dropped during the period was approximately 128 tons.*[19]

POETIC LICENSE

ONE OF THE SQUADRONS ... HAD A SKIPPER I LIKED - COMDR. JIM MORIN, ... A FRESH-FACED BOSTON IRISHMAN WITH A TWINKLE IN HIS EYES. IF YOU LOOKED CLOSE YOU SAW THE SQUINT WRINKLES AND THE FLAT STARE THAT CAME WHEN HE MEANT BUSINESS, AND YOU KNEW THAT THIS MAN HAD BEEN AROUND THE BASES. HE WAS 38 AND ON HIS LAST TOUR OF COMBAT. HE WAS [THE CO] FOR THE SILVER FOXES ... ALSO, HE HAD A SENSE OF HUMOR.

"ONE DAY WE WERE SITTING IN HIS READY ROOM WHEN THE TYPIST ... WHOSE JOB IT WAS TO TYPE COMMANDS WHICH THEN APPEARED ON A LARGE LUMINOUS SCREEN IN ALL THE READY ROOMS, BEGAN ONE OF HIS "UNPREDICTABLE VOYAGES OF DISCOVERY," AS JIM REFERRED TO HIS MADLY GARBLED TRANSMISSIONS. ... THE BOY STARTED OUT BRISKLY: *PILOT MAN YOUR.* THEN THERE WAS A PAUSE. THEN FOR SOME REASON, FOUR DOTS APPEARED, FOLLOWED BY A SINGLE LETTER . ANOTHER PAUSE. FOUR MORE DOTS. THE MAN SEEMED TO BE WORKING UP TO SOMETHING. SUDDENLY, IN A STACCATO, HE TYPED *PILOY.* THERE WAS A SOMEWHAT LONGER PAUSE AS THE TYPIST GATHERED HIMSELF FOR THE SUPREME EFFORT. IT CAME OUT IN A DYNAMIC SINGLE BURST: *PIHOTS MAN YOUR PLANQB.* CMDR. MORIN LOOKED AT ME. "I THINK HE'S TRYING TO TELL US SOMETHING," HE SAID GRAVELY. "I'M PRETTY SURE THAT'S IT."

I ASKED JIM MORIN ONCE IF HE WAS WORRIED ABOUT GOING INTO COMBAT ... WAS HE AFRAID OF THE HEAVY FLAK THERE? "AFRAID", MY GOOD MAN?" .... "I DON'T KNOW THE MEANING OF THE WORD. I'M UTTERLY FEARLESS AND ALWAYS HAVE BEEN." THEN HE SAID, A BIT GRIMLY, "SURE I'M SCARED OF THE FLAK. WE ALL ARE UNTIL WE GET IN IT. THEN WE'RE TOO BUSY." "HOW ABOUT SAMs?" "NO BIG THING," CMDR MORIN SAID. "BUT A FELLOW USUALLY BENDS HIS BIRD AROUND RATHER SMARTLY WHEN HE SEES ONE COMING UP."

FRANK HARVEY

We also hit Hanoi and Haiphong. During these first missions back over North Vietnam, my camouflaged planes took more hits than our sister squadron. I attributed that to our paint jobs so I had the camouflage paint removed.

One day I led a gaggle of A4s into downtown Hanoi. The North Vietnamese lighted us up with radar and we knew they would launch missiles quickly. Marine Captain Ed Huff, who was flying ahead of the strike group, quickly zeroed in on a radar beam and zapped the radar with an anti-radiation missile. The SAM-2's fire control system needed radar information to steer the missiles. Ed saved our bacon. I got him a Silver Star.[20]

One day I led a gaggle of A4s into downtown Hanoi. The North Vietnamese lighted us up with radar and we knew they would launch missiles quickly. Marine Captain Ed Huff, who was flying ahead of the strike group, quickly zeroed in on a radar beam and zapped the radar with an anti-radiation missile. The SAM-2's fire control system needed radar information to steer the missiles. Ed saved our bacon. I got him a Silver Star.

Jake Ward, the CO of one of the F4 squadrons, replaced CAG McDonald at this time. Then Bud Ingley[21] arrived to be the new XO for VA-155 because my detachment date was nearing.

We flew very heavily during this line period so Chuck Peters and I agreed that we would have no time for a formal Change of Command ceremony. We selected 29 June for the turnover, and Chuck promised me that two days before that date I would hang up my flight gear and fly no more missions. In 1965, a Squadron CO had been shot down on the day before he was to leave his ship for the States. Chuck remembered this and wanted to get me out of the sky. I guess lots of people heaved sighs of relief when I flew my last mission.[22]

Many in Vietnam found disengagement from combat to be a problem. Disengagement often triggered strange reactions. Some people became exceedingly nervous during the last days of their tours, worrying, of course, that they might be killed. That would affect their performance, and if they did not remain sharp and focused, they increased their chances of being killed on their last missions or getting others killed.[23] Others just became extremely cautious.

The day before our Change of Command ceremony, the North Vietnamese shot down Gene Smith.[24] The command immediately began a SAR effort because he had crashed about six or eight miles inland. The carriers had launched some big strikes at the time and few pilots and planes remained aboard to help out with the rescue attempt. We sent out two-plane sections, one from each A4 squadron, to aid in the rescue. When the last available pilot from my squadron flew off to help, I ran back to my stateroom and unpacked my flight gear. As soon as the aircraft that had flown with Gene returned, the deck crews refueled and rearmed them for an immediate launch. I got a quick briefing on the crash site and the kind of opposition to expect, and then with a wingie from the Blue Tails, I was launched.

As we showed up, the two pilots at the scene had run low on fuel. They quickly told me where to look and left. We ran in low and fast and passed over the burn mark on the ground where Gene's A4 had crashed at the end of its dive.

North Vietnamese soldiers swarmed all around the site. We saw no sign of Gene. We flew in and out a number of times, but saw no additional activity and heard nothing on the radio.

A little later, our relieving pilots called and relayed a message from Captain Bill Houser[25] of the *Connie*. Houser instructed me to remain out over the water and not to go inland again. No one saw Gene on the ground and no one saw Gene eject, so the Navy reported him later as killed in action.

With this gloomy note, Chuck Peters and I held our Change of Command ceremony in our Ready Room the next morning. The informality of the ceremony kept emotions low. After the preceding seventeen months or so, the emotions had been pretty well drained out of us anyway. That noon, after Chuck took command of the Silver Foxes, I flew off to the Philippines.

Only two of my original Silver Fox pilots remained behind with Chuck: LTJG Jim Hughes and LTJG Al "Albie" Hyde. LCDR Paul Moore, our stalwart replacement stayed on also. The other new Foxes included Bud Ingley, Tony Karpaitas, Bill Cook, W. P. Smith, Ed Huff, and the four nuggets; Stu Fitrell, Neil Holben, Glenn McGeorge, and young ENS Butterfield.

During my time as the Commanding Officer of VA-155 in 1965 and 1966, my most demanding time and probably my most productive time as a pilot leader, I had lost only one aviator KIA, Gene Smith, and one aviator MIA, Wendy Rivers.

1 JUNE 1966
VA-155

CO
CDR JAMES MORIN

XO/CO DESIGNATE
*CDR CHARLES PETERS

XO DESIGNATE
CDR BUD INGLEY

OPERATIONS OFFICER
LCDR TONY KARPAITAS

ADMINISTRATIVE OFFICER
#LCDR GENE SMITH

MAINTENANCE OFFICER
LCDR PAUL MOORE

PILOTS
*LT WIMER (BILL) P. COOK
CPT ED HUFF, USMC
LT W. P. SMITH

LTJG ALBIE HYDE
LTJG JIM HUGHES
&
4 NUGGETS
LTJG STU FITRELL
LTJG NEIL HOLBEN
LTJG GLENN MCGEORGE

ENS BUTTERFIELD

#KIA ON MORIN'S WATCH

*KIA LATER

CASUALTIES DURING THE MORIN LINE PERIODS

SILVER FOXES VA-155		
USS *CORAL SEA*		
LT ED DICKSON	KIA	7 FEB 65
LCDR WENDY RIVERS	MIA/POW	10 SEP 65
USS *CONSTELLATION*		
LCDR GENE SMITH	KIA	27 JUN 66

ONCE AGAIN, WHEN REPORTS ANNOUNCED THAT SILVER FOX ONE HAD BEEN SHOT DOWN, MANY PEOPLE THOUGHT THAT MORIN HAD BOUGHT IT UNTIL FOLLOW-UP REPORTS IDENTIFIED CHUCK PETERS.

Three Days

Later that afternoon, I boarded a chartered airliner at Clark Air Force Base in the Philippines and headed for California. On my arrival at Lemoore, word reached me that Chuck Peters had been shot down around Hanoi on 1 July 1966, twenty-four hours after our Change of Command.[26] After being hit, he had flown his uncontrollable A4 to the safety of the sea and ejected. After his parachute opened, others saw Chuck hanging limply in his harness. After landing in the water, no one saw him come to the surface. Looking back, I assume that Chuck had died before he entered the water or he was so badly hurt or wounded that he could not get out of his parachute harness and drowned when the parachute filled with water and pulled him under. Once again, when reports announced that Silver Fox One had been shot down, many people thought that Morin had bought it until follow-up reports identified Chuck Peters.[2] What a difference three days made!"

Δ

Missed Opportunities

The United States failed to take three actions by the time Jim Morin returned to the United States that would have changed the outcome of the war.

First, they did not land large forces in NVN below Hanoi. Such action would have caused the enemy to withdraw all or most of its troops from SVN. While the JCS considered such action, nothing occurred. Former NVA COL Bui Tin verified the threat such an invasion presented to North Vietnam.

> *Hanoi was terribly afraid that the United States would carry out such an attack. In early 1963, word went out in the North to build traps and pungi sticks[28] in a number of areas from Vinh to Nghe An Province south to Vinh Linh and drills were carried out every year. A massive landing would have had an effect totally different from an air raid.[29]*

Second, they did not concentrate their forces in the I Corps Area up against the DMZ along the 17th Parallel from which they could have launched major attacks to cut off the Ho Chi Minh Trail in Laos near the junction of the border between North Vietnam and South Vietnam.[30] North Vietnamese leaders feared any ground interdiction of the trail.[31]

Third, they did not blockade the port of Haiphong and bomb the ships bringing in supplies to North Vietnam.

Such actions would have sent clear, harsh, and effective messages to the North Vietnamese that the United States wanted the North Vietnamese out of South Vietnam rather than the tepid and ineffective messages that President Johnson and Secretary McNamara sent with their incremental increases of armed force and their severe restriction on the use of armed force.

Casualties During the Morin Line Periods

Other Squadrons		
USS *Coral Sea*		
RMCA Dwight G. Frakes	Drowned	24 Feb 65
LCDR Kenneth E. Hume	KIA	Mar 65
LT David A. Kardell	KIA	9 May 65
LT William M. Roark	KIA	4 Jun 65
CDR Peter Mongilardi, Jr.	KIA	25 Jun 65
LT Edward B. Shaw	KIA	Summer 65
LT Charles B. Goodwin	KIA	Summer 65
LCDR Harry E. Thomas	KIA	13 Aug 65
Andrew L. Furrer	KIA	
LCDR Robert Shumaker	MIA/ POW	
USS *Constellation*		
LT Harry J. Belnap	KIA	23 Jun 65
LTJG Virgil K. Cameron	MIA/ KIA	24 Jun 66

Sidebar Notes

Losses to AAA Fire
Nichols and Tillman, *Yankee Station*, 53.

Aviators
Nichols and Tillman, *Yankee Station*, 34-5.

Poetic License
Frank Harvey, "The Air War in Vietnam," *FLYING*, A special supplement September 1966, 50.

1 June 1966 VA-155
Morin, *Manuscript* and *Interview with Admiral Morin*, 20 October 2010, Tallahassee, FL.

Casualties During the Morin Line Periods
USS *Coral Sea Website* and USS *Constellation Website* (1 June 2010).
- LCDR Virgil King Cameron graduated from the U. S. Naval Academy in 1964, #32924 and died in 1966 at age 27. He most likely was a LTJG at the time the Navy declared him Missing in Action. He remained classified as MIA for eight years and during that time the Navy apparently promoted him to LT and then LCDR. RADM Morin said that was normal procedure for officers declared MIA. *1993 USNA Register of Alumni*, 393, and *Arlington National Cemetery Website* (2 July 2010). MIA, 7 September 1965.
- LT Bill Roark, also a graduate of the U. S. Naval Academy, had a destroyer named in his honor.

End Notes

[1] Gutzman, *Vietnam*, 12-13 and Nalty, *Vietnam War*, 101 (McNamara), 164 (Uongbi and Sea Dragon), 20 (December strength).
[2] Prior to the Battle of Ia Drang, the most famous battle that units of the 7th Cavalry fought was at Little Big Horn where Chief Sitting Bull annihilated Brevet Major General (Lt. Colonel) George Custer's small force.
[3] Ann M. Bradley, *Unpublished paper*, c. 1966.
[4] Colonel Dewey Waddell and Major Norm Wood (eds.), "Part I, The Tale of Two Bridges," *Air War-Vietnam* (NY: Arno Press, 1978), 50.
[5] *Ibid.*
[6] Nichols and Tillman, *Yankee Station*, 53.
[7] *Ibid, 49.*
[8] In the study of comparative ground forces in Army schools, American officers found that the Soviets always had more AAA units than the U. S. Army had. For all intents and purposes, the Army relied on the U. S. Air Force to protect its ground units from enemy air forces and with the advent of the ground-to-air missiles the Army had abandoned providing AAA weapons such as the track-mounted "quad-50" caliber AAA gun system and the "twin-40mm" AAA guns to its ground forces. In the 1950s, the Nike missile weapons systems replaced the high firing 90mm AAA guns that once protected American cities. Personal knowledge. JHB.
[9] Lt. Colonel Gordon Nelson and Major Norm Wood (eds.), "Part III, The Battle for the Skies over North Vietnam," *Air War-Vietnam* (NY: Arno Press, 1978), 236.
[10] Naval Air Pacific.
[11] Huff is deceased.
[12] Karpaitas retired as a Captain, USN.
[13] Bill Cook. LCDR Wimer P. Cook flying a A4-E for VA-155 died on 22 December 1967 over North Vietnam, apparently by his own bomb. *Coral Sea Website.*
[14] W. P. Smith retired as a Captain, USN.
[15] Morin, *Email to Bradley*, 14 July 2009.
[16] William Newell Small graduated from the U. S. Naval Academy in 1948. He retired as an Admiral in August 1985. Small, the highest ranking of Morin's peers, earned a Legion of Merit, a Joint Staff Commendation Medal, a Navy Commendation Medal, and a Navy Unit Commendation Ribbon. *1993 USNA Register of Alumni*, #199961, 293.
[17] Charlie Smith retired as a Captain, USN.
[18] Much has been written about the bridge. Frank Harvey wrote about LCDR Pete Garber, who flew A6s on the *Constellation*, and had a fixation on the bridge. Harvey wrote: "He was after the Thanh Hoa Bridge, which he said was the "big status symbol" target in North Vietnam... All in all, Garber said, just about every gung-ho pilot in both the Air Force and Navy had taken a crack at the Thanh Hoa Bridge...But they didn't all get off without getting hurt because the Than Hoa Bridge is surrounded by masses of big and little guns that sprouted like steel asparagus." Garber said to Harvey, "That bridge belongs to me." Frank Harvey, "The Air War in Vietnam," *FLYING*, A special supplement September 1966, 50.
[19] Waddell and Wood (eds.), *Air War-Vietnam, 46.*
[20] Morin, *Email to Bradley*, 21 May 2009.

[21] Ingley retired as a Captain, USN. Albie Hyde wrote the following about Ingley: "CDR Ingley, who was very new as XO, took over as CO, and he just didn't have it. Ingley was an OK naval officer, and a fair pilot but he was no leader. In a way the pilots who went to the east coast had a good deal because they didn't have to operate in a squadron that was slowly sliding down hill." Hyde, *Letter to Bradley.*

[22] Pulling someone out early was a good move. Peters might have been better not to have told Morin that he would stop flying when he did, but rather just grounded Morin without warning two days before his tour ended.

[23] "In the Mekong Delta, we had a sergeant assigned to our advisory team who just became more and more unhinged and useless as his rotation day neared. His reaction apparently was not uncommon. There were reports that commanders in some units pulled people out unannounced a day or so before they rotated to negate this sort of psychological." JHB.

[24] LCDR Gene Albert Smith is listed as killed in action on 27 June 1967. USS *Constellation Website.* (1 June 2010).

[25] Houser retired as a Rear Admiral.

[26] CDR Charles Henry Peters is listed as killed in action on 1 July 1966. USS *Constellation Website.* (1 June 2010).

[27] Albie Hyde wrote: "When it was time for the Skipper to be relieved it was like losing a father. Cdr Peters took over and was shot down the next day so we never knew whether he would be a good CO or not. He was a good naval officer, a poor pilot and an unknown quantity as a leader." Hyde, *Letter to Bradley.*

[28] Pungi sticks were sharpened pieces of wood or bamboo that the enemy placed in small and large camouflaged holes to wound troops who unknowingly stepped into the holes. While not fatal, the sticks could seriously injure a soldier's foot or feet or his leg or legs.

[29] Bui Tin, *From Enemy to Friend, a north vietnamese perspective on the war* (Annapolis, MD: Naval Institute Press, 2002), 159.

[30] *Ibid,* 86. Author referred to ideas presented after the war by GEN Bruce D. Palmer, Vice Chief of Staff and Acting Chief of Staff, U. S. Army.

[31] *Ibid,* 74-5.

Casualties During the Morin Line Periods - Other Squadrons

Journalist 2-C Robert West and Lieutenant (jg) Manuel Perez, THE STORY OF A CARRIER WEST, 1964-1965, UNITED STATES SHIP CORAL SEA listed Frakes, Furrer, Hume, Kardell, Mongilardi, Roarke, and Shumaker as casualties.

Sea Breeze, March 1965, 19, listed Frakes as a casualty.

Sea Breeze, January 1966 listed Frakes, Furrer, Goodwin, Hume, Kardell, Mongilardi, Roark, Shaw, and Shumaker as casualties.

Eight

SHORE DUTY AND MORE

Watching a bear

Hot War During the Cold War / Vietnam 1966-1971[1]

Important Events		James Morin
North Korean pilots began training NVN pilots to fly MiG 21 fighters.	1966	CDR Morin to BUPERS; appointed Recording Secretary to Awards Board.
Silver Fox Albie Hyde flew last combat mission.		Morin selected to be a CAG.
U. S. conducted first air strikes on Hanoi.		
Former Silver Fox CO Jack Harris died during the fire on the USS *Oriskany*.	**Oct 1966**	
U. S. Air Force conducted Operation BOLO and destroyed 7 MiGs in air-to-air combat.	Jan 1967	
Communist gunners downed 13 U. S. helicopters in one day.	1967	Morin expected assignment to Pacific Fleet and return to Southeast Asia. Assigned instead to Atlantic Fleet as CAG.
U. S. N. aircraft struck military targets in Hanoi.	1967	
Fires on USS *Forrestal* killed 134 crew members and destroyed 21 aircraft.	Jul 1967	
U. S. conducted first air strikes on Haiphong shipyards.	1967	
50,000 anti-war protestors marched on the Pentagon.	Oct 1967	
NVA and VC launched the TET Offensive in SVN. Enemy suffered a military defeat; won a propaganda victory.	**Jan 1968**	
NVA bombed Laos. One CIA helicopter crew chief shot down one NVN aircraft with an AK-47 assault rifle.	**1968**	**Morin joined fleet in Mediterranean; became CAG 7 on USS *Independence*.**
Chinese continued to operate radar sites in North Vietnam.	1968	
President Johnson suspended bombing north of the 20th Parallel.	Mar 1968	
Paris Peace Talks began to end Vietnam War.	**May 1968**	
	Jun 1968	Morin returned to U. S. and took command of Carrier Replacement Wing 4.
President Johnson halted all bombing raids over North Vietnam.	Oct 1968	
The U. S., the U. K., and the U. S. S. R. signed the "Nuclear Non-Proliferation Treaty."	1968	
President Richard Nixon took office.	Jan 1969	
B52s bombed Laos in secret raids.	Mar 1969	
101st Airborne Division fought the Battle of Hamburger Hill.	May 1969	
President Nixon ordered/approved the successful invasion of Cambodia.	Apr 1970	
U. S. fighter shot down one MiG 21, the only one in 1969-1970.	1970	Carrier Replacement Wing 4 became Light Attack Wing 1. Morin remained CAG.
U. S. continued protective air strikes into North Vietnam all year.	**1971**	**Morin left Light Attack Wing 1. Ended his flying career.**
Operation PROUD DEEP ALPHA conducted for five days in 3 SVN provinces below the 19th Parallel.	1971	

Cold War 1966-1971

The late 1960s saw several developments: the intensification of the War in Vietnam, the development of a virulent anti-war movement in the United States, and the negotiations with the Soviets to reduce the chance of nuclear war.

Δ

In 1966, GEN Westmoreland expanded the war of the big battalions in Vietnam, and new U. S. divisions arrived and began conducting larger operations against the Viet Cong in the South. In 1967, large American operations increased. In January 1968, during the traditional truce at TET, the Chinese New Year, the North Vietnamese and Viet Cong launched a countrywide offensive against South Vietnamese cities and bases, and the U. S. Embassy and Ton Son Nhut Airbase in Saigon. Enemy forces seized the ancient capital city of Hue in the Highlands.

The American press reported the TET Offensive as a massive enemy victory, but, in fact, South Vietnamese and American forces repelled most attacks except the one at Hue and crushed the Viet Cong who had surfaced from underground in hopes of a total victory. NVA GEN Giap believed the offensive was a terrible defeat. Moreover, the counteroffensive inflicted even more damage, "resulting in some of the greatest losses for our [NVA] side in all of 1968, 1969, and 1970." Former NVA COL Bui Tin reported:

> *Practically all our urban hideaways which took us years to build, were uncovered in the process and wiped out. Even rural bases were lost on a vast scale. Northern troops had to be brought south to fill in the huge gaps created by the losses. On the spot mobilization became extremely difficult. In not a few localities, so-called regional troops at the provincial and district levels – even so-called village militia – were actually fresh recruits form the northern provinces of Than Hoa, Nghe An, and Ha Tinh.*[2]

However, the TET Offensive unnerved American leaders and soon President Johnson decided not to run for re-election, fulfilling NVN goals to "shake up support in the rear," and "give a big boost to the American anti-war movement."[3] In addition, it started to turn the American people against the war, even though when American and Vietnamese troops recaptured Hue, they found that the enemy had murdered around 5,000 Vietnamese civilians, the greatest single atrocity of the war.

REFLECTION: SOVIET WAR PLANNING

MARSHAL N. A. BULGANIN, THE PREMIER OF THE SOVIET UNION FROM 1953 TO 1955, WROTE IN A BOOK PUBLISHED IN 1948 THAT "PLANNING FOR WAR SHOULD NOT BE LIMITED TO THE DRAWING UP OF OPERATIONAL MILITARY PLANS BUT SHOULD TAKE FULLY INTO ACCOUNT THE ECONOMIC AND PSYCHOLOGICAL ELEMENTS AS WELL."

ACCORDING TO HIM, "THE PRINCIPAL MISTAKE OF BOURGEOIS MILITARY LEADERS LIES IN THEIR OVERSTRESSING OF THE STRICTLY MILITARY ASPECTS OF WAR; AT BEST, THEY ARE ABLE TO EXTEND THEIR THINKING TO A CONSIDERATION OF WAR POTENTIALS. ACTUALLY, ACCORDING TO HIM, WAR PLANNING MUST COVER THE MILITARY, THE ECONOMIC, AND THE PSYCHOLOGICAL FIELDS, AND OPERATIONS MUST BE CORRESPONDINGLY UNDERTAKEN IN ALL THREE SPHERES."

"MODERN WAR HE BELIEVES IS BOTH A NATIONAL AND SOCIAL CONFLICT. PROPAGANDA IS A WEAPON AS EFFECTIVE AS CONVENTIONAL MILITARY MEASURES. POLITICAL TECHNIQUES CAN PARALYZE AN ARMY THROUGH DESTRUCTIVE PROPAGANDA AIMED AS THE ANNULMENT (RATHER THAN) THE ANNIHILATION OF HOSTILE FORCES. PSYCHOLOGICAL WARFARE CAN BE THE MAJOR FACTOR LEADING TO THE OVERTHROW OF THE DEFEATED NATION'S GOVERNMENT AND ITS REPLACEMENT BY A NEW REGIME SOCIALLY AND POLITICALLY ACCEPTABLE TO THE VICTOR."

1968
A Mistake or a Deliberate Attack?

USS *Liberty*
Six Day War
June 1968

The new developments changed the political situation: American leaders announced implementation of Vietnamization, a program to turn more of the war operations over to the Vietnamese; and, Peace Talks in Paris began in May 1968.

On the other side of the world, the United States faced another crisis as a six day war erupted between Israel and Egypt and Syria: the Israelis attacked and nearly sunk a U. S. intelligence ship, the USS *Liberty,* on 8 June 1968. The incident may have been as dangerous as the Cuban Missile Crisis.

The Israeli strike killed thirty-four officers and men, wounded 170 crew members, and severely damaged the ship.

Israel claimed that the attack was a mistake; some officials included ADM Thomas Moorer, the CNO, believed that the Israelis deliberately attacked the *Liberty.* Moorer believed that President Johnson covered up the attack in order to keep good relations with Israel. RADM Lawrence Geis, commanding the carriers in the Mediterranean, told a crew member that Johnson said to him: "I will not embarras our ally."

The U. S. did not embarrass Israel.

> Formal American investigations concluded that the attack was a mistake. The American investigations seem superficial in retrospect. Moreover, Congress never investigated the incident.
>
> The U. S. awarded a Medal of Honor to the ship's captain and awarded several other war decorations to officers, sailors, and National Security Agency officials.

Israel quickly closed the incident.

> Israeli investigations also concluded that the attack was a mistake.
>
> Israel paid reparations to the families of those killed and wounded and to the U. S. for damage to the ship.

Abrams changed the war on the ground.

Most of the survivors of the attack believe to this day that the attack was deliberate.[4]

Soon after the incident, Clark Clifford, a long-time Democratic Party leader, replaced McNamara as Secretary of Defense, and in July 1968, GEN Creighton Abrams replaced GEN Westmoreland as the Commanding General of USMACV. Abrams changed the war on the ground.[5]

In January 1969, President Richard Nixon replaced President Johnson. The Battle of Hamburger Hill and the My Lai Massacre dominated news in 1969. Little remembered the Viet Cong formed the Provisional Revolutionary Government of South

Vietnam and Communist governments in Bulgaria, Cambodia, Red China, Cuba, Czechoslovakia, East Germany, Hungary, Mongolia, North Korea, North Vietnam, Poland, Rumania, the U.S.S.R., and Yugoslavia recognized the VC government. In addition, one Communist-leaning Arab country, Syria, recognized the Viet Cong government.

In 1970, President Nixon ordered the invasion of Cambodia to disrupt NVA supply routes and bases. The highly successful invasion triggered protests in the United States.

On the political front, the U. S. Senate repealed the Gulf of Tonkin Resolution and later passed a resolution barring the use of American funds for operations in Cambodia, and the Supreme Court refused to hear a state challenge of the legality of the war.

The United States soon began withdrawing troops from South Vietnam, and in doing so the political leaders undermined the real success that GEN Abrams had achieved. Historian Lewis Sorley chronicled the success in *A Better War, The Unexamined Victories and the Final Tragedy of America's Last years in Vietnam:*

> *There came a time when the war was won. The fighting wasn't over, but the war was won. This achievement can probably be best dated in late 1970, after the Cambodian incursion in the spring of the year. By then the South Vietnamese countryside had been widely pacified, so much that the term "pacification" was not longer even used.*[6]

In 1971, ARVN troops began major operations and tried to cut the Ho Chi Minh Trail. Initially successful, they finally had to withdraw. In mid-year the New York *Times* published the stolen, secret *Pentagon Papers.*[7]

Δ

An anti-war movement developed early in the war in the United States and overseas. In 1965, the first "teach-in" was conducted at the University of Michigan to protest the war. "In 1966, veterans from World Wars I and II, along with veterans from the Korean and Vietnam wars staged a protest rally in New York City where discharge and separation papers were burned in protest of U. S. involvement in Vietnam."[8] That same year, President Charles DeGaulle, who had failed to solve the Vietnamese problem after World War II, condemned the American involvement and called for a withdrawal. In 1967, anti-war activists organized massive rallies in San Francisco and New York. The next year, Martin Luther King linked "the civil rights movement with the anti-war movement."[9] A prominent official, U. S. Attorney General Ramsey Clark, opposed the war while in office from

REFLECTION: THE MESSENGER

"...THE VILLAGE MESSENGER, AN ELDERLY MAN, KNOCKED HESITANTLY ON THE DOOR OF JOHN J. AND CAMILLE GEOGHEGAN. ALTHOUGH THE TELEGRAM WAS ADDRESSED TO MRS. BARBARA GEOGHEGAN, WIFE OF LIEUTENANT JOHN LANCE (JACK) GEOGHAGEN, THE MESSENGER KNEW WHAT IT SAID AND HE KNEW THAT JACK GEOGHAGEN WAS THE ONLY CHILD OF THE FAMILY.

AS THE GEOGHAGENS READ THE NEWS, THE MESSENGER BROKE DOWN, QUIVERING AND WEEPING AND ASKING OVER AND OVER AGAIN IF THERE WAS ANYTHING HE COULD DO TO HELP THEM. BEFORE THEY COULD DEAL WITH THEIR OWN GRIEF, THE GEOGHAGENS FIRST HAD TO DEAL WITH HIS; THEY HUGGED AND COMFORTED THE MESSENGER AND HELPED HIM PULL HIMSELF TOGETHER FOR THE LONG TRIP BACK TO TOWN THROUGH THE DEEPENING GLOOM."

THE NAVY ORDERED ME TO THE BUREAU OF NAVAL PERSONNEL (BUPERS) IN WASHINGTON, DC, AND WE MOVED FROM LEMOORE IN LATE JULY.

1967 to 1969, and he joined the anti-war movement after he left office. In 1967, 50,000 anti-war protestors marched on the Pentagon and in August 1968, anti-war protestors disrupted the Democratic Convention in Chicago and violence grew. The next year, 250,000 anti-war protestors assembled in Washington, DC.[10] On many campuses, professors led anti-war meetings and students disrupted meetings of anyone who supported the war.[11]

In 1970, students at Kent State University aggressively protested the war, and the Governor of Ohio called in the National Guard. During one confrontation, a few guardsmen opened fire and killed four students.

The anti-war movement affected some troop units in Vietnam and many veterans who returned from the war. Disaffected veterans organized Vietnam Veterans Against the War (VVAW) in 1967. In February 1971, the VVAW conducted an inquiry into the war focusing on alleged atrocities committed by American and allied troops.[12] The movement, skillfully exploited by the North Vietnamese to undermine U. S. war policy, also adversely affected some American POWs held in North Vietnam.

Δ

On the superpower stage, the United States and the Soviet Union continued to seek ways to limit nuclear armaments and decrease the chance of nuclear war.

In 1967, the two nations agreed to an "Outer Space Treaty," promising not to use nuclear weapons in orbit around the earth or on the moon and promising not to permit the "use of the moon or any celestial body for military purposes, including weapons testing of any kind."[13] In 1967, the United States signed the "Latin American Nuclear Free Zone Treaty," a "Multilateral agreement signed by 24 Latin American countries banning the manufacture, acquisition, testing, deployment, or use of nuclear weapons in Latin America."[14]

In 1968, the United States, the Soviet Union, and the United Kingdom signed the "Nuclear Non-Proliferation Treaty" which limited "the spread of military nuclear technology by the recognized nuclear-weapon states - U. S., U. S. S. R., U. K., France and China - to non-nuclear nations wishing to build or acquire nuclear weapons." Eventually, 187 states signed the treaty.[15] More importantly, talks began on the Strategic Arms Limitation Treaty in 1969.

All these events took place while Morin served in Washington, the Mediterranean, and the United States.

Ψ BUPERS

The Navy ordered me to the Bureau of Naval Personnel (BUPERS) in Washington, DC, and we moved from Lemoore in late July. My wife and children found this to be a hard move to make. They had to leave behind the families with whom they had established special bonds because of the wartime experiences they had shared together. All the wives and children had worried about the safety of their husbands and fathers, and all had shared a special grief when the enemy had shot down and killed or captured one of the pilots from Lemoore. Those families who had a member missing in action faced grim futures.

Ψ

While Commanding Officers' wives had no official rank, each one gained inescapable responsibilities while the officers and men serving under her husband fought in Southeast Asia. War made all the unofficial responsibilities tougher.

At Lemoore while we flew our missions into North Vietnam South Vietnam, and Laos, Betty had to provide a shoulder to lean on for some wives, give guidance to the other wives who faced tough situations, and provide constant encouragement to the younger wives - often newly married and new to the Navy. She had to go with the Base Commander and the Chaplain to see Wendy's wife after he was shot down. Fortunately, unlike some other CO's wives, she did not have to make a trip to report the death of a Silver Fox.

BUPERS assigned me to the Enlisted Plans Section. Before I really got into my new job, the Secretary of the Navy convened a new board which was to meet for six weeks to study the Navy Department's Awards and Decorations system. Appointed the Recording Secretary, I served with a distinguished group of men. The Board consisted of retired ADM George W. Anderson, Jr., a former Chief of Naval Operations, retired VADM William

Reflection: Silver Fox CO Jack Harris

CDR Jack Harris died during a fire on the USS *Oriskany.*

"At 0728 26 October [1966] fire broke out in a forward magazine and raged through 5 decks, claiming the lives of 44 sailors.

Many of those killed were veteran combat pilots who, a few hours earlier, had flown on raids over Vietnam.

Serious damage to the carrier ended ORISKANY's WestPac deployment and following limited repairs in Subic Bay (RP) she returned to San Diego on 03 Nov 1966.

Four months later ORISKANY departed the San Francisco Naval Shipyard and began preparations for a return to Yankee Station, departing for WestPac on 16 June 1967."

He spoke in generalities and, knowing that I had just returned from combat duty, asked if I had anything to say about the system. Did I have anything to say about the system?!!! You betcha!

R. Smedberg, III, a former Superintendent of the U. S. Naval Academy, retired Marine Corps LTG Frederick L. Wieseman, and Percival Brundage, a former member of President Dwight Eisenhower's staff. Anderson, a naval aviator who had flown off the new USS *Yorktown* in World War II, had earned an Army Distinguished Service Cross and a Bronze Star Medal and a slew of the high "service" awards often accumulated by senior officers.[16] Smedberg had earned a Silver Star and a Bronze Star in combat and several impressive service awards during his career. Wieseman held a Bronze Star and similar service awards. Anderson, Smedberg, and Wieseman had graduated from the Naval Academy.[17] I did not know much about Brundage.

The day the Board convened and was sworn in, ADM Anderson read his appointing letter from the Secretary of the Navy. He spoke in generalities and, knowing that I had just returned from combat duty, asked if I had anything to say about the system. Did I have anything to say about the system?!!! You betcha!

It was an interesting board. We visited the Atlantic and Pacific Fleet Headquarters and the Marine Headquarters. We corresponded with every Navy admiral and Marine general on active duty and with many retired admirals and generals. They all urged caution but admitted that we tended to be too tight. Presentations by the Army and Air Force on their systems confirmed what we line officers already knew. I was able to sell three ideas:

> 1. Add a medal, the Navy Achievement Medal, which up to that time was only a ribbon;
>
> 2. Add a third level unit citation, the Meritorious Unit Citation, to recognize those units which contributed to the effort but which did not engage in combat (I was thinking about the Service Force ships, the oilers, ammo ships, supply ships that worked hard to keep Task Force 77 going); and
>
> 3. Add a Combat Action Ribbon to recognize those young Marines and Navy medical corpsmen out in the trees who, if they were not decorated or wounded, would return wearing no more ribbons than a clerk back at Danang whose life was not in jeopardy daily.

That fall, the Navy selected me to be an Air Wing Commander, a CAG!

CAG

That fall, the Navy selected me to be an Air Wing Commander, a CAG!

Since my former CAG, Wes McDonald, served as the Aviation Commander detailer (assignment officer), I figured that I would get what I wanted. I found that two CAG slots at Lemoore and one at Miramar would come open in 1967. With my experience, I believed that it was only fitting that I be sent back to the Pacific Fleet. I was told, however, that Secretary of Defense McNamara had established limitations on combat tours and that I would not be going back into combat. Guys who deployed for their third, fourth, and fifth combat cruises later on probably wondered what happened to those limitations.

A year later, after being detached from BUPERS, I began my CAG training. Designated to command Air Wing 7, I was to sail to the Mediterranean in May 1968 in USS *Independence* for nine months. With these new orders, Betty wondered where my three years of shore duty went. A nine-month separation was not pleasant to contemplate, but at least I would not be getting shot at. My new Air Wing had six squadrons:

A year later, after being detached from BUPERS, I began my CAG training. Designated to command Air Wing 7, I was to sail to the Mediterranean in May 1968 in USS Independence for nine months.

o three A4 (attack) squadrons,
o two F4 (fighter) squadrons, and
o one RA5C (heavy attack) squadron.

To prepare for my new task, I refreshed in the A4 at NAS Cecil Field; then I started training in the F4 at Oceana and Key West. After all the years flying single-engine aircraft by myself, I now began flying this Cadillac, the Phantom, After day and night CQ in the F4, I went to NAS Albany, GA to check out the big RA5C.

To get an idea about what to expect in the new assignment, I corresponded with CAG 7, Ken Shugart, a former Lemoore A4 skipper.[18] I learned that the INDY'S cruise schedule had been modified so I when I went to BUPERS to pick up my ticket to fly to Europe, I could not convince the lady there that I should go to Athens. She had no schedule change and insisted in sending me to Rome.

WHEN I CALLED ROTA FROM TORREJON, I LEARNED THAT AN OLD FRIEND FROM MEMPHIS DAYS, CDR TED DAUM, THE SKIPPER OF VQ-2 AT ROTA, HAD CRASHED ON TAKE-OFF THAT MORNING AND DIED.

My itinerary took me to Madrid, then to Torrejon Air Force Base, and on to Naval Air Station Rota. When I called Rota from Torrejon, I learned that an old friend from Memphis days, CDR Ted Daum, the skipper of VQ-2 at Rota, had crashed on take-off that morning and died. His loss hit hard. Ted and his wife, Mary, were godparents for some of our children, and Betty and I were godparents for some of their five children.

I spent almost three days in Rota helping Mary Daum make the necessary decisions she had to make. Now I know it was the Good Lord who had the lady send me through Madrid instead of on to Athens. Fatal crashes are fact of life when you fly, but they devastate families. Most people do not know that generally more pilots in any given group die in air accidents than in combat. That is why the services pay us flight pay – more correctly, hazardous duty pay.[19]

Ψ

I flew on to Naples; then caught INDY's mail plane to my new home. Ken and I held our Change of Command ceremony on board the *Independence* at Augusta Bay, Sicily where eighteen years earlier I learned the meaning of "shipmate." Later, the ship got a new Captain, Spense Mathews,[20] who made sure that I got a lot of ship-handling time.

One evening at sea while I worked on paperwork in the Air Wing Office, I received a call to report to the bridge immediately. I ran up the ladders and walked on to the completely darkened bridge not knowing the reason for my summons. I found out quickly: the Captain called out, "The CAG has the CONN [control of the ship]." At the time, we were halfway through our approach to an oiler to refuel. My eyes had not night-adapted yet, and consequently I could hardly make out the lights on the oiler. Fortunately, the Navigator, CDR Don Hegrat,[21] another old friend from Memphis days and VF-14, kept giving me recom-

...THE CAPTAIN CALLED OUT, "THE CAG HAS THE CONN."

mendations which I took until I could see more in the dark. We came alongside smartly. Lucky me! CAPT Mathews would test me often again during the remainder of the cruise.

In the summer of 1968, we celebrated the twentieth anniversary of the organization of the Sixth Fleet. As part of the celebration, our Air Wing and the Air Wing from the USS *Shangri-La* put on an air show for a deck full of VIPs from NATO. Other than that, it was a long cruise.

WHEN WE RETURNED TO NORFOLK, THE NAVY NOTIFIED ME THAT I WOULD COMMAND CARRIER REPLACEMENT WING 4 AT NAS CECIL FIELD. AS THE CAG OF FOUR FLEET TRAINING SQUADRONS, I WOULD SUPERVISE THE TRAINING OF A4, A6, A7, AND F4 PILOTS BEGINNING IN JUNE 1968.

Ψ

When we returned to Norfolk, the Navy notified me that I would command Carrier Replacement Wing 4 at NAS Cecil Field. As the CAG of four Fleet Training Squadrons, I would supervise the training of A4, A6, A7, and F4 pilots beginning in June 1969.

To my dismay, CAPT Billy Phillips, my old nemesis from *Ranger* days, currently commanded the wing. CAPT Billy, who was not too thrilled that I would replace him, called around to all his friends trying to rally their support so that I, "an attack puke," would not get the coveted job. No one supported him. Moreover, I was still a commander, and that bothered him too. I do not know if he knew that I had been selected for Captain during the INDY cruise, but had not been promoted yet.

Dave Leue, the former XO/CO of VA-153, *The Blue Tail Flies*, took over from me as CAG 7, and I headed south to relieve the world's greatest fighter pilot as CAG 4. Before moving to Jacksonville in the summer of 1969, our oldest son, Jim, started Plebe summer at the Naval Academy.[22]

CAG 4 proved to be a great and prestigious job. The commanding officers of the squadrons had already commanded squadrons with the Fleet. The experienced officers made my job easy and fun. Back in those years, there were not many flying billets for Captains. I took advantage of my opportunity and checked out in the A6 and A7 and made it a point to fly all four types of the fleet aircraft in the same day just to record the feat

CAG 4 PROVED TO BE A GREAT AND PRESTIGIOUS JOB.

in my log book. In early 1970, after the Navy decided to go to the functional wing concept, it decommissioned Air Wing 4 and re-commissioned it as Light Attack Wing 1.

UNDER THE NEW ORGANIZATION, I ASSUMED RESPONSIBILITY FOR ALL THE A7 SQUADRONS IN THE ATLANTIC FLEET WHILE THEY WERE ASHORE.

Under the new organization, I assumed responsibility for all the A7 squadrons in the Atlantic Fleet while they were ashore. The A7s replaced all the A4s. By the time the "E" model of the A7 came along, our attack pilots finally flew a very accurate bomber that could carry a big ordnance load a long way. I missed the A4 for sentimental reasons, but our attack pilots needed a better light attack aircraft. Initially, we suffered with many logistical problems with the A7E. However, we slowly made progress introducing this excellent aircraft to the Fleet, thanks in great part to Bill Gortney, my friend from *Ranger* days who had retired and worked as the Factory Representative for Ling Tempco Vaught.

I remained the CAG through early 1971.

End Notes

[1] Gutzman, *Vietnam*, 12-16. Nichols and Tillman, *Yankee Station*, 154-158. Hyde, *Flight Log*.
[2] Bui Tin, *Enemy to Friend*, 63-4.
[3] *Ibid*, 61-2.
[4] For information about the USS *Liberty*, see Peter Hounam, *Operation CYANIDE, Why the Bombing of the USS Liberty Nearly Caused World War III* (London: Satin Publications, Ltd., 2003), "Thomas Moorer" and the "Liberty Incident," *Wikipedia*, and William D. Gerhard and Henry Millington, *Attack on a Sigint Collector, the U.S.S. Liberty (UNITED STATES CRYPTOLOGIC HISTORY, Special Crisis Collection, Volume 1*, NATIONAL SECURITY AGENCY/CENTRAL SECURITY SERVICE, 1981. (18 July 2010).
[5] For the dates of the various actions and events during the Vietnam War, see Gutzman, *Vietnam*, 12-16.
[6]Lewis Sorley, *A Better War, The Unexamined Victories and the Final Tragedy of America's Last Years in Vietnam* (NY: Harcourt Brace & Company, 1999), 217.
[7]·Gutzman, *Vietnam*, 16.
[8] *Ibid*, 420.
[9] *Ibid*, 13.
[10] *Ibid*, 15 (250,000 protestors).
[11] Personal knowledge of demonstrations at Rice University 1968-1970, JHB.
[12] Gutzman, *Vietnam*, 420 and 435.
[13] "Outer Space Treaty (1967)," http://www.atomic archive.com /Treaties/Treaty4.shtml.
[14] "Latin America Nuclear Free Zone Treaty (1967)," http://www. atomicarchive.com /Treaties/ Treaty5.shtml.
[15] "Nuclear Non-Proliferation Treaty (1968)," http://www.atomic archive.com/Treaties/Treaty6.shtml.
[16] Many Army officers considered that the various Distinguished Service Medals awarded by the different services were the "good conduct medals" given to flag and general officers serving in the Pentagon or awarded at retirement. Similarly, some considered that the Legion of Merit was the "good conduct medal" for retiring Army colonels. Personal knowledge. JHB.
[17] Admiral Anderson graduated in the Class of 1927, Lt. General Wieseman in the Class of 1931, and Vice Admiral Smedberg in the Class of 1926. See *1993 USNA Register of Alumni*, 210, 212, and 223. For Anderson's naval service, see Karl Schuon, *U. S. Naval Biographical Dictionary* (NY: Franklin Watts, Inc., 1944), 2-3.
[18] Kenneth Laverne Shugart, Jr. retired as a Rear Admiral in 1983. He died in 1985 at age fifty-nine. Shugart graduated eighteenth in his class from the Naval Academy in 1948. He earned four Distinguished Flying Crosses , a Bronze Star, and several service awards. *1993 USNA Register of Alumni*, #19437, 290.
[19] A study of the Class of 1927 from the Naval Academy shows that fifteen graduates died in air accidents in time of peace, three died in crashes related to combat action, one was shot down, and twelve died in naval surface action. JHB Paper.
[20] Matthews retired as a Rear Admiral.
[21] Hegrat retired as a Commander.

Sidebar Notes

Reflection: Soviet War Planning
Stephan T. Possony, *A Century of Conflict, Communist Techniques of World Revolution, 1848-1950* (Chicago: Henry Regnery, 1953), 353.

1968 A MISTAKE OR A DELIBERATE ATTACK?
See End Note 4 and Appendix E.

Reflection: The Messenger
Lt. Gen. Harold G. Moore (Ret.) and Joseph L. Galloway, *We Were Soldiers Once and Young, The Battle that Changed the War in Vietnam* (NY: HarperPerrenial, 1993), 382-3.

Reflection: Silver Fox CO Jack Harris
"USS ORISKANY (CVA-34)," *The Virtual Wall, Vietnam Veterans Memorial*. http://virtualwall.org /units/oriskany.htm.

Photo Credits

Watching the Bear
No reference. (24 March 2011).

[22] James B. Morin, Jr. graduated ninth in his class in June 1973. *1993 USNA Register of Alumni*, #40129, 440. He retired after a full career in the grade of Captain, USN.

Nine

SHIP COMMAND MIDDLE EAST

USS LaSalle

Hot War During the Cold War - Vietnam 1972-1973[1]

Important Events		James Morin
	Oct 1971	**CAPT Morin took command of the USS *LaSalle*.**
	Jan 1972	Morin moved *LaSalle* to Portsmouth Shipyard.
President Nixon visited China.	21 Feb 1972	
NVA began massive Easter Campaign to overrun SVN.	**30 Mar 1972**	
Operation LINEBACKER I began. U. S. launched massive air strikes against NVN.	6 Apr 1972	
B52s attacked Hanoi and Haiphong for the first time.	**16 Apr 1972**	
NVN patrol boats and MiGs attacked U. S. 7th Fleet; damaged one destroyer.	19 Apr 1972	
President Nixon announced the mining of Haiphong harbor.	8 May 1972	
Seventh aircraft carrier arrived on Yankee Station.	15 May 1972	
President Nixon removed most of the bombing restrictions over NVN.	23 May 1972	
President Nixon and General Secretary Breshnev signed Anti Ballistic Missile (ABM) Treaty and Interim Agreement on Strategic Offensive Weapons.	26 May 1972	
Watergate Break in occurred.	26 Jun 1972	
	7 Jul 1972	**Morin departed GITMO, Cuba on *LaSalle* for Middle East.**
	28 Jul 1972	"Pollywog" Morin crossed the equator.
	8 Aug 1972	*LaSalle* arrived in Mozambique.
Last U. S. troop unit departed SVN.	11 Aug 1972	
	23 Aug 1972	**Morin arrived in Bahrain.**
U. S. Air Force began all-out air campaign against NVN.	18 Dec 1972	
B52s pounded Hanoi for an hour.	26 Dec 1972	
	9 Jan 1973	**Morin departed Bahrain for the U. S. and command of the USS *Franklin D. Roosevelt*.**
Paris Peace Agreement Signed, ending Vietnam War for the United States.	**Jan 1973**	

Cold War Middle East 1972-1973

The Cold War reached the Middle East in 1953. President Dwight Eisenhower permitted the U. S. Central Intelligence Agency to overthrow the Communist-leaning government of Premier Mohammed Mossadegh and install Shah Mohammad Reza Pahlavi as the Shah of Iran. The Shah became a stalwart ally of the United States and the West and remained so for many years.

Δ

Things wobbled a bit under President Kennedy. Army MAJ Alexander Haig, after earning a Master's Degree at Georgetown University in Washington, D. C., joined the International Plans & Policy Group at the Pentagon. Eventually, the young combat veteran from Korea found himself assigned to the "Middle Eastern Desk" and dealing with issues related to Iran.[1] Haig wrote about the situation he found at his new job:

1. "The first two years of the Kennedy Administration were a period of trouble and misunderstanding in United States relations with Iran." A State Department report ordered by President Kennedy in February 1961 stated that recent elections in Iran had been "rigged"... and that the Shah "is capable of making ... a switch [to neutralism]."

2. The Central Intelligence Agency (CIA) told the President that any reduction of U. S. support for the shah could turn him to seek economic and possible military aid from the USSR.

3. "During his campaign for the presidency, Kennedy had referred to the possibility of revolution in Iran if the shah did not undertake reforms to control corruption and improve the life of the people."

4. "...the new administration suggested to the shah that future American aid would be linked to policy changes designed to fight corruption, shrink the size of the armed forces, encourage economic growth, and bring a dramatically large measure of social justice to the Iranian people."[3]

5. The Kennedy administration feared losing Iran.

6. Khrushchev had told Kennedy in Vienna that Iran would eventually fall to the Soviet Union.

SEA POWER: ALFRED THAYER MAHAN

"TALK OF EMPIRE, TRADE, AND NATIONAL GREATNESS CAME TOGETHER IN 1890 IN THE PUBLICATION OF A MONUMENTALLY IMPORTANT BOOK, ***THE INFLUENCE OF SEAPOWER UPON HISTORY***, BY CAPTAIN ALFRED THAYER MAHAN.

MAHAN, WHO WAS ATTACHED TO THE NAVAL WAR COLLEGE, WAS MORE COMFORTABLE AROUND BOOKS THAN ON SHIPS. ALTHOUGH HE HAD SERVED THROUGHOUT THE WORLD, HE HAD THE LOOK, TEMPERMENT, AND INCLINATION OF A COLLEGE DON.

HIS MASTERPIECE SET FORWARD THE SIMPLE THESIS THAT NAVAL POWER WAS THE KEY TO NATIONAL GREATNESS.

TAKING GREECE, ROME, AND ENGLAND AS EXAMPLES, HE ATTEMPTED TO DEMONSTRATE THAT COUNTRIES RISE TO WORLD POWER THROUGH EXPANDING THEIR FOREIGN COMMERCE AND PROTECTING THAT COMMERCE WITH A STRONG NAVY.

WITHOUT A POWERFUL NAVY, MAHAN EMPHASIZED, A NATION CAN NEVER ENJOY FULL PROSPERITY AND SECURITY. WITHOUT A STRONG NAVY, IN SHORT, NO NATION COULD EVER HOPE TO BE A WORLD POWER."

7. The State Department recommended continued aid to Iran.[4]

After Haig arrived in Iran, he learned that most of the Iranian officers "were far more worried about being invaded by Iraq than by the U.S.S.R." So was the Shah.

Subsequently, the President formed an interdepartmental group to study the issue of aid to Iran. The Pentagon, in turn, formed a team to devise "a plan for the military defense of Iran," Haig joined officers from other branches of the armed services on the team. After Haig arrived in Iran, he learned that most of the Iranian officers "were far more worried about being invaded by Iraq than by the U.S.S.R." So was the Shah.

Once in the field, Haig found that, in addition to border clashes between the Iranians and the Iraqis, the Kurds had begun a guerrilla war to establish an independent Kurdish state, something the Soviets had posited as a means to destabilize the region and create problems for the United States.[5]

The American team spent five or six weeks touring Iran to determine how best to defend the nation from a Soviet attack. The team's final report called for:

1. "the addition of 10,500 troops above the 150,000 already authorized;"

2. "construction of new bases and airfields;"

3. "construction of early-warning radar stations along the Soviet border;"

4. "enhanced training for the Iranian armed forces;" and

5. "restoration of ... two frigates to the assistance budget."[6]

Haig added that the President held "skeptical views on the shah and the situation in Iran," and Secretary of Defense McNamara wanted to reduce aid to Iran.

Δ

When Morin sailed for the Middle East, Richard Nixon had been president for three years. Nixon moved to lessen tensions with the Communist bloc in 1972 and initiated a period of détente with the Soviet Union and expanded trade with China.

Morin visited African countries before the Soviets moved into Africa and before Iranians deposed the Shah and upset the balance of power in the Middle East.

Ψ

Captain

In early 1971, the Navy notified me that I had been selected to be Captain of a ship. The pleasant surprise told me that the Navy still expected bigger and better things of me. Most Naval Aviators finish their operational flying after squadron command, except for the fortunate few that kept on flying as Wing Commanders. My extra two years as CAG 4 and Commander, Light Attack Wing One were icing on the cake for me. When a pilot's flying days come to a halt, it is a shock. My additional years gave me the opportunity to prepare myself for the inevitable. I turned my Wing over to Doug Mow[7] and the Morins moved to Norfolk where the Fleet Commander designated me the Commanding Officer of the USS *LaSalle* (LPD-3), then undergoing overhaul at the Naval Shipyard Philadelphia. We found that we had a fun place to live in, if we ignored the location.

LaSalle, an amphibious vessel, displaced 15,000 tons. She could carry about 800 marines, had a helicopter flight deck, and could ballast down so that landing craft could enter and leave through a stern gate. To prepare for my command, I commenced training at Amphibious Force, Atlantic Fleet Headquarters, attended a number of required schools, and spent a week at sea on another LPD. Then, I headed to Philadelphia where I took the opportunity to inspect the *LaSalle* in dry dock. After the Shipyard completed the overhaul in October 1971, I took command, and shortly thereafter, we sailed to our new home port of Norfolk.

Ψ

I had a very interesting Executive Officer, CDR Tom Clark.[8] Tom had enlisted in the Navy before World War II, and by the end of the war he had attained the rank of Chief Boatswain's Mate. I suspected he made a pretty salty one.

LaSalle

"Being an amphibious landing ship, the LaSalle had enough room to carry about 800 marines with all their gear. It also had two "mike" boats, landing craft meant to carry the marines up to a beach, drop its ramp, and allow the marines to run out on to the beach.

On our new mission as flagship, we carried but two marines, the Admiral's Chief of Staff, and the Admiral's driver/body guard. So the ship had a lot of empty space — to be used if we ever had to evacuate anyone from a hostile environment.

The ship had a gate on its stern which could be lowered. The ship would ballast down allowing [it] to take on water and the stern would sink below the water. The "mike" boats were stored in the well-deck and when the time came, they'd be floated and launched from the stern of the ship, fully equipped.

Again, because we carried so few marines, there was still plenty of room in the well-deck for other things."

Rex Harrison

1972
Nuclear Arsenals

"Soviet and [U. S.] weapons systems were far from symmetrical...

The Soviet Union had continued its development and deployment of heavy ballistic missiles and had overtaken the U. S. lead in land-based ICBMs. During the SALT I years alone Soviet ICBMs rose from around 1,000 to around 1,500, and they were being deployed at the rate of 200 annually. Soviet submarine-based launchers had quadrupled. The huge payload capacity of some Soviet missiles ("throw-weight") was seen as a possible threat to U. S. land-based strategic missiles even in heavily protected ("hardened") launch-sites.

The [U. S.] had not increased its deployment of strategic missiles since 1967 (when its ICBMs numbered 1,054 and its SLBMs 656), but it was conducting a vigorous program of equipping missiles with "Multiple independently-targeted Re-entry Vehicles (MIRV). A MIRV missile permitted a missile to carry a number of warheads directed at separate targets. MIRVs thus gave the United States a lead in numbers of warheads. The [U. S] also retained a lead in long-range bombers.

The Soviet Union had a limited ABM system around Moscow; the [U. S.] had shifted from its earlier plan for a "thin" ABM defense of certain American cities and instead began to deploy ABMs at two land-based ICBM missile sites to protect its retaliatory forces. (The full program envisaged 12 ABM complexes)."

After the war, he moved from Chief to Warrant Officer Bosun, and later the Navy selected him to be a Limited Duty Officer (Deck). Although he did not see much shore duty in his line of work, Clark managed to go to night school whenever he could. With additional education, he applied to be an Unrestricted Line Officer. The Navy accepted his application and he eventually commanded his own Landing Ship Tank (LST). I do not believe that Tom Clark liked aviators. He struck me as the kind of officer who would say: "I don't fly your planes, so don't sail my ships." When we left Philadelphia and steamed eighty miles or so down the Delaware River to the Atlantic and when we entered Hampton Roads at Norfolk, he hovered nearby so I would not have a mid-air collision with another ship or put *LaSalle* in the mud.

A few weeks later, we sailed for the Naval Station at Guantanamo Bay in Cuba (GITMO). There, the Fleet Training Group put us under intensive training to bring all of us in the ship's company up to a decent level of combat readiness. Each morning at first light, we got underway, and then ran drills all day until we returned at dusk. One morning, I waited to get underway because our pilot had not come aboard. We had a full day scheduled so I told Clark, "Let's get underway!" With a spring line forward, I brought the stern out, then twisted the ship parallel to the pier, rang up "all ahead one third," and steamed smartly out into the bay. Standing off to the side, the XO nodded his approval. Another time, we had to practice refueling underway, day and night. Making the approach to the oiler, then coming alongside, reminded me a lot of flying in formation. Clark watched me like a hawk but saw I could handle it. A few more evolutions and I knew he had begrudgingly given me an "up" on my safe solo check. He learned that some aviators could be pretty good ship handlers.

We were scheduled to return to Norfolk before Christmas. Before leaving the training area, my crew wanted to hold "swim call" off the coast of Cuba, but I hesitated because of my concern about sharks and about the amount of time we would lose.

Instead, I had the well deck flooded and held swim call inside the ship while heading for Norfolk at sixteen knots.

After we returned to Norfolk, the Navy decided to send the *LaSalle* to the Persian Gulf to replace the USS *Valcour*, the flagship for the Commander, Middle East Forces. *Valcour*, a very old seaplane tender, had seen better days. The Navy planned to decommission and scrap her when she returned to the United States.[9]

AFTER WE RETURNED TO NORFOLK, THE NAVY DECIDED TO SEND THE LASALLE TO THE PERSIAN GULF TO REPLACE THE USS VALCOUR, THE FLAGSHIP FOR THE COMMANDER, MIDDLE EAST FORCES.

At the same time, the Navy planned to make major renovations on the *LaSalle*. Consequently, in January 1972, I moved my ship to the Naval Shipyard Portsmouth, VA and entered dry dock there. While in dry dock, I spent time explaining our upcoming mission to the officers and men of the *LaSalle*. We were to perform a dual military and diplomatic mission - show the flag and promote goodwill. In doing so, the Navy expected us to promote understanding and mutual respect between our country and the various countries of the Middle East.

USS *VALCOUR*

Soon workmen began doubling the air conditioning capacity and installing special air filters to keep out blowing dust and sand. Others built a helicopter hanger on the forward part of the flight deck and installed a large canvas awning across the forward half of the flight deck to be used for ceremonies. Specialists then installed updated and sensitive satellite receiving equipment, new electronic equipment for the Admiral's staff, and a closed-circuit television station. *LaSalle* was the first small ship to be equipped with her own closed-circuit TV station. Finally, the painters came aboard and painted the *LaSalle* a bright white, covering its Navy grey.

During this time, the Commander of the Middle East Forces notified me that the USS *Valcour* had organized a band, and that, therefore, I had to organize one to play in most of the ports we were to visit. Unknown to me, some sailors had found old musical instruments in one of the staterooms they had painted, and these – an old set of drums, an amplifier, a guitar, a trombone, etc. – soon found their way into the hands of sailors who could play them. One night while they were "jamming," I

A WEEK BEFORE SAILING, MY WIFE UNDERWENT A SERIOUS OPERATION. ANY OTHER OFFICER ON THE *LASALLE* WOULD HAVE BEEN LEFT BEHIND UNDER THESE CIRCUMSTANCES, BUT NOT THE CAPTAIN.

stopped by and told them that they were my new band. Then, I granted them "open-gangway" - they would not have to stand watches while in port - if they did not charge for performances in the ports we expected to visit. One sailor said, "You'd have thought Christmas was occurring all over again." The new bandsmen bought new equipment, supplies, and lights, and the newly named *Greenfields Band* gave its first concert before we left Norfolk. There would be many more.[10]

After four months in the shipyard, we refloated the *LaSalle*, refueled her, and returned to the Naval Station Norfolk where we remained for about a month before getting underway for the Persian Gulf. While there, stevedores loaded into our well deck and cargo deck the household goods and cars of the families that opted to move to Bahrain.

A week before sailing, my wife underwent a serious operation. Any other officer on the *LaSalle* would have been left behind under these circumstances, but not the Captain. The Commander, Amphibious Forces, Atlantic did arrange for another Captain to take the ship to Guantanamo to commence refresher training, and that gave me a little time with Betty. But soon I had to head for GITMO and eight months at sea, leaving Betty at the worst possible time. She had to recuperate from her surgery without me.

I flew to GITMO where I resumed command and completed the last portion of refresher training. One of my crewmen noted:

> *When he returned to the ship at GITMO, he never wore his problems on his sleeve nor did he let it affect his performance or how he treated his shipmates. He had a very positive outlook on life. He did, however, have a soft spot in his heart for all the dependents we left behind.*[11]

Ψ

Soon departure day, 7 July 1972, arrived. We left Cuba and headed east for our first port of call on the long trip to Bahrain – Dakar, Senegal. We steamed under an overcast sky all the way across the Atlantic, navigating by dead reckoning [DR] because my navigator, John Sadler,[12] could not get star shots for a fix. We expected that very late one evening we would spot the loom of a light house on the African coast, if, of course, our DR proved correct. Right on time, we picked up the reflection in the night sky. Sadler had hit it right on the nose.

The next morning, 21 July 1972, we entered the harbor through a small passage in the mole surrounding it, and rendered a twenty-one gun salute. The French maintained a naval base there, and we moored at one their piers. In the two days in port, I made official calls on key people at the U. S. Embassy, the French Navy, and the Senegalese Government and attended associated functions. We found that our Ship's Store became a popular attraction for the American Embassy wives at this port and every other port thereafter. Soaps, tissues, detergents, mops, toothpaste, etc. were in high demand. Fortunately, we had stocked up well with household supplies to take to Bahrain for our families who would move there.

After leaving port, we took two days to reach Monrovia, Liberia, where we stayed for a day and a half. Children of Liberian military personnel came aboard and entertained us. I presented one young singer with my *LaSalle* ball cap, which made a big hit. As in Dakar, I made the usual round of calls and toured the city.

On 24 July 1972, we arrived off Abidjan, Ivory Coast.[13] To get to the city from the sea, we had to navigate the narrow Vridi Canal. I boned up on canals the night before and we made an uneventful passage the next day. However, we encountered some difficulty going alongside the pier in the port because a strong current was setting us on to the pier. There were no tugs to help us. Small harbor craft, however, took our mooring lines ashore.

We found Abidjan to be a beautiful city. The U. S. Ambassador, John Root, requested that we host a dinner aboard that night with dancing on the flight deck for the foreign embassy people. We paid for the evening because the Ambassador said

1969-1972
SALT I

"SALT I, THE FIRST SERIES OF STRATEGIC ARMS LIMITATION TALKS, EXTENDED FROM NOVEMBER 1969 TO MAY 1972.

DURING THE PERIOD THE UNITED STATES AND THE SOVIET UNION NEGOTIATED THE FIRST AGREEMENTS TO PLACE LIMITS AND RESTRAINTS ON SOME OF THEIR CENTRAL AND MOST IMPORTANT ARMAMENTS. IN A TREATY ON THE LIMITATION OF ANTI-BALLISTIC MISSILE SYSTEMS, THEY MOVED TO END AN EMERGING COMPETITION IN DEFENSIVE SYSTEMS THAT THREATENED TO SPUR OFFENSIVE COMPETITION TO STILL GREATER HEIGHTS.

IN AN INTERIM AGREEMENT ON CERTAIN MEASURES WITH RESPECT TO THE LIMITATION OF STRATEGIC OFFENSIVE ARMS, THE TWO NATIONS TOOK THE FIRST STEPS TO CHECK THE RIVALRY IN THEIR MOST POWERFUL LAND AND SUBMARINE-BASED OFFENSIVE NUCLEAR WEAPONS."

WHEN I ARRIVED, THE IVORIANS PROVIDED ME FULL HONORS. I TROOPED THE LINE. MY HOSTS THEN PRESENTED ME WITH A NICE PLAQUE. LOOKING BACK, I AM AWED BY THE COURTESIES THE IVORIANS PAID TO ME, A NAVY CAPTAIN,

that he did not have funds for such events. But on the next two nights, the Embassy and the Ivorian Minister of the Armed Forces wined and dined us ashore.

During the short visit, I was asked to inspect the Ivorian Naval Personnel at their base. A Lieutenant served as the Chief of Naval Operations. When I arrived, the Ivorians provided me full honors, and I trooped the line. My hosts then presented me with a nice plaque. Looking back, I am awed by the courtesies the Ivorians paid to me, a Navy Captain, but remember that they paid respect to the United States which my ship, crew, and I represented.

We got underway at noon on 27 July 1972 without tugs. Moving off the pier and getting out into the river proved difficult. We had to twist to get headed fair. Ambassador Root rode out to sea with us. As the first U. S. ship to visit in many years, he said we had made a most favorable impression. We flew him back ashore in our helo. Then, a few hours later, we set the watch to keep a sharp lookout for the arrival of Davey Jones as we neared the Equator.

Early the next morning, two Shellbacks awakened me and took me under guard to the mess decks for breakfast of *Sea Crater Grits*, *Seaweed Au Jus*, and *Simmered Pollywogs*. All other Pollywogs were fed similar fare before heading to the flight deck for our initiation ceremony. For the ceremonies, they gave me the honor of being the first to crawl on my belly through the slop shoot composed of three-day old garbage laced with foul smelling aqueous liquid. Then hauled before Neptune Rex and his court, I had to kiss the Royal Baby's greasy belly, run the gauntlet, and get ceremoniously dunked in the pool rigged for "baptism." All the other Pollywogs followed. All went well; no one got hurt. That afternoon, we cleaned up the ship and then had holiday routine. At 1710 that evening, we crossed a point not visited by many, zero degrees Longitude and zero degrees Latitude.

A young shipmate recalled the events in more detail.

At eight bells, the Jolly Roger was hoisted and Davey Jones gave his final warning. We were doomed!

All the pollywogs were expected to dress appropriately, meaning all your clothes were to be worn inside out and backwards. To begin some 300 pollywogs gathered on the foc'sle to await their summons before a Royal Court. The few shellbacks aboard had collected garbage for about a week and prepared an obstacle course fit for this Royal Court.

Each pollywog had to crawl in front of the court, made up of Davey Jones, the Royal Whore, Royal Surgeon, Neptunus Rex, Royal Judge, Royal Baby, Defense counsel, Master-at-Arms, and the Royal Dentist. Each pollywog had charges written up against him, and plead as each might, the court found them all guilty.

Looking down the line of pollywogs ahead gave you a little reason for hope. The first form on indignity was having your face wallowed in the Royal Baby's gut, ripe with some greasy, foul smelling muck. Looking further down the line, each shellback waited with the shellalie (cut up pieces of fire hose about three feet long) and farther was the gauntlet of misery.

To our surprise though, at the front of the line was Captain Morin. Yes, even with years and years of service, our captain was still a lowly pollywog. We couldn't see much as were getting beaten with shellalies and having overly ripe eggs smashed on our heads or down the backs of our shirts.[14]

My young sailor added that my being there reassured the men that they would make it through all right, and he reported with some pride that everyone became a Golden Shellback because we had crossed at zero degree latitude and zero degree longitude.

To get to our next port of call, Lourenço Marques, Mozambique, we had to sail around the southern tip of Africa. As we progressed south from the Equator, the weather started to cool and the sea conditions slowly worsened as we encountered twenty-five to thirty knot winds on our bow. Soon we encountered winter off South Africa. On the eastern side of Africa we sailed into spring, and by the end of our twelve-day trip, summer weather returned.

Before we got to Mozambique, however, we experienced an engineering problem on 31 July 1972. The engineering crew

1972
DÉTENTE

"AS PRESIDENT, RICHARD NIXON RADICALLY REDEFINED AMERICA'S RELATIONSHIP WITH ITS TWO FOREMOST ADVERSARIES, CHINA AND THE SOVIET UNION. IN A REMARKABLE TURNABOUT FROM HIS RECORD OF STAUNCH ANTI-COMMUNISM, HE OPENED RELATIONS WITH CHINA AND BEGAN STRATEGIC ARMS LIMITATION TALKS WITH THE SOVIET UNION.

THE GOAL OF DÉTENTE (THE EASING OF TENSIONS BETWEEN NATIONS) WAS TO CONTINUE TO RESIST AND DETER SOVIET ADVENTURISM WHILE STRIVING FOR MORE CONSTRUCTIVE RELATIONS WITH THE COMMUNIST WORLD.

NIXON BELIEVED THAT IT WAS NECESSARY TO CURB THE ARMS RACE, IMPROVE GREAT POWER RELATIONSHIPS, AND LEARN TO CO-EXIST WITH COMMUNIST REGIMES.

THE NIXON ADMINISTRATION SOUGHT TO USE THE CHINESE AND SOVIET NEED FOR WESTERN TRADE AND TECHNOLOGY AS A WAY TO EXTRACT FOREIGN POLICY CONCESSIONS."

Tight Ship

"Captain Morin ran a tight ship but wasn't over-bearing.

He expected his men to learn and conduct themselves according to the military code-of-conduct. Meaning, he wanted his men to learn to become responsible for themselves as the military demanded: get up, take care of your assigned areas, muster on time, dress according to the day's requirement, show proper military courtesies, stand your assigned watches, complete your divisional responsibilities, study for advancement, and train on your assigned fire watch and general quarters station every day.

You learned through your leading petty officer (LPO) in your division that Captain Morin expected his crew to show proper military respect, no matter who the person might be.

We learned to show proper respect for our superior petty officers, chief petty officers, and commissioned officers. Every morning we were expected to salute sharply each officer the first time we saw them that day."

discovered a crack in the superheater header of Number 2 boiler, a problem common to the type of boilers we had. To continue, we had to go to one boiler operation.

I immediately notified Fleet Headquarters of our problem. The boiler breakdown concerned me: if anything happened to our remaining boiler, we would go dead in the water in some very poor sea conditions. Earlier on our trip from Cuba to Senegal, while steaming on one boiler, we had to secure it (shut it down) because of inattention to feed water levels. It had been eerie to go dead in the water, but at least at that time, we had favorable sea conditions.

We did not have the capability of repairing the superheater header. I hoped that we could make it to Cape Town, South Africa for repairs. I did not know that our government had forbidden any U. S. ships to stop at Cape Town because of the racial situation in South Africa. Fleet Headquarters soon informed me to continue on and that the shipyard personnel would meet us in Mozambique.

As we swung around the southern tip of Africa one night, I watched the lights of Cape Town from my bridge chair. The ship rolled badly with the rough seas on our beam. I slept very little for a few nights as we encountered heavy shipping traffic and passed by fishing boats everywhere. Once we turned north, the ship rode better because the wind and the heavy seas were behind us.

At noon on 8 August 1972, we arrived off the mouth of the river that led to Lourenço Marques. [15] There, a Portuguese pilot came aboard to take us up the river. To insure that we had sufficient back-up power in case of an emergency, I had lit off the Number 2 boiler. We would put up with the steam leak for the trip up the river.

The trip was not relaxing because the pilot said that the marking buoys were not reliably positioned, the channel was narrow, and we could see shoal water close aboard. We found during our visits in that part of the world that most of the ports were poorly marked. A two-knot current again complicated our entry into the harbor and our going alongside a commercial pier.

Once tied up, we began the usual rounds of official calls with our Consulate and the Portuguese military. The next day, we toured the beautiful, European-designed city.

As promised, help arrived. Eleven civilian technicians from the Newport News Shipyard came aboard to repair our boiler while we sailed to Bahrain. Former Machinist Mate John Tomaio remembered the work that was done.

> *The [Boiler Tenders took time] to do firesides (cleaning the outsides of the tubes inside the boiler) as well as watersides (cleaning the insides of the boiler tubes) on the boiler tubes and take apart the damaged superheater tubes (higher up toward the stack) in prep for a faster repair when we finally did get access to a repair facility.*
>
> *They worked around the clock in less than terrific conditions. Point of fact, the conditions were horrible.*[16]

Because of the nasty work, I went down to the lowest level to visit the men. I sat on the bilge line with my feet ankle-deep in the bilge water, and opened a bottle of brandy and shared it with the sweaty sailors. Sometimes, you need to do something special for your men. Tomaio added: "Everyone knew he had broken several hard rules [drinking aboard ship; drinking with enlisted sailors, etc.] ...After that he had the hardest working bunch of BTs in the fleet. It was a greatly appreciated gesture and he became endeared by more than just those BTs. We all came to appreciate him a little more after that."[17]

On 11 August, we got underway again. During our trip northward, "spring like" weather returned. It was wonderful. As we approached French-controlled Madagascar, we sailed close to the Comoroes Islands to view those beautiful tropical islands. French officials, however, in their inimitable way, registered an official complaint with our State Department because they believed that we had passed too close to their territories. We had. After Atlantic Fleet Headquarters queried me about the com-

Tight Ship

"When the Captain entered a compartment, no matter the time of day, the first to see him was to shout out, "Attention on deck!" The group of sailors in the compartment were to quickly ascertain who was the most senior, and that person was to come forward and offer his assistance to the Captain."

These were the lessons well learned, because during the remainder of my career, the same was expected. [He] taught his crew and officers well.

It's how he taught us that is of interest. I never saw Captain Morin actually stop and rake anybody over the coals about anything, but he would make sure you were given proper instruction from within your division.

As for Captain Morin himself, as he walked down a passageway, he exuded confidence. ...He dressed to perfection and he performed his responsibilities with confidence and precision. He exemplified the finest of naval traditions, exactly what he asked of his men and we all took notice. So as it were, he taught by example and a finer example I never saw."

Rex Harrison

plaint, I reported that we had not been inside three miles. I heard nothing further, but the incident may explain why I cannot visit France! I am kidding.

At least, I thought, the admiral could have gotten twenty-one Bahrainis on camels to fire their rifles in reply or do something to respond to our courtesy.

Home Port

Approaching the Straits of Hormuz, we watched a constant stream of large tankers exiting the Persian Gulf. Once in the Gulf, we found our visibility reduced to less than three miles because of dust and haze. Consequently, we could not relax. On the very hot and clear morning of 23 August, we stood off the Bahrain light ship. The Commander, Middle East Force, RADM Marmaduke Bayne, had directed me to render a twenty-one gun salute to the relatively new nation of Bahrain when we entered the port. I had asked RADM Bayne the day before, "Where was the battery that would return our salute?" He told me that there was no battery. After that, I had the *chutzpah* to quote Navy Regulations to him to remind him that we did not render a salute unless the host country could return it. He answered: "DO IT!" I found out later that a lot of things were done differently in that part of the world. At least, I thought, the admiral could have gotten twenty-one Bahrainis on camels to fire their rifles in reply or do something to respond to our courtesy.

Once again we encountered a narrow channel into a port. This one, however, had two ninety-degree turns. The British pilot who joined us did not have experience handling a ship such as the *LaSalle*. Consequently, he did not use enough rudder to start a turn, and then he would use too much. Initially, I coached him, but finally I moved over and stood by the helmsman and gave the helmsman the rudder commands needed to get the results that the British pilot wanted.

We had made the ship spotless for our arrival and the crew manned the rail for our entry into the port just before noon. With the temperature at 116 degrees Fahrenheit, my choker white uniform wilted quickly as I stood in the sun rendering a hand salute while the crew fired the twenty-one gun salute. By the time we docked, I looked as if I had been dunked. With the first few lines over, I gave the CONN to my Operations Officer and ran below to dry off and change uniform. Dressed in a new uniform, I returned quickly to the quarterdeck to render honors to RADM Bayne when he stepped aboard his new Flagship. His Staff quickly shifted from the *Valcour* to the more spacious and air conditioned flag spaces on the *LaSalle*. They loved it. It was like going from a Ford Pinto to a Lincoln Town Car for them. We had arrived at our new home port.

RADM MARMADUKE BAYNE, A SUBMARINER AND A REMARKABLE MAN, HAD BEEN EXTENDED - OVERTOURED - IN HIS JOB AS COMMANDER, MIDDLE EAST FORCE BECAUSE OF HIS RECOGNIZED EXPERTISE AND KNOWLEDGE OF THAT PART OF THE WORLD.

Ψ

RADM Marmaduke Bayne, a submariner and a remarkable man, had been extended - overtoured - in his job as Commander, Middle East Force because of his recognized expertise and knowledge of that part of the world. When State Department officers had questions about the Gulf or the Emirates, they solicited his advice. It was a pleasure and an education to work for him as his Flag Captain. "Duke" Bayne - I never called him that - went on to serve as the Commandant of the National War College and become the first President of the Defense University. Unfortunately, cancer cut his career short, but after surgery he lived in retirement to age eighty-four.[18]

We remained in port for the next four weeks during the hottest time of the year. Besides the oppressive heat, the humidity ranged above ninety percent. The combination was overwhelming. Our white paint and double air conditioning saved us and kept us cool. My new problem was getting the young sailors to close the hatches whenever they went out on a weather deck.

During this time we stayed busy. We offloaded the household effects and the cars of my married crew members who had opted to move their families to Bahrain. Social activities kept us active. The ship yard workers completed the repairs on Boiler Number 2 and then found cracked piping in Boiler Number 1 when they cleaned it. They stayed and repaired it and then left for the U. S. (Thank goodness that boiler hung together for as long as it did because it took us around Africa and then all the way to Bahrain.]

On the third day in port, RADM Bayne took me on my official call on the Amir of Bahrain, His Highness Sheik Isa Bin Sulman al Khalifa, at Government House.

I have often asked myself, what we would have done if we had had problems sooner with the boiler? In addition to these activities, military and governmental dignitaries called to visit the Admiral in his new Flagship. Those visits required the crew to keep the ship in tip-top condition at all times. We also hosted several formal dinners aboard the *LaSalle*, and, in turn, attended several in the city.

On the third day in port, RADM Bayne took me on my official call on the Amir of Bahrain, His Highness Sheik Isa Bin Sulman al Khalifa, at Government House. The Amir was about five feet tall and quite rotund. We had coffee and exchanged pleasantries. Then, I met the Prime Minister and other officials. All wore traditional Arab dress.

The Amir graciously authorized a helicopter tour of the twenty-mile by five-mile island, the south half of which he reserved as his hunting preserve. I took the tour that afternoon. I discovered the small island to be gravelly desert outside of the capital city of Manama. I also found that the intense sun, haze, sand, and ocean blinded me if I did not wear sunglasses. Several other visits by American and foreign military officers stationed in Saudi Arabia and from CENTO[19] Nations followed. The Admiral hosted many dinners for these groups aboard the *LaSalle*.

On the evening of 28 August, the Amir held a reception in our honor.

We also had some fun. Texans who ran the British Petroleum refinery on the island - fed by a pipeline from Saudi Arabia - invited us out for a tour and lunch. While there, they gave us a presentation about the upcoming oil crisis. They told us to tell everyone at home to expect a fuel supply emergency soon and to prepare themselves for it.

Ψ

On the evening of 28 August, the Amir held a reception in

our honor. RADM Bayne briefed me on the proper courtesies to show and how the visit and the dinner would be run. The Amir sat on his throne at the far end of a huge reception hall which was about the size of a basketball court. High back chairs lined both sides. I believe they called it a *majlis*, the area where every Arab ruler could hear the petitions of his subjects. One of the Amir's entourage introduced me, and I had to walk the entire length of the floor up to the Amir. There, his protocol secretary introduced me to the Amir, we exchanged a few words, and then the protocol secretary directed me to a chair. The game as I understood it was to walk to the chairs farthest from the throne to sit. Then, someone would intercept you and lead you to a chair closer to the throne to recognize your status. Very Biblical. "He who humbleth himself shall be honored, he who ... etc."

After all the guests had been introduced to the Amir, he rose, and led us to the dinner. We passed through a beautiful gardened area to the formal dining room where we were seated around a U-shaped table. Out of deference to the women and Westerners in attendance, we sat on chairs and had eating utensils. In many places on the table I saw platters four feet in diameter filled with mounds of golden rice and roasted goats complete with their heads, eyes, lips, hoofs, and all. The joke – perhaps protocol would be a better word was that if the Amir honored you by offering you the eyes, you were required to eat them. Luckily, that did not happen: the goats were for show.

Waiters who resembled the cast of thousands in the movie *Lawrence of Arabia* served us. With leathery skin burned black, they wore long robes, no shoes, bandoliers of bullets crossing their chests, and curved daggers at their belts. We had an excellent meal. Once the official party left the room, according to Arab tradition, all the lesser officials - the guards, waiters, etc. - dove into the goats and rice. When they finished, their wives and children ate. Locals called these functions "goat grabs." You could see greasy hand prints of past feasters on the walls.

We returned to the *majlis* for smokes, no alcohol. We stood around and the cast of *Lawrence of Arabia* returned to serve us

Lawrence of Arabia

"Thomas Edward (T. E.) Lawrence was born on August 16, 1888 in North Wales ... [He] became famous for his exploits as British Military liaison to the Arab Revolt during the First World War. Lawrence had been fascinated by archaeology since childhood. After graduating with honors from Oxford in 1910, he served as an assistant at a British Museum excavation in Iraq (then known as Mesopotamia)...

When war broke out ... in 1914, Lawrence spent a brief period in the Geographical Section of the General Staff in London, and was then posted to the Military Intelligence Department in Cairo.

In 1916, the Arabs rebelled against the Turkish empire. Lawrence was sent to Mecca on a fact-finding mission, ultimately becoming the British liaison officer to the Arabs...

After the war, Lawrence served in the British Delegation at the Paris Peace Conference, where he promoted the cause of Arab independence. Despite his efforts Syria, Palestine, and Iraq were mandated to France and Britain. Lawrence returned to England exhausted and disappointed."

WE ALSO FACED SOME CULTURAL PROBLEMS SUCH AS THE LOCAL CALENDAR. WE SHIFTED TO THE ARAB WEEK.

very strong coffee in thimble-sized cups. The waiters stood by while you drank your sip of coffee and then took your cup when you finished. They then placed the cup at the bottom of the pile of cups and moved on to the next group. No one knew how many times each cup was used. Best not to worry about it, I guess! After coffee, the servants brought around rose-scented water to pour on your hands. This was a nice touch, but then we stood around with dripping hands. Then along came this gent with glowing charcoal in a silver incense burner. RADM Bayne had not briefed me on this one! I figured this was my chance to dry my hands over the charcoal. That proved to be a social blunder. I learned that you were supposed to sprinkle a little frankincense on the charcoal and then waft the smoke towards you so that you could enjoy the aroma.

During my time in Bahrain, I always enjoyed visiting the market in town, the Suq. It found it very interesting. Besides the market, I visited the regular "downtown" section of shops. After I got my local driver's license, I drove around the city and the port and familiarized myself with all the key points of interest. I also explored the ship channel in my gig. I took it out a number of times to check the buoys against my chart. At low tide, I could see the sand bars that lined the channel.

Ψ

...THE INEVITABLE "BAHRAINI BELLY" GOT ME FOR ABOUT TEN VERY UNPLEASANT DAYS.

We faced stomach problems in the Middle East with contaminated local drinking water and sometimes with improperly washed fruits and vegetables. As best as I tried to dodge the problem, the inevitable "Bahraini Belly" got me for about ten very unpleasant days. I had taken care not to drink local water off the ship, but I suspect that some of the local ice cubes got me.

We also faced some cultural problems such as the local calendar. We shifted to the Arab week. Their Saturday is our Monday, so we shifted our divine services to Friday, which became our Sunday. Although confusing at first, we adjusted. TGIF for us became TGIW (Wednesday).

Grand Tour

FOLLOWING THE OVERTHROW OF THE SHAH, ALL THE ROYAL FAMILY MEMBERS, GOVERNMENT OFFICIALS, AND SENIOR MILITARY OFFICERS FLED THE COUNTRY. SHAFIG WENT TO PARIS. THERE, AYATOLLAH KHOMEINI'S GOON SQUADS SOON ASSASSINATED HIM.

Finally, after our hectic and fascinating four-week social whirl, we got underway for our first grand tour of the Gulf and the Indian Ocean.

In the Gulf, there were numerous oil rigs, and at night we could clearly see their burn-offs, which made perfect navigational references for us.

We made our first port of call Kharg Island, an Irani island off their coast. The weather was excellent. The following morning, we anchored off Kharg, and I went ashore to call on His Royal Highness, Prince Shafig, the Governor of Kharg. The Shah's nephew, the prince held many posts besides the Governorship: Commander in the Imperial Irani Navy, Commander of the Helicopter Squadron, and Commander of the Hovercraft Squadron. A man in his early thirties, a human dynamo, he spoke "American" and wore U. S. Naval Aviator's wings because he had gone through flight training at Pensacola. Having spent nearly two years in the U. S., he often would roll his eyes and say, "Ah-American women."

Prince Shafig held a reception and lunch for us and presented me with a beautiful silver cigarette box. After lunch, he invited me for a ride in a hovercraft which he piloted. The Prince took me for a wild ride, slewing around corners at fifty knots and bumping over waves.

That evening, RADM Bayne hosted a formal dinner on board. During the evening, Prince Shafig requested that the Admiral permit him to helo qualify on a ship at sea. Later, the Admiral cornered me and told me that the Prince would fly out the next day after we got underway. I told him that NATOPS (our standard operating procedures) did not permit such action until we briefed a pilot on signals, communications, procedures, safety, etc., and until we did that, I could not allow the Prince to fly onto my ship. Again, RADM Bayne said: "DO IT!" The Prince and another pilot made eighteen landings, and we had no problems.

Following the overthrow of the Shah, all the royal family members, government officials, and senior military officers fled the country. Shafig went to Paris. There, Ayatollah Khomeini's goon squads soon assassinated him.

Ψ

OUR GUNNERS FIRED SMARTLY EVERY FIVE SECONDS WITHOUT MISSING A COUNT. THE OMANIS ANSWERED WITH THEIR 16TH CENTURY MUZZLE-LOADING CANNONS EMPLACED IN THE 16TH CENTURY PORTUGUESE FORT THAT OVERLOOKED THE SEA.

Following this interlude, we soon sailed for Muscat, Oman, outside the Straits of Hormuz. I did not sleep that night because of the traffic. Rather than get run over by a huge tanker, I just got in line with them to steam out of the Gulf.

We arrived off Muscat on the morning of 24 September 1972. We found the harbor to be no more than a cove so we stayed out a ways in deep water. Once in place, we fired a twenty-one gun salute to Oman. Our gunners fired smartly every five seconds without missing a count. The Omanis answered with their 16th Century muzzle-loading cannons emplaced in the 16th Century Portuguese fort that overlooked the sea. Unlike our timing, the Omanis fired about every sixty seconds and their cannon belched flame, smoke, and paper wadding. We maintained our hand salutes for a long time that day! Afterwards, I did not mention to the Omanis that they had run up our flag upside down.

Muscat, and probably the rest of Oman, has to be the worst place in the world. It was the hottest place, built up against steep granite hills that surround the cove. I saw no vegetation, just crowded housing and sullen faces. The American Consul General told me that it cooled down to 100 degrees Fahrenheit at night.

RADM Bayne and I were to make an official call on the Sultan, a young man who had deposed his tyrannical father. I had recently seen the young ruler's photo in *National Geographic.* The Admiral told me that we would go ashore in full dress whites - that meant with medals, sword, and white gloves - and directed me to be on deck at a certain time. When he showed up, he said, "Leave your sword on board." I presumptuously told him that full dress whites meant with sword, and if he wanted no sword, then I should remove my medals and replace them with ribbons. Once again, he said: "Morin, DO IT!" So I removed my sword. I learned then that no one was allowed in the Sultan's presence carrying a weapon of any kind.

We went ashore in RADM Bayne's barge. An official car picked us up at dockside and drove us to the Palace. The Sultan met us in his office where we drank coffee and dipped our fingers into a bowl of very sweet stewed dates.

After our visit, our driver took us to the U. S. Consulate to pay a courtesy call on the American Chargé de Affairs, Mr. Quinlan. The Chargé had arranged a nice lunch for us. After the calls, we returned to the *LaSalle* to receive the return call of the Sultan made by his Prime Minister. We rendered full honors to the Prime Minister. Following that visit, we stood by for the return call of Mr. Quinlan. RADM Bayne directed me to fire a thirteen-gun salute. I reminded him that a Chargé only rated a nine-gun salute. The Admiral said, *"Morin!!"* We gave Mr. Quinlan thirteen guns!

That evening, we returned to shore and drove to the home of Dr. and Mrs. Bosch, Christian missionaries who had lived in Oman almost all of their adult lives. Dr. Bosch was a renowned archeologist and he had filled his home with artifacts. The home was unusual. It was built into the side of a granite hill above the city and the back walls of the rooms at the rear of the house were rough granite.

The following day, I toured Ft. Murani from which the Omanis had fired their salute. No one had done anything to modernize the fort after it was built. A regiment of Omani soldiers occupied it, and by our standards lived in some pretty Spartan conditions. I saw the old cannons used for the salute. The troops had to pull them back from their firing positions to reload them; then, they had to run them back into battery to fire. Considering the physical effort involved in handling the pieces, firing one shot every sixty seconds by the two cannons seemed pretty efficient in hindsight.

I SAW THE OLD CANNONS USED FOR THE SALUTE. THE TROOPS HAD TO PULL THEM BACK FROM THEIR FIRING POSITIONS TO RELOAD THEM; THEN, THEY HAD THEN RUN THEM BACK INTO BATTERY TO FIRE. CONSIDERING THE PHYSICAL EFFORT INVOLVED IN HANDLING THE PIECES, FIRING ONE SHOT EVERY SIXTY SECONDS BY THE TWO CANNONS SEEMED PRETTY EFFICIENT IN HINDSIGHT.

Zeigler's welcome made a big hit and set an excellent tone for the rest of the formal meal.

After the tour of the fort, we hosted a formal dinner that evening aboard the *LaSalle.* The Prime Minister represented the Sultan. A few Omani Government ministers attended, as did Dr. and Mrs. Bosch, and Mr. Quinlan. We also hosted some British officers who ran the Omani Army and Navy. I had a young officer aboard who spoke fluent Arabic. ENS Conway Ziegler, a Naval ROTC graduate of Princeton, had joined us just before we left the United States. After the guests sat down for dinner, I had Ziegler appear on our closed-circuit television system and welcome our guests in their native language. He spoke live from the Admiral's cabin. Zeigler's welcome made a big hit and set an excellent tone for the rest of the formal meal. That night, one of the Omani ministers seated next to me, said: "He speaks very good Arabic, but he has a Lebanese accent." Apparently, he picked up his accent while he studied in Beirut.[20] I had Conway on the quarterdeck when the guests departed. The guests complimented him profusely as they left the ship, but were surprised when they saw his red hair, blue eyes, and freckles.

Conway Zeigler played an important role in all our social affairs with Arabic-speaking guests aboard the *LaSalle*, and our guests always appreciated his welcome and were fascinated by his hair, eyes, and freckles.

Ψ

Early on the morning of 1 October 1972, we entered the harbor off the city of Victoria in the Seychelles and anchored. We rendered honors as we sailed in.

The following day, we got underway for the Seychelles Islands where we were to represent the United States at the celebration of 150th anniversary of England taking over the islands. RADM Bayne returned to Bahrain, but his Chief of Staff, CAPT Edward Baty, with the rest of the Admiral's staff accompanied us. We found the Indian Ocean to be smooth as glass during the trip, with absolutely no wind. With the smooth sea, we saw the phenomenon known as "green flash" as the sun disappeared at sunset. Green Flash developed because of the refraction of the sun light as the sun dips below the curvature of the earth. If you blink, you will miss it, but I saw the phenomenon innumerable times.

Early on the morning of 1 October 1972, we entered the harbor off the city of Victoria in the Seychelles and anchored. We rendered honors as we sailed in. The British destroyer HMS *Lowestoft* and the French frigate *Brestois* had preceded us into the port and had anchored. With the arrival of the *LaSalle*, I became the Senior Officer Present Afloat (SOPA). Naval tradition required that all the junior captains in port call on the SOPA, so soon after we anchored the French Captain, a Lieutenant Commander, arrived in full dress to pay his respects. We gave him full side honors, and then I escorted him to my cabin to chat. I offered him tea, coffee, or fruit juice to drink because by regulation I could not have liquor aboard. After a short visit, he returned to his frigate to await my return call. When I arrived at his ship, the crew piped me aboard and I went to the French Captain's cabin. There, we enjoyed some excellent cognac. I returned to the *LaSalle* to await the call of the British Commanding Officer. After the British Captain called, I boated over to his destroyer and enjoyed a couple of excellent gin and tonics. Foreign officers knew about the U. S. Navy's rules about alcohol, but in cases such as these, American captains faced terribly embarrassing situations. Later, Secretary of the Navy John Lehman changed the 1914 no-alcohol rule and permitted American captains to serve alcoholic beverages in port to visiting officials where it would be an embarrassment not to do so.

CAPT JAMES MORIN

Princess Margaret, the Queen's sister, and Lord Snowden, the princess' husband, represented Queen Elizabeth for the festivities. We attended the opening ceremonies that started the celebration in the local sports stadium. That evening, CAPT Baty, the British Captain, the French Captain, and I, all decked out in our dinner dress uniforms, attended a reception for Princess Margaret at the British Governor's huge residence. The old home was very open, and it had beautiful wooden floors and paneled walls.

PRINCESS MARGARET, THE QUEEN'S SISTER, AND LORD SNOWDEN, THE PRINCESS' HUSBAND, REPRESENTED QUEEN ELIZABETH FOR THE FESTIVITIES.

We socialized with the local British and Seychellois until we heard the music – "God Save the Queen." Princess Margaret,

Later, I talked with Lord Snowden. He asked to tour the LaSalle. He came aboard the next morning. During his time on board he asked a number of questions about our capabilities. He left impressed.

on the arm of the Governor, walked down a long curving staircase into the room. When she spotted the four naval officers, she smiled and made a bee-line for us. The Governor, however, deftly grabbed her elbow and steered her away so he could make the appropriate introductions to all the guests. Ultimately, he introduced the Princess to us, and we exchanged small talk for a few minutes. Later, I talked with Lord Snowden. He asked to tour the *LaSalle*. He came aboard the next morning. During his time on board he asked a number of questions about our capabilities. He left impressed.

During the next few days, we toured the tropical paradise and visited the U. S. Air Force Tracking Station there. The Americans were pleased to have a U. S. Navy ship in port. Many of them toured the ship, purchased items in our ship's store, and had lunch or dinner on board.

On 5 October 1972, we got underway and sailed to Mombassa, Kenya to make our next port visit. We arrived on the 8th and found that the only available pier could not accommodate our ship. The locals directed us to make up fore and aft between two mooring buoys in the river. We faced a tough maneuver because of the river's current, a brisk wind, and the fact that we had to fit in between the ships already moored in the river. When moored, we found that we were no more than fifty to sixty yards from a heavily treed river bank. Once made up, I called on the Mayor, the Harbor Master, and the Commander of the Kenyan Navy, a British Royal Navy Commander. Then, we toured Mombassa.

The next morning I left for Tsava National Park. I stayed for two days at a very fancy hotel/motel that was built above a large watering hole. At night, we watched the wild animals come in to drink water. The following day, we loaded into a bus and drove out on the plain to see wild animals up close. Our driver stopped frequently so we could take pictures, but he would move on quickly when he noticed that an elephant or a rhino was getting antsy. I enjoyed the break and being off the ship, but I could not relax fully because I did not know what was going on aboard the *LaSalle*.

In the morning of 12 October 1972, we left Mombassa for Port Louis, Mauritius. Mauritius is another small island off Madagascar. En route we experienced another serious steam leak which required wrapping up a boiler for repairs. This delayed us about a half day. As we proceeded well into the southern hemisphere, we bucked heavy seas and winds.

We arrived off Port Louis about noon on the 16th and made up to a pier. Once docked, I made the usual round of calls: the Prime Minister, the Lord Mayor, and the U. S. Ambassador, Mr. Brewer. That night Mr. Brewer held a reception and dinner in our honor at his home. We enjoyed the evening. The hospitable embassy staff, in turn, entertained my officers and men. The staff obviously enjoyed our visit because they did not get to see many American warships.

CAPT Baty, on behalf of RADM Bayne, hosted a formal dinner on board the following evening. The guests resembled the United Nations: the Prime Minister, Mr. Ramgoolam; Ambassador Brewer; the Lord Mayor, the British High Commissioner; the Ambassadors of France, India, and Pakistan; and the Soviet Union's Chargé.

1972
WATERGATE

WATERGATE WAS THE NAME GIVEN TO THE BREAK INTO THE DEMOCRATIC PARTY NATIONAL COMMITTEE OFFICES IN THE WATERGATE HOTEL BUILDING IN WASHINGTON, D. C. ON 26 JUNE 1972.

THE POLICE ARRESTED FIVE BURGLARS WHO ATTEMPTED TO PLANT MICROPHONES IN THE OFFICES.

DURING THE INVESTIGATION OF THE INCIDENT, PRESIDENT NIXON'S AIDES, H. R. HALDEMAN AND JOHN ERLICHMAN WERE LINKED TO THE BURGLARY AND ITS COVER-UP. THEY RESIGNED.

LATER, PRESIDENT NIXON WAS LINKED TO THE INCIDENT AND THE COVER-UP. HE EVENTUALLY RESIGNED IN 1974.

Ψ

The most important day to Britain's Royal Navy is Trafalgar Day, 21 October, the day that Lord Nelson defeated the French fleet. British naval officers hold formal dining-ins annually to celebrate the Royal Navy's most illustrious victory. While in Mauritius, the Captain of the British Communications Station, HMS *Mauritius*, invited me to attend as an honored guest. Our own dining-ins follow the British tradition, but they are mild reflections of what turn out to be pretty wild parties in the Royal Navy.

TRAFALGAR DAY

THE MOST IMPORTANT DAY TO BRITAIN'S ROYAL NAVY IS TRAFALGAR DAY, 21 OCTOBER, THE DAY THAT LORD NELSON DEFEATED THE FRENCH FLEET.

Royal Navy dining-ins begin sensibly: Royal Pipers start the program; the presentation of the beef follows; and the required joke telling precedes the serving of the formal meal. Once the meal ends, waiters distribute cigars to all along with many de-

Fittingly, the High Commissioner extolled the heroism of Lord Nelson and the Royal Navy. When he finished, my host asked me to say a few words. I cannot recall what I said, but never at a loss for words, I said something.

canters of port wine. With port on hand, the toasting begins: first the Queen and then the Royal Navy. With an American present, the toasts continued: the President and then the U. S. Navy. With those toasts, I had to return the honor and toast the Queen and the Royal Navy again. Then, more toasts followed. Eventually, the guest speaker – the British High Commissioner at this event – addressed the officers. Fittingly, the High Commissioner extolled the heroism of Lord Nelson and the Royal Navy. When he finished, my host asked me to say a few words. I cannot recall what I said, but never at a loss for words, I said something.

Then the games and the rough part began. The officers had drunk lots of port wine. The games got physical: officers lost sleeves or tore seams out of their dinner jackets and trousers. As the senior officer present, I was not required to participate in the mayhem but was talked into a short race across the dining room floor. I agreed to this simple task. But first, each participant in the race had to put his forehead on a short broomstick and go around it several times before heading for the hurdle. Not wishing to shame the U. S. Navy in this contest, I did my turns and started. Once airborne over the first hurdle, I did a slow roll from the effects of the rotations around the broomstick. If I had not drunk so much port, I probably would have killed myself.

Meanwhile, someone had brought in some explosive devices that resembled five-inch-long firecrackers. Every now and then, an officer would set one off in the dining room. They were very loud. Someone gave me one to hold and light before throwing it and told me that the fuzes burned for seven seconds before exploding and that the device and the fuzes were reliable. I was not sure about the information, but I did light one off.

After midnight, the party started slowing down. There was one firecracker left, and the group decided to go out on the front lawn and light it off. The British Lieutenant Commander who served as our liaison officer for the port visit lit the fuze and put a bucket over the firecracker. The device exploded with a loud bang and in the high humidity a dense cloud of smoke followed the explosion. The British officer ran into the smoke and began laughing loudly. We followed to see what had happened and found that the bucket had not budged but the bottom was nowhere in sight. I looked up expecting to see the saw-like disc descending on us to cut us to ribbons. I think it is still in low orbit around the earth. And to think that I held one of the things in my hand and lighted and threw it.

We toured the island in the following days. It was very lush and hilly. Sugar cane plantations covered most of the open ground.

On 22 October 1972, we got underway for Karachi, Pakistan. Our tube problems in a boiler continued to plague us, so we found ourselves steaming on one boiler again. Once the boiler in question cooled down, crew members entered it and found a ruptured side wall tube. They removed it and blocked it off.

We then satisfactorily hydro-tested the system. Because of our problems with the power plant, I never was confident that the *LaSalle* would ever be free from potential engineering casualties in the steam plant. Some of our problems stemmed from the design of our power plant; others stemmed from human error.

When we reached Karachi, a pilot came aboard to take us up the channel where we went pierside. As we moved into the port, we saw that many of the port facilities had been damaged in the short Indian-Pakistani war that had just ended. This made our visit more important than most because of our military ties to anti-Communist Pakistan. RADM Bayne rejoined his flagship. The usual calls followed, and that night the Admiral hosted a dinner aboard the *LaSalle*. The next day, I visited Karachi, which teemed with people.

I met two Pakistani Naval Officers who had been at the Naval Post Graduate School with me in 1957. They asked me to get U. S. missiles for Pakistan. They must have thought that I had influence. Influence? I could not keep my boilers running without problems!

We sailed for Bahrain on 2 November and arrived there on the morning of the 5th. We found the weather becoming very pleasant and enjoyed our next ten days in Bahrain before we

WATERGATE CATALYST

"WATERGATE WAS NOT SIMPLY AN ISOLATED EVENT IN WHICH A PRESIDENT, CORRUPTED BY POWER, WAS DETECTED IN HIGH CRIMES AND MISDEMEANORS AND JUSTLY PUNISHED BY BEING DEPRIVED OF HIS OFFICE. IT WAS A CULMINATION OF A LONG AND BITTER STRUGGLE FOR CONTROL OF GOVERNMENT AND THE HEARTS AND MINDS OF THE AMERICAN PEOPLE, BETWEEN LEFT AND RIGHT, BETWEEN CONGRESS AND THE EXECUTIVE BRANCH, BETWEEN THE INTELLIGENTSIA AND THE COMMONALITY, AND, ABOVE ALL, BETWEEN RICHARD NIXON AND THOSE WHO, MANY YEARS BEFORE, HAD INVESTED THEIR OWN POLITICAL VALIDITY AND SELF-ESTEEM IN THE INNOCENCE OF ALGER HISS.

THE CATALYST ... WAS THE VIETNAM WAR, WHICH SPLIT THE COUNTRY INTO TWO CAMPS AND CREATED A CONFRONTATION ... WHICH TESTED TO THE UTMOST THE POWER OF OUR INSTITUTIONS TO PRESERVE THE CONSTITUTION."

ALEXANDER HAIG

We got underway on 20 November to conduct formation steaming evolutions. The language barrier presented a problem. With ships maneuvering and signals being missed or misunderstood, I gave up trying to get any sleep.

left on 16 November for some sea time with Commander, Destroyer Squadron 4. With the Squadron's three destroyers we headed off to the Irani Naval Base of Bandar Abbas from which we were participate in the annual CENTO exercise with the Iranis, Pakistanis, and British. We anchored off Bandar Abbas with fourteen other ships on the 18th to participate in a pre-exercise conference.

The Irani Naval Commander-in-Chief, RADM Attai, and four other admirals joined us for dinner that night. Admiral Attai was very interested in the *LaSalle* as a command ship. Consequently, I gave him a tour, and he was amazed at all the space we had below decks for boats and vehicles. I was surprised that he did not get the Shah to buy him one like it because the Iranians were buying everything else at the time.

Ψ

We got underway on 20 November to conduct formation steaming evolutions. The language barrier presented a problem. With ships maneuvering and signals being missed or misunderstood, I gave up trying to get any sleep. As we exercised, we worked our way south to another Irani Naval Base, Bandar Jask, where we planned to meet for a final exercise brief. We entered the port 600 yards apart in column, executed a smart ninety-degree turn to port, and dropped anchors on signal. It was a formation anchoring. At slow speed, the last ninety-degree turn proved hard for the *LaSalle* to do smartly. The destroyers and frigates turned quickly, and they had reserve power for slow speed maneuvering. I was still in my turn when the signal came to "Let go." I got pretty close to the little ship ahead of me because his radius of turn was so much shorter than mine. I put our anchor just below the water line, signaled "anchored," and continued to back and fill to move to my designated spot. That evening, all Captains came to the *LaSalle* for the final phase briefing.

We got underway again the next morning, the 24th. For the

next two days of the exercise, our escorts protected the *LaSalle* and the HMS *Plumleaf* from simulated attacks by submarines. We reentered Bandar Abbas on 26 November and again executed a formation anchoring. This time before I got the signal to execute the final ninety-degree turn I had my rudder over to get my bow moving. The next morning, I attended the post-exercise critique at Irani Naval Headquarters and then attended a sumptuous lunch. *LaSalle* was to get underway at 1300 hours. Because of the lunch, I knew that I could not get back to the ship by that time so I signaled my XO to get underway on time. After lunch, RADM Bayne and I flew back and landed on the ship about 1400. We arrived in Bahrain on 28 November, ending our grand tour. A quiet period followed: *LaSalle* was to remain in port for the remainder of the year.

Return

Upon our return the Navy notified RADM Bayne that he had been selected for promotion to Vice Admiral and ordered him to return to Washington to be the head of the National Defense University. New orders also arrived detaching me from command of the *LaSalle*, effective January 1973, and after training, assigning me to command the aircraft carrier, USS *Franklin D. Roosevelt* (CVA-42) home ported in Mayport, FL near Jacksonville Beach.

A round of dinners to honor the Admiral and Mrs. Bayne followed at the Palace, various embassies, and at many private residences. The well-respected couple had been instrumental in fostering good relations in a part of the world where the United States needed friends. When RADM R. J. Hanks, USN[21] arrived, he faced a substantial challenge: building on what RADM Bayne had accomplished.

During this port period, Prince Hamed, the Amir's twenty-one-year-old son, and a Brigadier General in the Bahraini Defense Force (BDF) invited all the ship's officers to a dinner for men only at BDF Headquarters. At the military compound out-

1972 Situation in the Middle East

"As for the Middle East situation, tensions continued in 1972 because of the radical actions of Arab guerrillas and Israel's frequent military operations against Syria and Lebanon, although a … cease-fire had existed since the Middle East War five years before. In 1972, there were such developments as (i) the cooling-off of relations between the Soviet Union and Egypt, (ii) increased fluidity in the domestic situation in Egypt and (iii) increased Soviet aid to Iraq, Syria and Turkey. There were no signs of concrete progress toward peace in the Arab-Israeli conflict."

"The Soviet Union began to show moves to promote close relations with its southern neighbors by taking such measures as Premier Kosygin's visit to Iraq (from April 6 to 10, 1972), the conclusion of the Soviet-Iraqi treaty of friendship and cooperation, and Chairman Podgorny's visit to Turkey (from April 11 to 17, 1972). Despite Egyptian President Sadat's visit to the Soviet Union immediately after these events (from April 27 to 29), these Soviet moves were considered as something different from its past Egypt-oriented Middle East policy. In contrast to the subsequent cooling-off of its relations with Egypt, Iraqi President Bakr's official visit to the Soviet Union (in September), which promoted Soviet-Iraqi cooperation in the economic and military fields, and the Soviet … moves for a rapprochement with Syria were conspicuous, it is believed that the Soviet Union will continue making efforts to increase its influence over the Middle East, while taking into consideration … Egypt's declining diplomatic influence."

1972
Situation in Africa

"... [China]... put emphasis on Africa in its aid diplomacy. In November 1972, [China] concluded an agreement on economic and technical cooperation with Nigeria. It concluded a similar agreement with Zaire in January 1973 and gave loans totaling $100 million. It also gave loans on favorable terms and steady technical cooperation to other Black African countries. It is considered that China's aid diplomacy, coupled with its way of doing things which is attracting the attention of Africans, will become a factor that will cause China's influence in Africa to increase in the future. It is believed, however, that China will not intensify its political maneuvering because of past experiences.

The [U. S. and the U. S. S. R.] also have great interest in Africa, and it is believed that they will continue to give the region considerable aid in the future.

... many African countries welcomed aid from any country from the standpoint of the principle of nonaligned neutrality, and tried to maintain their independence by being careful not to become too dependent upon aid from any particular country or group of countries.

Reports indicated that national liberation movements in Portuguese Guinea, Mozambique and Angola intensified. Amilear Cabral, a leader of the liberation movement in Portuguese Guinea, was assassinated (in January 1973) at a time when it was expected that Portuguese Guinea would soon declare its independence, and developments in the situation in the area assumed greater importance."

side the city where we were to have dinner, we saw several Jordanian Army Officers who trained and officered the BDF. They attended in civilian clothes because their presence was considered to be a secret. All the guests ate under tents on a beautifully mild, clear evening. After dinner we sat on seats off the parade ground and watched the BDF military band – dressed in kilts and playing bagpipes – march and play. They were very good. At sunset, in a very British style, the band played "retreat" and the color party lowered the colors.

After dinner, the fun began. Amid lots of laughing and yelling in Arabic the men formed a ring around a fire. Then, one person would sing a refrain in Arabic, and everyone else would laugh and appreciate whatever was sung. Next, the circle danced around the fire much in the Greek or Jewish traditions. The other guests pulled each of us into the circle and linked arms, and we tried to figure out how to dance the next tricky step. When the circle stopped, the Amir's son called out another name, and that person sang another refrain. Then, the circle danced again. The dancing continued well into darkness. We may have been the butt of their songs: I do not know, but we had fun.

We celebrated New Year's Eve in the British Officers' Club at Jufair, a communications center. Britain was pulling all her military forces out of the Middle East at the time, and Jufair was last British military facility in Bahrain. After we welcomed 1973 in, the Americans decided to hold the first Bahraini Touch Football Classic at the base soccer field at 1000 hours New Year's Day. A sufficient number of Americans showed up to organize two good teams. The classic began. We played for over an hour on the gravel field, and I was pleased that I had not pulled a hamstring or twisted an ankle. Everyone was tired, and the group decided that next team to score a touchdown would win. I received a punt, and as I ran out of bounds, one of my opponents pushed me towards a bench on the sidelines. I leaped over the bench, put my foot down in the only puddle of water on the field, and heard something break. Teammates helped me to my feet. I felt okay and walked over to the car and returned to the ship. The Hospital Corpsman took x-rays, but did not see any breaks. That afternoon, I attended a reception. The standing around became very painful. When the Staff Doctor arrived, I told him about my problem. We immediately went back to the ship, and the doctor looked at the x-rays. Finding a break, he put my leg in a cast.

That evening, CAPT Howard Crosby, USN, my relief, reported aboard.[22] We had a week to work together before our scheduled Change of Command ceremony. As best I could, I hobbled around the ship on crutches as we inspected every space. Going down ladders was murder. The day before the ceremony, we held an inspection of all ranks on deck. I let my Executive Officer inspect with the new Captain. I sat in a chair too pooped to do much more. That night, the officers and men honored me at a farewell dinner, and the next morning I relinquished command.

Ψ

I left the *LaSalle* about noon on 9 January 1973. Still hobbled, I arrived at the airport where the Station Manager of Middle East Airlines, whom I had met at various social gatherings in Bahrain, took good care of me and helped me to my airplane. Once airborne that evening and headed to Beirut, I looked down and saw all the "burn-offs" out in the desert. I had seen burn-offs often while steaming in the Persian Gulf so they were an appropriate last remembrance of my tour in *LaSalle* and that exotic part of the world.

I LEFT THE LASALLE ABOUT NOON ON 9 JANUARY 1973. STILL HOBBLED, I ARRIVED AT THE AIRPORT WHERE THE STATION MANAGER OF MIDDLE EAST AIRLINES, WHOM I HAD MET A VARIOUS SOCIAL GATHERINGS IN BAHRAIN, TOOK GOOD CARE OF ME...

Sidebar Notes

Sea Power: Alfred Thayer Mahan
Martin, *America and Its Peoples*, II, 558.

LaSalle
Rex Harrison, *Email to John Bradley* (24 August 2008), 2.

1972 Nuclear Arsenals and 1969-1972 SALT I
"Strategic Arms Limitation Treaty I," http://www.atomarchive.com/Treaties/Treaty8.shtml (11 July 2010).

1972 Détente
Martin, *America and Its Peoples*, II, 881.

Tight Ship
Harrison, *Email to Bradley*, 2.

Lawrence of Arabia
http://www.lucidcafe.com/library/95aug/lawrence.html (14 July 2010).

1972 Watergate
Martin, *America and Its Peoples*, II, 873-76.

Watergate Catalyst
Haig, *Inner Circles*, 328.

1972 Situation in the Middle East
"Situation in the Middle East and Africa," http://.mofa.go.jp/policy/other/bluebook/1972/1972-1-13htm.

1972 Situation in the Africa
"Situation in the Middle East and Africa," http://.mofa.go.jp/policy/other/bluebook/1972/1972-1-13htm.

End Notes

[1] Gutzman, *Vietnam*, 16-17.
[2] Alexander M. Haig, Jr. with Charles McCarry, *Inner Circles, How America Changed the World, A Memoir* (hereafter, *Inner Circles*) (NY: Warner Books, 1992), 84.
[3] *Ibid.*
[4] *Ibid*, 85. For 2 thru 5.
[5] *Ibid*, 86 for 6-8.
[6] Christopher Andrew and Vasili Mitrokin, *The Sword and the Shield, The Mitrokin Archive and the Secret History of the KGB* (NY: Perseus Books, 199), 181 and "KGB Subversion and Disinformation Plan," in *Berlin Crisis, wikipedia.org*, 4-7 (19 April 2010).
[7] *Ibid*, 89.
[8] Mow retired as a Rear Admiral in 1981. He graduated from the U. S. Naval Academy in 1950. During his career he earned a Silver Star, three Distinguished Flying Crosses, three Bronze Stars, and high service awards. *1993 USNA Register of Alumni*, #21874, 306.
[9] Clark retired as a Commander.
[10] *"Designated flagship for Commander, Middle East Force, Valcour departed Norfolk in August 1949 for the first of sixteen deployments to the Middle East. She returned to Norfolk in March 1950 and conducted a second tour as Middle East Force flagship between September 1950 and March 1951. In May 1951, while departing Norfolk for independent ship exercises, she suffered a steering casualty and veered across the bow of the collier Thomas Tracey. The ensuing collision ruptured an aviation gasoline fuel tank and started a raging fire that took the lives of 36 men. After a major firefighting and salvage operation, she was brought back into port the following day. Valcour then underwent an extensive overhaul, during which air conditioning was installed and her 5"/38 gun was removed to compensate for the added weight. Between 1952 and 1965 Valcour deployed every year to the Middle East as one of a trio of ships that served alternately as flagship for Commander Middle East Force. Through 1961 Valcour followed a highly predictable schedule, departing Norfolk in January, relieving USS Duxbury Bay (AVP-38) upon arrival on station, being relieved by USS Greenwich Bay (AVP-41), and returning to Norfolk in August. Highlights of this service included the boarding, salvage, and return to its crew of the burning and abandoned Italian tanker Argea Prima in May 1955 and a visit to the Seychelles Islands in 1960. She was the first U. S. Navy ship to call there in 48 years. In around 1960 Valcour received some conspicuous equipment upgrades, including a tripod mast with a newer air search radar and a tall communications antenna which, with its deckhouse, replaced the quadruple 40mm gun mount on her fantail. She completed her fifteenth Middle East cruise in March 1965. In a 1965 force realignment, Valcour's two running mates were ordered decommissioned and Valcour was selected to be the sole Middle East flagship. As such, she was reclassified AGF-1 in December 1965 and departed the United States for her new home port of Bahrain in April 1966. Though designated the permanent Middle East Force flagship in 1971, in January 1972 she was selected for inactivation." U. S. Navy Historical Center Website*, 26 September 2009.
[11] *Ibid.* The band eventually played successfully in seventeen countries – GITMO, Cuba; Dakar, Senegal; Monrovia, Liberia; Abidjan, Ivory

Coast; Lourenco Marques, Mozambique; Mombasa, Kenya; Mogadishu, French Somalia; Djibouti, Ethiopia, Port Victoria, Seychelles; Port Louis, Mauritania; Muscat, Oman; Manama, Bahrain (home port); Kharg Island, Iran; Bandar Abbas, Iran; Karachi, Pakistan; Colombo, Sri Lanka; Singapore; and Diego Garcia.

[12] Rex Harrison, *Email to John Bradley* (24 August 2008), 2. Harrison added that Morin said; "The wives of our seagoing officers and men have never received the recognition and credit they so richly deserve."

[13] Sadler retired as a Commander.

[14] The Ivory Coast is now the Republic of Côte d'Ivoire. The national language is French.

[15] Harrison, *Email to Bradley*, 2.

[16] Following Portugal's granting of independence in 1974, the city was renamed Maputo. The port is the capital of Mozambique and the largest city in the country.

[17] John Tomaio, *Email to John Bradley*, 25 August 2008.

[18] *Ibid.*

[19] *Washington National Cemetery Website,* consulted 27 September 2009. Born in Norfolk, Virginia, Bayne graduated from the University of Tennessee in 1942. "He served on two submarines in the Pacific during World War II and later commanded two submarines... as well as a submarine division and submarine flotilla. He also served as an aide to the commander in chief of the Atlantic Fleet [and] executive assistant to the Secretary of the Navy." He commanded U. S. Middle East Forces from 1969 to 1972.

[20] CENTO - Central Treaty Organization. The original organization, the Middle East Treaty Organization (METO) or the "Baghdad Pact," was formed in 1955 to oppose Soviet expansion and influence in the Middle East. Britain, Iran, Iraq, Pakistan, and Turkey were initial members. The first headquarters were in Baghdad. The U. S. became an associate member in 1958. The same year, Iraq withdrew, and the organization was renamed CENTO. In 1974, Turkey invaded Cypress, and Britain withdrew its forces. In 1979, the Iranian Revolution ousted the Shah, and CENTO collapsed. CENTO never proved terribly effective.

[21] He also studied in Egypt, but apparently that stay did not affect his Arabic accent.

[22] Most likely Robert Jack Hanks, Class of 1946, U. S. Naval Academy, #17855. *1993 USNA Register of Alumni,* 280. Hanks retired as a Rear Admiral in 1977. This probably was his last assignment.

[23] Possibly Captain Howard Sidney Crosby, U. S. Naval Academy, Class of 1948, #20190. *1993 USNA Register of Alumni*, 294. He retired in 1979.

Ten

DANGEROUS TIMES

USS Constellation[1]

ARTICLE 94[2]
MUTINY OR SEDITION

UNITED STATES UNIFORM CODE OF MILITARY JUSTICE, 1951

(a) ANY **PERSON SUBJECT TO THIS CODE-**

(1) WHO WITH THE INTENT TO USURP OR OVERRIDE LAWFUL MILITARY AUTHORITY REFUSES IN CONCERT WITH ANY OTHER PERSON OR PERSONS, TO OBEY ORDERS OR OTHERWISE DO HIS DUTY OR CREATES ANY VIOLATION OR DISTURBANCE IS GUILTY OF
MUTINY;

(2) WHO WITH THE INTENT TO CAUSE THE OVERTHROW OR DESTRUCTION OF LAWFUL CIVIL AUTHORITY, CREATES, IN CONCERT WITH ANY OTHER PERSON OR PERSONS, REVOLT, VIOLENCE, OR OTHER DISTURBANCE AGAINST SUCH AUTHORITY IS GUILTY OF
SEDITION;

(3) WHO FAILS TO DO HIS UTMOST TO PREVENT AND SUPPRESS AN OFFENCE OF MUTINY OR SEDITION BEING COMMITTED IN HIS PRESENCE, OR FAILS TO TAKE ALL REASONABLE MEANS TO INFORM HIS SUPERIOR OR COMMANDING OFFICER OF AN OFFENCE OF MUTINY OR SEDITION WHICH HE KNOWS OR HAS REASON TO BELIEVE IS TAKING PLACE, IS GUILTY OF A FAILURE TO SUPPRESS OR REPORT A MUTINY OR SEDITION.

(b) A PERSON WHO IS FOUND GUILTY OF ATTEMPTED MUTINY, MUTINY, SEDITION, OR FAILURE TO SUPPRESS OR REPORT A MUTINY OR SEDITION SHALL BE PUNISHED BY
DEATH
OR SUCH OTHER PUNISHMENT AS A COURT-MARTIAL MAY DIRECT.

Strategic Vulnerability

Had the leaders of the Soviet Union realized the opportunity they had, they could have dealt the United States grievous blows during the years from 1971 to 1973 because drugs, racial conflicts, and the last tempestuous months of the Vietnam War seriously undermined the discipline and operational capabilities of the U. S. Armed Forces.

U. S. Army

The Army had its hands full in South Vietnam in the late 1960s and early 1970s. Troops in many units – often rear area units – became ill-disciplined because of drug use and the influence of the anti-war movement on many of the soldiers, often draftees, who deployed to South Vietnam. As a result, some troops "fragged" their officers, especially those deemed to be too tough and demanding. In one case, a black soldier shot at two officers who demanded that he turn down the volume on his boom-box that played inside a bunker. The soldier missed the officer he fired at and killed the other officer standing outside the entrance to the bunker.[3] Overall, the Army claimed that soldiers killed at least 1,011 officers and NCOs during the Vietnam War.[4]

Outside groups – often Marxist - worked hard to destroy the discipline and operational capability of the troops in Vietnam. The following information from a paper associated with the National Lawyers Guild which Philippine Authorities seized in 1972 in Olongapo highlighted the goals of the outsiders:

> *'GI Work' has as its long range goal the weakening and finally the destruction of the military arm, to be replaced by a people's army. It is aimed at reaching the point at which combat troops will not fight in Vietnam or elsewhere... various actions incapacitate the ruling class war machine and finally GIs join the liberation struggle against the ruling class... but the anti-imperialist class struggle goal is the overall strategic guide.*[5]

Additional comments provide a deeper understanding of the efforts of the Guild:

> *The weakening and eventual destruction of the military arm will proceed in many stages. The first task is to arouse in those GIs whose class backgrounds and positions in the military make them our allies against imperialism. We must show these low-ranking [enlisted men] how they are being used, and explain to them what imperialism is and how it works. Large and grow-*

1972-1973
NATIONAL LEADERS

PRESIDENT
RICHARD NIXON

SECRETARY OF DEFENSE
MELVIN LAIRD

JOINT CHIEFS OF STAFF

CHAIRMAN
ADM THOMAS MOORER

CHIEF OF STAFF, AIR FORCE
GEN JOHN D. RYAN
GEN GEORGE S. BROWN
(EFFECTIVE 1 AUG 73)

CHIEF OF STAFF, ARMY
GEN WILLIAM WESTMORELAND
(RETIRED 30 JUN 72)
GEN BRUCE PALMER, JR.
(ACTING)
GEN CREIGHTON ABRAMS
(EFFECTIVE 12 OCT 72)

CHIEF OF NAVAL OPERATIONS
ADM ELMO ZUMWALT

ing organizations of anti-war GIs should result from this arousal. Destroying the military arm should be the goal of these organizations. Eventually this arm will be replaced by a people's army which will lead the struggle to build a socialist society. At this time, however, the goal is entirely destructive.[6]

Few Army troops served in the Olongapo area, but Navy and Marine personnel visited the area frequently and a large number of Air Force personnel worked less than fifty miles away at Clark Air Force Base.

1971 ACCESSIONS TO U. S. ARMY (ENLISTMENTS AND DRAFTEES)

MONTH	ALL ACCESSIONS	PERCENT CAT 4
JAN	36,133	25.4%
FEB	32,553	27.1%
MAR	26,182	27.1%
APR	19,900	28.2%
MAY	20,899	28.4%
JUN	23,883	25.3%
JUL	13,230	25.9%
AUG	14,290	24.3%
SEP	16,268	23.2%
OCT	10,962	6.5%
NOV	17,218	14.7%
DEC	14,567	21.4%

In Europe, Army commanders faced drug and racial problems without the help of seasoned company officers and experienced NCOs because so many of those leaders had deployed to Vietnam. Consequently, drug rings operated in the line units, often managed by junior NCOs and by black soldiers. The drug leaders allegedly threatened any of the junior NCOs and soldiers who "ratted" them out and bribed other Army personnel to keep quiet about their activities. The Army identified, tried, and convicted many of them and sent them to the U. S. Disciplinary Barracks at Ft. Leavenworth, a military prison, to serve their sentences. Once at Ft. Leavenworth, many of the convicted drug dealers and users had their convictions overturned upon judicial review, and others appealed their sentences and got them overturned. In 1977, thirty or more black soldiers whose convictions had been overturned faced re-trial. Organized into a holding detachment, many of these soldiers became troublemakers, often threatening others on the post, and thus they continued to be disciplinary problems to the Army.[7]

The fact that many of these troops had been drafted into the Army did not help the matter. In addition, the lowering of recruiting standards caused by Secretary McNamara's infamous plan to recruit 100,000 less intelligent Mental Category 4 men into the armed services further aggravated the disciplinary situation. Table 1 shows the number of monthly additions to the Army and the percentage of Category 4 additions during each month in 1971.[8]

Even among the cadets at the U. S. Military Academy disciplinary problems developed from 1970 to 1977. Many cadets openly resented the cadet officers who maintained required standards; many cadet officers bent to peer pressure and did not demand that cadets perform their duties properly or did not gig them for failing to do so. At one graduation, a large number of the graduating class booed the First Captain – a former enlisted soldier – when his name was called out to receive his diploma because they considered him to be "too military." Several cadets, some with famous fathers who had graduated from West Point, became openly anti-military and anti-war. Most resigned; one joined the *Village Voice* in New York City and continued to attack West Point and the military.

The great majority of cadets put on civilian clothes as soon as possible when they left the post so that they could hide their military affiliation. Even cadets going away from the post on official business and on athletic trips shed their uniforms as soon as they could. Some, reportedly, bought and wore wigs to hide their short haircuts when they left on pass or leave. Many young officers facilitated this unmilitary behavior. There also were underlying racial issues. Reportedly, some black cadets would not support the Honor Code when a black cadet apparently violated its tenets. On athletic team tables in the Mess Hall, quite often ten black cadets would group themselves at separate tables. Black cadet leaders and black commissioned officers would not break up such clusters.

Weak discipline and non-military behavior existed in part because many officers serving as instructors, some of whom had returned from Vietnam, set poor examples: they dressed poorly; they wore long hair; they espoused anti-war attitudes; and they looked down on their fellow officers who served as Tactical Officers. One wrote a book - albeit a fictional account - severely criticizing his division's leaders in Vietnam, including the division commander who had fought with the 101st Airborne Division in World War II, commanded a regiment at the Battle of the Bulge, commanded a regiment during the Korean War, and had won a Distinguished Service Cross. The anti-war press loved the book, as did many of his peers and many cadets.

Everything exploded in 1976 when a cheating scandal erupted and touched hundreds of cadets who deliberately violated the Honor Code and then worked with Army lawyers to salvage their careers and keep from being discharged from the Military Academy. Eventually, the Superintendent permitted many to return after sitting out a year rather than throwing them out of the Academy permanently and let most, if not all, of the First Classmen (seniors) involved in the scandal graduate.[9]

Even troops from the 82nd Airborne Division who supported summer training at West Point looked pretty shoddy in the early 1970s. Troops coming from less distinguished Army units, the Army Reserves, and the National Guard looked disgraceful compared to similar troops before the Vietnam War. Only the Rangers reflected pride in their service with their sharp uniforms, short haircuts, and personal demeanor.[10]

U. S. Navy

The Navy obviously also had a serious problem as shown by comments in the underground paper – *Kitty Litter* - that periodically appeared on the USS *Kitty Hawk*. The broadside for August 1972 when the carrier operated off on North Vietnam began:

> *On June 5, 1972, the Kitty Hawk's skipper, Captain Owen L. Oberg was relieved by Captain Marland W.*

WEAKEST PRESIDENT IN AMERICAN HISTORY

"...WHEN RICHARD NIXON RESIGNED, THE NORTH VIETNAMESE KNEW THEY WOULD WIN, ...

PHAM VAN DONG SAID OF GERALD FORD,

"HE'S THE WEAKEST PRESIDENT IN U. S. HISTORY."

BUI TIN

TIN ADDED: "WE TESTED FORD'S RESOLVE BY ATTACKING PHUOC LONG. WHEN FORD KEPT AMERICAN B-52'S IN THEIR HANGERS, OUR LEADERSHIP DECIDED ON A BIG OFFENSIVE AGAINST SOUTH VIETNAM."

CIVIL RIGHTS DIFFICULT TIMES AT HOME

1950
PRESIDENT TRUMAN DESEGREGATED THE ARMED FORCES.

1954
BROWN V. TOPEKA DESEGREGATED PUBLIC SCHOOLS.

1955-1956
BUS BOYCOTT IN BIRMINGHAM, AL.

1960
SIT-IN MOVEMENT BEGAN IN GREENSBORO, NC.

1961
FREEDOM RIDES BEGAN.

1963
MEDGAR EVERS ASSASSINATED IN MS.

1964
CIVIL RIGHTS ACT ENACTED.
WAR ON POVERTY BEGAN.

1965
MARTIN LUTHER KING LED DEMONSTRATIONS IN SELMA, AL.
VOTING RIGHTS ACT ENACTED.
RACIAL RIOTS ERUPTED IN WATTS, CA.

1966
BLACK PANTHER PARTY ORGANIZED.

1967
RACE RIOTS OCCURRED IN 127 CITIES.

1968
MARTIN LUTHER KING CONDEMNED U. S. INVOLVEMENT IN VIETNAM.

KING ASSASSINATED IN MEMPHIS, TN.
RIOTS FOLLOWED IN 168 CITIES.

> *"Doc" Townsend. Thus began an ordeal that has become the talk of the ship, at least among the lower caste of the Navy's structure.*
>
> *Full details of Captain Townsend's personality are not available, since the Captain himself is frequently unavailable. He hides in his cabin or on the bridge and is never seen without his Marine lackey. Captain Oberg (now Rear Admiral), while a super-patriot and war hawk, was always available for anyone in the crew to talk to.*

CAPT Townsend countered these comments when he wrote to the Special House Armed Services Subcommittee on 11 December 1972; "Most Commanding Officers have responsibilities which preclude daily visits among the crew. However, visits with the crew could in no way be described as "rare." For the last two months I have made an effort to meet with different numbers of crewman [sic] on the forecastle. I am accompanied by a Marine orderly whenever I go about the ship. This is a matter of convenience and he assists me in the details of keeping the bridge informed of my location."[11]

Kitty Litter continued:

> *Captain Townsend has shown himself to be one of the original "law 'n order kids," both at mast and on the 1MC [public address system]. He has imposed such arcane and ridiculous punishments as 8 days bread and water for such tiny and meaningless "crimes" as unauthorized absence.*
>
> *In two consecutive masts, Captain Townsend has put several black men into the brig for fighting with white men, while dismissing "with a warning" a white man who spoke in such a demeaning manner to a black man and then punched him in the stomach!*
>
> *Needless to say, this angered the black men on board, as well as many white men (one of whom is the legal officer), his reason for the dismissal: the man was performing his job in the only way he knew how. This is the same quality of argument as "I like blacks okay, as long as they know their place." Racism, it appears is all right if the racist is in a position of power.*

The author apparently knew about the internal, confidential goings-on at CAPT Townsend's many disciplinary sessions or more likely he made up the allegations from the reports of disgruntled sailors or from the scuttlebutt about the various incidents and the disciplinary actions that developed on the ship. However, the Commanding Officer presented more substantive information about so-called discriminatory disciplinary actions: "These incidents occurred in early June 1972 when the blacks, both of them with bad records, were punished for throwing a man down an escalator. The white was only warned since he had a clear record. Also the incident arose from placing the white, who was non-rated, in a supervisory position."

[H.A.S.C. No. 93-13]

HEARINGS

BEFORE THE

SPECIAL SUBCOMMITTEE ON DISCIPLINARY PROBLEMS IN THE U.S. NAVY

OF THE

COMMITTEE ON ARMED SERVICES
HOUSE OF REPRESENTATIVES
NINETY-SECOND CONGRESS
SECOND SESSION

HEARINGS HELD NOVEMBER 20, 21, 22, 24, 25, DECEMBER 6, 7, 8, 9, 11, 12, 15, 18, 1972

(Printed October 1973)

The Captain continued about an alleged "slugging" incident. *Kitty Litter* reported that "a white mess cook refused to give a black two sandwiches instead of the usual one. The black swore and called him a "honky," and the mess cook slugged him." In this confrontation, the CAPT Townsend wrote, "The black did not call the white a honky nor did the mess cook slug the black." Seems as if the *Kitty Litter* author mixed up his information and his information was incorrect.[12]

The charge of racism could have been right on target, but unfortunately some black servicemen began to use the charge in this era to cover up their own inappropriate conduct or the inappropriate conduct of other black servicemen. And based on the mixed up information reported in *Kitty Litter*, such charges appeared to be creations of uninformed and discontented black sailors.

The *Kitty Litter* author added an anti-war blast to his commentary:

> *In direct contrast to Captain Oberg, who only came on the 1MC in order to tell the crew of important events which are out of the ordinary, Captain Townsend talks every day. Usually his speeches are nothing more than propaganda machines, telling us of the day's wanton destruction and compunctionless murder, adding after each separate description of individual atrocities, "which is good," Good for whom? Certainly not for the people we kill every day in "protective reaction strikes."*
>
> *Probably the only person these actions are "good" for is the Captain and other lifers who want "feathers in their caps" for murdering their fellow human beings.*

The broadside continued:

> *In order, not to make the wrong impression, it must be stressed that the Captain, no matter how misguided he is, is not an ogre. He has relaxed the haircut regulations on the ship (much to the chagrin of Kitty Hawks number one jingo Morris Peele) and said that people can have their hair as they want within reason as long it is neatly combed and not "fuzzy." After this incident of blatant racism, he spent several hours talking with black sailors on the forecastle, finally adopting a least one of their suggestions, the formation of a Human Resources council to replace the token Minority Affairs Committee.*

These comments contradicted the one about CAPT Townsend

ZUMWALT

"ZUMWALT HAD US ALL CONFUSED. HE SET UP IN BUPERS AN OFFICER WHOM ANYONE COULD CALL TO MAKE A COMPLAINT ABOUT HIS UNIT, TOTALLY BYPASSING THE CHAIN OF COMMAND.

I HAD A SAILOR WHO WORKED ON THE FLIGHT DECK WHO COMPLAINED THAT OUR TANKER AIRCRAFT... HAD TO DUMP THE EXCESS FUEL TO GET DOWN TO LANDING WEIGHT IF NO ONE NEEDED FUEL, WASTING FUEL. THE SAILOR WROTE TEDDY KENNEDY AND CONTACTED THIS FLUNKY IN BUPERS. BANG! I GOT A CONGRESSIONAL INQUIRY WHICH MEANT THAT I HAD TO RESPOND TO TEDDY'S OFFICE AND BUPERS WITHIN 24 HOURS AND EXPLAIN WHY WE WERE THROWING FUEL AWAY, PARTICULARLY DURING THE '73 FUEL CRISIS. AS IF I DIDN'T HAVE ENOUGH TO DO!

AS COS WE NEEDED Z TO BACK US UP NOT EMASCULATE US. BEING A CO WAS A HARD ENOUGH JOB WITHOUT THE CHANGES THAT Z MADE. HE CHANGED HAIR CUT STANDARDS, ALLOWED BEARDS, LET SAILORS GO TO THE EXCHANGE IN THEIR DIRTY WORKING DUNGAREES, ETC. ALL OF US IN LEADERSHIP POSITIONS, PETTY OFFICERS, CPOS, OFFICERS DIDN'T KNOW WHERE THE HELL WE STOOD."

JAMES B. MORIN, 2009

being unavailable. The *Kitty Litter* author concluded:

> *The Captain may or may not be an "Arnheiter", but he certainly needs to realign his priorities. Instead of cheering the Kitty Hawk's disregard for life, he should be racking his brain (even to the point of asking "inferior" enlisted men their opinions) to find ways to make life a little better on the Hawk.*
>
> *He must obey his orders, unfortunately, and continue with this stupid war, but he is not tied down to Oberg's paranoid anti-communism. Is it too much to ask that we be given near-human living conditions?*

Did the author in *Kitty Litter* contact or work with the National Lawyers Guild or its associated Pacific Counseling Center personnel working at the GI Center in Olongapo or the GI Center in Japan? A Navy report stated that: "Over the last two years in particular, considerable evidence has come to the Navy's attention indicating that projects and activities under the sponsorship of the National Lawyers Guild (NLG) in conjunction with the Pacific Counseling Service involve counseling and other activity prejudicial to good order and discipline within the Navy." Did sailors who rioted on the *Kitty Hawk* and the *Constellation* visit either of these Centers?[13]

The most vulnerable enlisted sailors whom the outsiders could target were the Mental Category 4 enlistees. In 1971, they were many fewer than the accessions into the army, but the Category 4 enlistees (Table 2) still made up an important percentage of new sailors.[14]

Δ

The Chief of Naval Operations, ADM Elmo Zumwalt, contributed to problems in the Navy by undermining the responsibilities of his subordinate commanders with his Z-grams. Issuing instructions from Washington about myriad minor issues - haircuts, beards, uniforms, pass policies, etc. - even while focusing on important issues such as race relations, did not sit well with most ship captains and operating line officers. While well-intentioned and aimed at maintaining traditional discipline, Zumwalt contributed to ill-discipline in the Navy.

Excerpts from Z-Gram 57 illustrated Zumwalt's approach.[15] He wrote:

> *Those demeaning or abrasive regulations generally referred to in the fleet as "Mickey Mouse" or "Chicken" regs have, in my judgment done almost as much to cause dissatisfaction among our personnel as have extended family separation and low pay scales. ...I desire to eliminate many of the most abrasive policies, standardize others which are inconsistently enforced, and provide some general guidance*

which reflects my conviction that if we are to place importance and responsibility of "the person" in proper perspective in the more efficient Navy we are seeking, the worth and personal dignity of the individual must be forcefully reaffirmed. The policy changes below are effective immediately and will be amplified by more detailed implementing directives to be issued separately.

This message contradicted the need for teamwork. Military units function best as teams where individuals sacrifice their individuality for the betterment and operational effectiveness of their team. Stressing individuality will encourage those who do want to do things their way in little matters to want to do things their way in important matters.

Zumwalt continued:

...In the case of haircuts, sideburns, and contemporary clothing styles, my view is that we must learn to adapt to changing fashions. I will not countenance the rights or privileges of any officers or enlisted men being abrogated in any way because they chose to grow sideburns or neatly trimmed beards or moustaches or because preferences in neat clothing styles are at variance with the taste of their seniors. Nor will I countenance any personnel being in any way penalized during the time they are growing beards, mustaches, or sideburns.

This provided a license to steal. If a sailor did not want to shave for a day or two or was criticized for being poorly shaved or unshaven at any time, he could tell his Petty Officer that he was growing a beard. Then, he could shave when he wanted to without any disciplinary action being taken for his deliberate manipulation of the instructions. Moreover, Petty Officers would have to hold corrections in suspension because they could not determine if a sailor was growing a beard or was just being lazy and trying to beat the system.

Zumwalt also wrote:

I view the prohibition against wearing clean, neat working uniforms or dungarees to and from work as unwarranted and I now direct that it be suspended for the convenience of all concerned.

Two probable results undermined this requirement. First, if a sailor worked in an engine room or performed heavy, dirty, manual tasks during the work day, he would not leave work wearing a "clean, neat working" uniform. Perhaps even worse, civilians seeing sailors going to work in less-than impressive dungarees and returning from work in dirty or heavily soiled dungarees hardly would be impressed by the appearance of the sailors. Soldiers, sailors, airmen, and marines needed to look sharp in public and not look like garbage collectors, oilfield

1971 ACCESSIONS TO U. S. NAVY

(ENLISTMENTS ONLY)

MONTH	ALL ACCESSIONS	PERCENT CAT 4
JAN	8,113	14.3%
FEB	7,545	17.1%
MAR	6,593	16.8%
APR	4,545	18.1%
MAY	3,581	16.3%
JUN	7,368	20.0%
JUL	8,100	14.5%
AUG	10,109	16.7%
SEP	7,877	13.7%
OCT	7,601	14.1%
NOV	6,136	16.2%
DEC	4,545	21.9%

THE MOST VULNERABLE ENLISTED SAILORS WHOM THE OUTSIDERS COULD TARGET WERE THE MENTAL CATEGORY 4 ENLISTEES.

"I THINK THE TROUBLE STEMS FROM THE SERVICES HAVING LOWERED THEIR STANDARDS TO PLEASE A FEW."

workers, or stevedores. Marine leaders seemed to remember that; some of the other service leaders did not.

In all organizations viewpoints differ and when change occurs, viewpoints often differ greatly. An experienced black petty officer who earned valor awards in Vietnam wrote the Chairman of the House Armed Service Committee in 1972 as follows:

> *I can truthfully say that during my twenty-one and a half years of Naval Service that I have never seen a black serviceman treated any differently than their white counterparts.*
>
> *I think the trouble stems from the services having lowered their standards to please a few. Some examples are as follows: Allowing Black favoritism, the Black Power sign to be used instead of the traditional military salute and the liberal reforms have caused a breakdown in military discipline. These are prime reasons we are experiencing these problems in the military. We are also accepting Blacks into the Armed Forces that are not motivated in the right direction, many of whom are only seeking to take home pay every two weeks or monthly but none of the responsibility that goes with it.*[16]

A master chief petty officer with seventeen years of service wrote the following:

> *It is not only my duty but my responsibility to speak out on what I consider to be misjudgment and unfairness in evaluating Admiral Zumwalt and his now famous Z-grams. Change is always difficult... Admiral Zumwalt has for the most stood alone in effecting changes that are designed to make everyone in the system both young and old feel they are vital part of the system...*
>
> *No authority or responsibility has been lessened. In fact, just the contrary has evolved. Z-grams have challenged and in many cases cause many sailors both officers and enlisted to accept a new area of responsibility and challenge never believed possible under the traditional system. Not enough thought has been given to what problems the Navy could of been facing had we not had Admiral Zumwalt and what I consider his courageous leadership.*[17]
>
> *There is no permissiveness in the Navy. However, I will say there are persons in the Navy who refuse to meet the responsibilities of leadership, using permissiveness as an excuse for their shortcomings...*
>
> *It is an honor for me to follow such a giant of a man.*

Another master chief petty officer wrote the committee a long letter. He said:

> *I am concerned for our Navy and afraid of what we are headed for if our course is not changed. Loyalty and discipline are our most valued possessions. There is too much laxity of late on the part of officers, chief petty officers and petty officers in rendering military courtesies, and the most flagrant fault is the lack of courage (guts, if you will) to counsel juniors on military discipline.*
>
> *Lack of discipline is a lack of combat efficiency and this is quite evident when just recently a U. S. Navy aircraft carrier, while engaged in combat had to be pulled off line to quell a racial disturbance, and another carrier going out on exercise had 130 men go on strike. The cracks in discipline are widening. At this point you may be saying to yourself why is he writing to me; he should be acting on the very thing he is complaining about. Believe me, sir, there are many concerned men and women who have been trying to retain some order and discipline by seeing to it that their troops (the individuals that work for us) are kept squared away. But what about those others who run wild while we are trying to do our job?*
>
> *These recent events of racial disturbance are not a total result of discrimination, and I do not deny that racial discrimination does not exist in the Navy. But what has been fuel for the fires is lack of discipline and this lack of discipline is encouraged by the unshined belt buckles, shoes, unkept hair and generally slovenly appearance which creates sloppy thinking, sloppy work, and the result is the battles are lost. Too many people desire to win the "good guy award" or if they do use their authority they are rebuked by their seniors for being too hard nosed or discriminatory toward a certain ethnic group.*
>
> *All major catastrophes in the loss of discipline in all organizations have been preceded by a general slackness in command. The old saying "that a taut ship (or station) is a happy ship" is still true. The reason is that on a taut ship the officers and men know where they stand and what is expected of them. There are no soft billets in a taut outfit! The officers and men are on the job and require others to be on the job. Chiselers and transgressors are promptly punished while their offenses are still minor...*[18]

The master chief ended his letter with this conclusion: "Discipline can be tautened. When this is correctly done, less punishment will be required and there will be more fun and zest in being a sailor and we'll be combat ready."[19]

1974-1975 NATIONAL LEADERS

PRESIDENT
GERALD FORD

SECRETARY OF DEFENSE
JAMES SCHLESINGER (1973-1975)
DONALD RUMSFELD (1975-1977)

JOINT CHIEFS OF STAFF

CHAIRMAN
GEN GEORGE S. BROWN, USAF

CHIEF OF STAFF, AIR FORCE
GEN DAVID C. JONES

CHIEF OF STAFF, ARMY
GEN FREDERICK C. WEYAND

CHIEF OF NAVAL OPERATIONS
ADM JAMES L. HOLLOWAY, III

USS *Kitty Hawk*

11 Oct 1972
Racial Incident
Enlisted Men's Club
Subic Bay, Philippines

Kitty Hawk Incident

ADM Zumwalt provided the following summary of the racial incident aboard the carrier, USS *Kitty Hawk*, to the Subcommittee of the House Armed Services Committee:

> *In the early morning of October 11 at the enlisted men's club in Subic Bay, a fight occurred which involved two blacks and a white enlisted man. The incident was quickly quelled by a Marine security force from the Subic Bay Naval station. However the original participants were joined by 30 to 40 others, both black and white, and a melee followed until the security force returned to the club. As a result of this incident, a number of blacks were placed in a bus and returned to Kitty Hawk, which was in its last day of a short up-keep and repair period. The security force arrested nine men – five blacks and four whites – as a result of the incident at the club, and charges were forwarded to the commanding officer, Kitty Hawk, on those personnel attached to the ship.*
>
> *The ship sailed the next morning, October 12, to return to the Gulf of Tonkin. Late on October 12 an investigating officer called the men involved in the previous evening's incident individually to his office to determine the facts of the case.*
>
> *About 8:15 p.m. one of the black men was called by the investigator to make a statement. He arrived at the office with nine other black shipmates. He declined to make a statement as was his right under the UCMJ* [Uniform Code of Military Justice]. *The group then departed and proceeded to the after mess deck. Enroute they were joined by 15 to 20 blacks. After about 10 minutes the men dispersed and proceeded to the forward mess deck. En route seven or eight of the group allegedly assaulted a number of white sailors.*
>
> *The human relations assistant met with the group of blacks shortly afterwards and calm appeared to be restored. The group subsequently dispersed. Within 30 minutes, at about 9:30 pm, a group of 30 to 40 blacks again assembled on the after mess decks and acted disorderly. The commanding officer, executive officer, and the commanding officer's designated human relations advisors met with several groups in an attempt to hear the grievances and terminate the disorder. Despite these efforts, the disorder continued for several hours.*
>
> *About 3:30 a.m., October 13, the commanding officer and the executive officer met with a group of black personnel on the forecastle in a session which lasted over an hour. When the commanding officer returned to the*

bridge to direct commencement of flight operations, the blacks dispersed. It is said that about 100 blacks roamed the ship during the night, and 27 of them have been identified as participating in assaults and rioting and have been so charged.

A total of 46 persons received medical treatment, 40 white and 6 black. Three of the men were injured seriously and required medical evacuation.

USS *Hassayampa*

16 Oct 1972
Racial Incident
Subic Bay, Philippines

Zumwalt concluded that the principal contributing factors for the incident were:

1. The indefinite and extended deployment of the carrier;

2. The resultant separations of the crew from their families;

3. The overcrowded conditions on the ship;

4. The "racially polarized liberty conditions in Olongapo;"

5. The belief of some of the young black sailors that "there was racial prejudice on Kitty Hawk;" and

6. The belief of some that "there had been injustice in handling of mast cases."[20]

Hassayampa Incident

ADM Zumwalt continued his report and summarized the racial incident on the fleet oiler, *Hassayampa*, during the Vietnam War.

He testified that the incident started "when a group of 10 blacks assembled in a living compartment while the ship was in Subic Bay [Philippines] and stated they would not sail with the ship unless reimbursed for money allegedly stolen from their wallets the previous evening." He added, "The command denied their request and 15 minutes later, at about 2:45 p.m. on October 16, several blacks assaulted a group of whites, resulting in minor injuries to four whites. The fight terminated shortly afterwards, but the antagonism from the incident remained."

Hassayampa's Captain took two actions. First, he "decided that eight blacks would be disembarked prior to sailing that evening in order to insure their safety from reprisals." Second, "he offered other blacks the opportunity to join the groups, and at 8:15 p.m., a total of 11 blacks left the ship and were placed in protective custody at Subic Bay." Subsequently, "The ship got underway immediately."[21]

Constellation Incident

USS *CONSTELLATION*

WITH USS *HASSAYAMPA* AND USS *RICHARD B. ANDERSON*, C. 1964

3 NOV 1972
RACIAL INCIDENT
ABOARD SHIP
OFF SAN DIEGO, CA

The CNO then summarized the racial incident that took place on the USS *Constellation*. CAPT John D. "Jake" Ward, commanded the ship while it sailed off of San Diego. Jim Morin's former shipmate and CAG reported:

> *This incident started about 9 a.m. on November 3 when black sailors began gathering in the after mess decks to discuss grievances. The executive officer met with two representatives of the group and then announced to the crew that the human relations council would meet at 9 p.m. and any grievances should be aired at that time.*
>
> *By about 11:45 a.m., approximately, 60 black sailors had congregated. About 2:50 p.m. a report of a potential riot on the mess decks brought a 30-man Marine reaction force, but since there was no apparent physical violence, the reaction force left. About 3 p.m., representatives of the ship's human relations council arrived on the mess decks and heard the group's grievances.*
>
> *About 4:30 p.m., the group asked that the commanding officer come to the mess decks to hear their grievances. They stated they would not leave until they had been heard by him. The commanding officer refused, due to ongoing air operations and the unruliness of the group, but offered to meet with representatives of the group on an individual basis.*
>
> *Representatives of the group did meet with the executive officer. During the period 5:30 p.m. to 3 a.m., November 4, the group essentially remained on the mess decks and discussed their grievances with the human relations council members. Two representatives of the group met with the commanding officer on the bridge at approximately midnight. The commanding officers told them of the proper channels by which to express grievances, discussed certain grievances, and agreed to meet with individuals and/or representatives at any time.*
>
> *The representatives left the bridge and subsequently told the group that the commanding officer refused to talk with them.*
>
> *The ship remained relatively quiet, but the commanding officer decided to return to port the next day, November 4, and to form a beach detachment composed of the element which had been involved throughout the night and other known or potential troublemakers. The word was passed that those sailors who wished to join the beach detachment could voluntarily do so and would not be regarded as in a disciplinary status.*

Constellation arrived in port at San Diego on the morning of November 4. The black group remained on the mess decks until approximately 11:30 a.m. and would not muster and participate in ship's work. A total of about 130 men, including 12 whites, left the ship at approximately 11:30 a.m. and formed the beach detachment. The detachment was placed under the cognizance of an officer assisted by several petty officers and members of the human relations council. The Constellation returned to sea at 8 on November 5.

On Monday, November 6, the beach detachment met with representatives from Constellation, NAS North Island, and NAS Lemoore. Meetings were conducted throughout the day but with no satisfactory resolution of the basic issues.

CAPT Ward met with the sailors of the beach detachment, but resolved nothing. Ward then gave the sailors liberty until the 8 a.m. the following morning, 9 November. By that time, eight members of the beach detachment had boarded the carrier; 120 remained on the pier and refused to board the ship. The local authorities transferred the 120 to their barracks at NAS North Island. There, they divided up the group and sent one group to NAS Miramar just north of San Diego, one to Naval Station Imperial Beach near Los Angeles, and kept one at NAS North Island.[22] Disciplinary proceedings followed.

ADM Zumwalt provided reasons for the incident:

The ship's employment is pertinent in this case. After a long overseas deployment and a brief standdown, commencing on August 1 and extending through September 28, ship's personnel worked long hours on repair and maintenance tasks in order to prepare for refresher training on early October. During the first 30 days after return to San Diego, there was significant change in personnel, with about 700 men either departing for discharge or new duty, or reporting aboard. The ship is scheduled to return to the western Pacific in January.

Δ

In addition to these racial disturbances, the Navy suffered "seventy-four incidents of actual and apparently intentional damage to U. S. Pacific Fleet ships during the period January 1 to December 11, 1972." The Navy determined that:

1. Eight were "intentional by confessions or convictions;" and

2. Forty-eight "were of such nature to indicate the act was deliberate but no investigation or confession was obtained."

1 JAN-11 DEC 1972 INCIDENTS OF INTENTIONAL DAMAGE PACIFIC FLEET CARRIERS

Carrier	Number of Incidents	Samples of Incidents
USS *America* (CVA 66)	2	Flooding of aircraft from sprinkler system
USS *Constellation* (CVA 64)	3	Contamination of boiler feed tanks
USS *Coral Sea* (CVA 43)	3	Wires cut on aircraft
USS *Hancock* (CVA 19)	8	Bomb arming wires tampered with
USS *Midway* (CV 41)	2	Loss of fuel
USS *Oriskany* (CVA 34)	1	15,000 gallons of oil spilled
USS *Ranger* (CVA 61)	7	Parted wires in launch valve of catapult

"IT IS A MISTAKE TO DEAL WITH SELF APPOINTED LEADERS OF GROUPS. TO DO SO WEAKENS THE CHAIN OF COMMAND..."

Eighteen cases remained under investigation at the time or courts-martial were pending. In one case, the Navy charged a sailor with sabotage.

Many of these incidents happened on aircraft carriers as shown in the table in the preceding sidebar, *Incidents of Intentional Damage Pacific Fleet Carriers.*[23]

The incidents occurred during time of war when the carriers were operating against North Vietnam or were enroute to or from the war zone.

Δ

Guidelines

The U. S. Marine Corps suffered a major racial incident on Okinawa, and from it, senior leaders developed guidelines for commanders faced with similar incidents in the future.

The Guidelines stated:

> *1. It is a mistake to deal with self appointed leaders of groups. To do so weakens the chain of command and tends to give command sanction to their unauthorized leadership. Instead each individual, except those "officially" in charge, should only be allowed to speak for himself.*
>
> *2. In any case possible, unit officers and [Senior Non-Commissioned Officers] should be used to control breaches of discipline.*
>
> *3. Never address racially segregated groups about racial matters. This includes a group that would have a small number of sympathetic persons from other racial groups present. No matter how fair the commander is, his words will be misinterpreted by the other groups. Furthermore there is understanding to be gained by the questions and responses that come from a fully integrated audience.*
>
> *4. Do not apprehend persons, particularly minority group leaders, in public places such as mess halls, clubs and theatres if they are identified and can be apprehended later, unless they constitute a clear present threat to the command in that public place.*
>
> *5. Assure that sufficient force is present and available before personally giving an order for an unauthorized group to disperse. Force once committed must be swift and capable of complete control of the situation. Force that is too little and too late is counter-productive and will lead to further violence.*

"ASSURE THAT SUFFICIENT FORCE IS PRESENT AND AVAILABLE BEFORE PERSONALLY GIVING AN ORDER FOR AN UNAUTHORIZED GROUP TO DISPERSE."

6. Refuse any demand that you or your subordinate commanders appear before an illegal gathering to address them. Instead deal with them individually in your office the following day.

7. Complete identification of participants in illegal racial activities is paramount. It is the mask of anonymity that emboldens persons to illegal activity and violence.

8. An alert, well trained, readily available reaction force is the best insurance against mob violence aboard your installation.[24]

ADM ELMO ZUMWALT

No Guidelines

Before CAPT Jim Morin took command of the USS *Franklin D. Roosevelt*, neither CNO Zumwalt or the commanders of the Atlantic Fleet or Sixth Fleet informed him about the recent racial incidents in the Pacific Fleet, and no senior naval commander provided him any guidelines to follow in case he faced a serious racial incident.

Sidebar Notes

1972-1973 and 1974-1975
National Leaders
Various sources.

1971 Accessions to U. S. Army (Enlistments and Draftees)
HASC Hearings-1972, Table 1, 984.

Weakest President in American History
Lewis Sorley, *A Better War, The Unexamined Victories and Final Tragedy of America's Last Years in Vietnam* (NY: Harcourt Brace & Company, 1999), 374.

Civil Rights
Difficult Times at Home
Martin, *America and Its Peoples*, 825 and 844.

Zumwalt
James Morin, *Email to John Bradley*, 11 July 2009.

1971 Accessions to U. S. Navy (Enlistments Only)
HASC Hearings-1972, Table 1, 985.

USS *Kitty Hawk*
U. S. Navy photo. www.navysite.de (14 July 2010).

USS *Hassayampa*
U. S. Navy photo. No details. http://www.angelwind.com/hassayampa (16 July 2010).

USS Constellation (with USS *Hassayampa*)
U. S. Navy Photo No details. http://www.angelwind.com/hassayampa (16 July 2010).

1 Jan-11 Dec 1972
Incidents of Intentional Damage, Pacific Fleet Carriers, *HASC Hearings-1972*, 11-13 (*Constellation*).

Photo of ADM Elmo Zumwalt
Wikipedia.

End Notes

[1] U. S. Navy photo.
[2] **Article 94, Mutiny or Sedition, Uniform Code of Military Justice, 1951**
Manual for Courts-Martial, United States, 1951, 442-3. There are two elements of proof need for conviction: "(1) That the accused created violence or a disturbance, or that, he refused, in concert with another person or persons, to obey orders or otherwise do his duty; and (b) that he did so with intent to usurp or override lawful military authority." 326.
[3] The shooter killed Major Robert Degan. Personal knowledge, JHB.
[4] Martin, *America and Its Peoples*, II, 796.
[5] PROBLEMS IN THE U. S. NAVY of the COMMITTEE ON ARMED SERVICES, HOUSE OF REPRESENTATIVES, NINETY-SECOND CONGRESS, SECOND SESSION. HEARINGS HELD NOVEMBER 20, 21, 22, 24, 25, DECEMBER 6, 7, 8, 9, 12, 15, 18, 1972. (Printed October 1973). (hereafter, *HASC Hearings-1972*), 1092.
[6] *Ibid.*
[7] Personal knowledge, JHB.
[8] *HASC Hearings-1972*, Table 1, 984.
[9] Personal knowledge, JHB.
[10] Personal observation, JHB.
[11] For *Kitty Litter*, see *HASC Hearings-1972*, 530.
[12] *HASC Hearings-1972*, 531.
[13] *Ibid*, 1091-92.
[14] *Ibid*, Table 1, 985.
[15] *Ibid*, 1042.
[16] *Ibid*, 1064-65.
[17] *Ibid*, 1065.
[18] *Ibid*, 1067.
[19] *Ibid*, 1067-68.
[20] *Ibid*, 10-11 (*Kitty Hawk*).
[21] *Ibid*, 11 (*Hassayampa)*.
[22] *Ibid*, 11-13 (*Constellation)*.
[23] *Ibid*, 912-13.
[24] *Ibid*, 191.

Eleven

HARD WORK AND NOT MUCH FUN

USS Franklin D. Roosevelt, c. 1970-1973[1]

USS *Franklin D. Roosevelt*[2]

Important Events		**James Morin**
CAPT Apollo Souchek took command of the USS *Franklin D. Roosevelt.*	Oct 1945	
CAPT Harry D. Felt (later CINCPAC) took command of the USS *Franklin D. Roosevelt.*	Aug 1948	
CAPT George W. Anderson (later CNO) took command of the USS *Franklin D. Roosevelt.*	Jul 1952	
CAPT Jack Youngblade took command of the USS *Franklin D. Roosevelt.*	Sep 1971	
	Oct 1971	**CAPT Morin took command of the USS *LaSalle*** **Began Middle East operations.**
Racial Incident, USS *Kitty Hawk.*	Oct 1972	
Racial Incident, USS *Hassayampa.*	Oct 1972	
Racial Incident, USS *Constellation.*	Nov 1972	
Subcommittee of House Armed Services Committee conducted an inquiry into racial incidents.	**Dec 1972**	
	9 Jan 1973	Morin turned over command of the *LaSalle* and returned to the U. S.
	12 Apr 1973	**Morin took command of the USS *Franklin D. Roosevelt.***
	Jul 1973	**Morin dealt with racial incident aboard the USS *Franklin D. Roosevelt.***
	Jul 1973	RADM Lawrence Geis began Article 32 Investigation of racial incident aboard the USS *Franklin D. Roosevelt.*
	Aug 1973	Morin briefed President of Southern Christian Leadership Council about the racial incident aboard the USS *Franklin D. Roosevelt.*
	Sep 1973	Morin departed FL with the USS *Franklin D. Roosevelt* for six-month cruise in the Mediterranean.

ROSIE 1973

The Navy gave me thirty day's leave before I had to report into COMNAVAIRLANT Headquarters in Norfolk.[2] I used the time to rest up because I was not very mobile anyway. By the time I went on leave, the doctors had put me in a walking cast.

Ψ

Hard Work

My orders to the USS *Franklin Roosevelt* meant another move for the family. Because of my wife's need for continued medical check-ups and the fact that in the eighteen months that I would command the *FDR* the ship would be in its home port only a fraction of the time, I requested that we be allowed to remain in quarters at the Naval Station Norfolk. The rules were that if you were not stationed at Norfolk, you did not rate quarters there. I understood that, but I felt I had extenuating circumstances. I finally met with the Commandant of the Naval District, a rear admiral, and pleaded my case. He cited the regulations and denied my request. I then saw the Atlantic Fleet Commander, ADM Ralph Cousins,[3] who had been on the *Coral Sea* during Viet Nam. He asked me, "Why, won't they let you stay there?" I replied: "Regs." He said he would take care of it, and he did. My family stayed in Norfolk.

After my leave ended, I reported into AIRLANT to be briefed by each Division on the Staff about information relating to *FDR*: operations, personnel, material condition, supply, training, safety, etc. Then, I attended some required schools. All of this took about six weeks. Finally, I flew down to Mayport and spent a few weeks at sea with the Commanding Officer, CAPT Jack Youngblade. Youngblade, a former enlisted sailor, had graduated from the Naval Academy in 1949 and earned his wings in

AFTER MY LEAVE ENDED, I REPORTED INTO AIRLANT TO BE BRIEFED BY EACH DIVISION ON THE STAFF ABOUT INFORMATION RELATING TO FDR: OPERATIONS, PERSONNEL, MATERIAL CONDITION, SUPPLY, TRAINING, SAFETY, ETC. THEN, I ATTENDED SOME REQUIRED SCHOOLS. ALL OF THIS TOOK ABOUT SIX WEEKS.

FINALLY, I FLEW DOWN TO MAYPORT AND SPENT A FEW WEEKS AT SEA WITH THE COMMANDING OFFICER, CAPT JACK YOUNGBLADE.

CAPT JAMES MORIN

1950, two years after I did. A well-decorated heavy attack pilot, he had taken command of the ROSIE in September 1971.[4]

Once aboard, I observed flight operations, ship handling, and various drills, and met the officers and crew. I no longer wore a cast, but I could only walk around the ship or stand around for about an hour before my leg would swell up and I would have to sit down and elevate it. Still not fully mobile, I had a hard time going down eight deck's worth of ladders to visit the steam plant. RADM Joe Moorer,[5] the Carrier Division Commander on board, notified the Admiral at AIRLANT that he doubted my physical fitness for command and recommended a month or two delay. CAPT Youngblade had orders to AIRLANT to be the Admiral's Chief of Staff and the Admiral wanted him to take over his new job ASAP. Consequently, he directed that the Change of Command be held as planned.

WENDY RIVERS HAD RETURNED TO THE UNITED STATES AFTER SEVEN AND A HALF YEARS AS A POW. HE AND MANY OF THE SILVER FOXES FROM CORAL SEA DAYS CAME TO MY CHANGE OF COMMAND CEREMONY.

Wendy Rivers had returned to the United States after seven and a half years as a POW. He and many of the Silver Foxes from *Coral Sea* days came to my change of command ceremony. One Fox remarked, "it turns out our performance in 1965 did not adversely affect his career path after all."[6] The eight Silver Foxes who came to the ceremony and I wanted to hear what Wendy had to say. Wendy told me that he had heard me fly by him, but before I came by a second time some farmers with pitchforks grabbed him and put him and his parachute in a nearby hootch under guard until we called off the SAR effort. Wendy was still very underweight and gaunt looking, but he was in high spirits after seven years of privation.

We had a great reunion of the Foxes and stayed up pretty late. The Foxes told me that it was a neat feeling to see me become a Captain of a carrier because for them a Carrier Captain was almost a grandfather figure.

On a glorious spring morning on 12 April 1973, I took command of *FDR. ROSIE*, a sister ship of the *Midway* and *Coral Sea*, had been commissioned on my eighteenth birthday, 27 October 1945, and she was showing her age. Ironically, the Navy had planned to commission the carrier the *Coral Sea*, but when

President Roosevelt died, the leadership decided to name it for him. We ran the guts out of all our carriers and *ROSIE*, the first of the post-World War II carriers, was getting tired.

Ψ

THIS BEGAN WHAT TURNED OUT TO BE TWENTY-TWO MONTHS OF HARD WORK AND NOT MUCH FUN IN COMMAND OF A BIG SHIP AT THE TIME IN OUR NATION'S HISTORY FILLED WITH SOCIAL UNREST AND RACIAL STRIFE.

Change of Command Day was our last day to socialize because the very next morning I took the *ROSIE* to sea for five days of Carrier Qualifications (CQ) for the Training Command and the F4 training squadron. This began what turned out to be twenty-two months of hard work and not much fun in command of a big ship at the time in our nation's history filled with social unrest and racial strife. In addition, ADM Elmo Zumwalt, Chief of Naval Operations, attempted to loosen up our standards of conduct and discipline which made it very tough for all Commanding Officers.

After CQ, we returned to port to onload the Air Wing Squadrons and the Fleet Training Group personnel from Guantanamo Bay, Cuba. The Fleet Trainers would ride with us for several weeks of intensive training and drilling for the Air Wing and the ship. The following weeks blurred together as we conducted regular flight operations with refresher training. We did not get back into port very often or for very long. I soon found that as Carrier Captain, I did not have sufficient hours in a day to complete all my tasks. I did not just sit in my bridge chair, feet up, monitoring the ship's movement and flight operations, as some people imagined. I did that sometimes, but I had thousands of other details to handle and countless problems to solve.

I never realized the inordinate amount of time I would have to spend conducting Captain's Mast (disciplinary hearings) for errant crew members. On a ship with over 4,500 sailors on board, even a small percentage of troublemakers and those who could not stay out of trouble amounted to a large number of people referred to the Mast. I would have sessions that lasted four or five hours. This tied up a lot of officers and NCOs who sent their people to the Mast and the members of the crew who

I NEVER ENJOYED MAST WHERE I HAD TO LISTEN TO AND REASON WITH SOME EIGHTEEN-YEAR OLD WHOSE PARENTS HAD NOT TAUGHT HIM ANY MANNERS OR SENSE OF RESPONSIBILITY AND WHO OFTEN DID NOT REALIZE THE CHALLENGES AND REQUIREMENTS HE WOULD FACE IN THE NAVY.

served as witnesses, and it took up too much of my time which I could have better used walking the ship and talking with my sailors at their work stations. They deserved my attention and I needed to see them as they performed their duties. Finally, I decided to hold Mast one hour a day just to whittle down the docket.

We really did not have many disciplinary problems at sea. Most of the cases developed when the sailors had been in port for awhile and got into all sorts of trouble ashore. Whenever we pulled out and went to sea, there would be a stack of Mast cases to be processed. Unlike the Army where company commanders – and not the colonels commanding brigades who were my peers – handled the disciplinary problems similar to mine – I had to process all the cases.

I never enjoyed Mast where I had to listen to and reason with some eighteen-year-old whose parents had not taught him any manners or sense of responsibility and who often did not realize the challenges and requirements he would face in the Navy. All of a sudden, he found officers and petty officers telling him when to get up, how he must dress, and how he must wear his hair. And then these supervisors demanded that he follow their orders and work fifteen to eighteen hours a day under some tough conditions. Moreover, the young sailor found himself crowded into a berthing compartment with little privacy and little storage space for his personal effects. Rather than agonize over the few sailors who could not cut the mustard, I constantly thanked the Lord for the thousands of other young men on the ship who did their jobs and stayed out of trouble.

After the Fleet Training Group burned us out, AIRLANT sent out a team to conduct our Operational Readiness Evaluation (ORE). We participated in an exercise with the USS *Independence* after that. Following a short break, we found ourselves back at sea with the British carrier, HMS *Ark Royal* and the French carrier, *Foch.*[7] Whenever the British Admiral assumed tactical command, he loved to keep the formation changing and making turns. The foreign carriers did not have many aircraft aboard so they did not have the problem I had with many aircraft tied down on the flight deck. I had to turn slowly, limiting the number of degrees of heel so that I did not break an aircraft free of its tie-downs. ROSIE would just begin to react with her rudder, when the British Admiral would put another signal in the air to do something else. At 0200 hours in the morning, I had to keep my eyes on every ship close to me. It was hair-raising. This exercise lasted ten days before we returned to Mayport on 8 June 1973 for a month in port. I had to get some needed repairs done on the ship and had to let the crew get some time ashore. My family drove down from Norfolk for the last two weeks in June, and we enjoyed some time together.

We got underway again on 6 July 1973 for three weeks of CQ and Air Wing training while en route to the Puerto Rican operating area for a missile exercise, culminating in an Atlantic Fleet Readiness Exercise.

The Racial Incident

The military is a microcosm of society, and in this period of social and racial problems in the United States, these same problems erupted in the Navy. Many young people entering the service had never worked hard in their lives, and found Navy rules and regimentation not to their liking. Some of the young sailors fought the "system" and were constantly in trouble. Others wrote their Congressmen to complain or wrote to Pers P office in the Bureau of Naval Personnel. ADM Zumwalt had established Pers P under a Captain to hear complaints from Navy personnel all over the world. Consequently and unfortunately, a disgruntled sailor could completely bypass the chain of command to lodge his complaint with Pers P when he could have asked his Petty Officer, his Chief, his Division Officer, or even his Captain to deal with his problem or complaint. I certainly did not appreciate getting a call or a letter from Pers P asking me "what about such and such?" I would respond, "What are you talking about? I know nothing about it!" The man

1973
Racial Breakout of Crew
USS *Franklin D. Roosevelt*

Commissioned Officers			
White	Black	Other	Total
192	2	2	196
98%	1%	1%	

Chief Petty Officers			
White	Black	Other	Total
139	4	13	156
89%	3%	8%	

Petty Officers			
White	Black	Other	Total
944	57	56	1,057
89%	5%	5%	

Enlisted Non-Rated			
White	Black	Other	Total
1,059	247	54	1,360
78%	18%	4%	

All Hands			
White	Black	Other	Total
2,353	310	125	2,788
84%	11%	4.5%	

OUR VERY SENIOR ADMIRALS DID NOT REALIZE THAT THE NAVY WAS DIFFERENT THAN THE NAVY IN WHICH THEY HAD CAPTAINED BIG SHIPS. THE FACE OF THE NATION WAS CHANGING; THE FACE OF THE NAVY WAS CHANGING.

in command of a capital ship has enough on his hands without talking to some desk-bound staff puke, who if he were good enough, would not have been desk-bound in the first place. Congressional inquiries proved even more difficult to deal with because often the sailor - or often his parents or his wife - blew a problem all out of proportion.

In 1971-72, USS *Constellation*, captained by my old shipmate from 1965-66, Jake Ward, and the USS *Kitty Hawk*, captained by Doc Townsend, with whom I had gone through flight training, suffered racial incidents which gained national notoriety. Both these fine officers, given command of our newest West Coast carriers, most probably would have attained Flag Rank, but victimized by these racial incidents, they retired after investigators and their convening authorities concluded that the carrier captains had not handled their situations properly.

Our very senior Admirals did not realize that the Navy was different than the Navy in which they had captained big ships. The face of the nation was changing; the face of the Navy was changing. Regardless of what educators said about more and better high school graduates, many of the high school graduates entering the services did not compare well with their predecessors.

At the line level in the Navy, we found sailors to be less educated and more ill-mannered from lack of proper home-training. Moreover, they had less sense of personal responsibility and had little sense of obligation to their country. The style of leadership needed to handle these men had to be different than that used ten, twenty, or thirty years before.

The Navy high command offered up Jake Ward and Doc Townsend as sacrificial lambs to show that the Navy would not tolerate racial insensitivity. No one found them to be racially insensitive, but senior Navy officials criticized the way they handled their incidents. There was no "right way" to handle them. The two captains did what they had to do to defuse frightening situations that bordered on mutiny. They faced no-win situations starting out, and they paid the price.

Ψ

Just before midnight on 15 July 1973, while conducting night flight operations, my Executive Officer, CDR Eugene M. Clemens, notified me that a group of four black sailors had assaulted and beaten up two white sailors in separate incidents. I immediately ordered my XO to inform every Department Head and the Air Wing Commander about the assaults, and have every Chief Petty Officer and Officer [the khaki] not on watch to walk the ship all night to prevent any other incidents. Nothing else occurred that night.

The following evening, the XO conducted a "rap session" with all the black sailors on the mess decks. About 10:30 p.m., the XO called me to say that the conversation was going around in circles and that he was getting nowhere. I told him that as soon as I had completed the recovery of the aircraft that were coming in at the time I would come down and address the crew.

When I got below, I asked for comments. Four complaints stood out:

> 1. A black man cannot get advanced in the Navy;
>
> 2. The lines of communication were such that a black sailor could not talk to his Division Officer when he wanted to;
>
> 3. Blacks were underrepresented in the Operations Department in the radarman, electronic technician, intelligence specialist, and photographer mate ratings; and
>
> 4. There was no formal On the Job Training (OJT) program on the ship by which blacks could get out of the menial tasks of pushing airplanes, chipping paint, and cleaning bilges.

After listening to their complaints, I told them that any man could get advanced in the Navy, but that a sailor had to be motivated to do so and had to study and pass exams to get ahead. No one asked the color of the exam taker. Second, I said there would be an OJT program for those who sincerely wanted to

Precedents for Racial Incidents

1. Recent influx of new personnel
2. Existence of black organizations
3. Concentration of blacks in units
4. Voluntary de facto segregation
5. Erosion of local command authority
6. Lack of faith by blacks in command responsiveness
7. Dual systems of Justice
8. Lack of preparedness with the command
9. Interrelationships between drug and racial problems
10. Ineffective human relations program
11. Expansion of suspicion between young black and white [personnel]
12. Adoption of militant leaders
13. Increase in racial assaults
14. Testing of the system
15. False sense of security
16. Outward black unity
17. Attempted organization of white reaction
18. Tension increases
19. Arming of personnel
20. Plans for outbreaks of violence

HASC Report, 180

Generalized Incident Model

The Congressional Hearing Report contained a study of other racial incidents. The study determined that there were four discrete phases in the incidents:

(1) A Precedent Phase,
(2) An Escalation of Tension Phase,
(3) A Catalytic Phase, and
(4) A Turbulence Phase.

The Report listed a number of common actions in previous riots, but the list was not found in the report. Key information about the Phases is quoted below.

Precedent Phase

"Prior to an incident there existed a specific set of precedent conditions which created the environment in which the incident occurred."

Escalation of Tension Phase

"This phase would normally consist of a series of assaults wherein the assailants would not be identified nor punished. Interwoven into this series of assaults would be a set of other racially oriented incidents. These could consist of, but not limited to, meeting of militant whites or blacks, disruptive demonstrations of ethnic pride...the demand for the command to meet with groups of blacks or whites...increased evidence of conflict between the command and lower enlisted ranks.

Primary evidence of this conflict might be found in minor confrontations between groups of blacks and the duty and security representatives of the command.

During the phase, inspection of barracks may uncover a variety of weapons such as chains, ball bats, or pieces of steel in the hands of the troops."

better themselves. Third, I would ensure that the lines of communications were open all the way to me, if necessary, using the chain of command.

I added that if they wanted to gain the high technology ratings, they would have to go to school and to do so they would have to show their ability to handle math and mechanical skills by scoring high on the requisite exams. Black or white, if a sailor made the score cut, he would go to school. With no further comments from the crew, I returned to the bridge for the rest of night operations.

During this event, several black Officers, Chief Petty Officers, and Petty Officers stood around the mess decks. Not one of them said, "I made it!" They let me stew in my own vital body fluids that night.

Ψ

The heavy pace of the exercise continued. On the night of 17 July 1973, black sailors beat up six more white sailors in the darkened passageways or in the heads. Once again, I broke out all the khaki to patrol the ship all night. This second occurrence changed everything from an isolated instance on the 15th to something more serious. I immediately prepared and sent a message to the Chief of Naval Operations in Washington, the Atlantic Fleet Commander in Norfolk, and my boss at AIRLANT.

RADM Joe Moorer, the Carrier Division Commander, missed all the FUN. He had transferred to one of the destroyers to watch a missile shoot.

Now, word reached me that my white sailors were incensed. Reports arrived that some of the engineering crew had cut up pipes to carry around with them for self-defense. My time bomb was ticking. I told my Minority Affairs Advisor, a black Chief Hospital Corpsman, to get out among the black sailors to find out what the hell was going on.

After flight operations that night, I maneuvered the ROSIE alongside an oiler for four hours to refuel. By the time that evolution ended, morning dawned. I had not been to bed for forty-eight hours, and I was getting tired.

About 8:00 a.m., as I was having breakfast in my chair on the bridge, my Minority Affairs Advisor reported that the word going around the ship was that the black sailors planned to stage a sit-down on the flight deck forward. I informed my XO and my Chief Master-at-Arms, J. Lively, of the report and directed them to keep moving about the ship and keep me informed of developments.

Around 10:30 a.m., I started seeing groups of two and three black sailors coming out of the island and walking forward on the flight deck. By 11, a group of sixty to sixty-four had gathered on the bow. We planned to begin flight operations at noon and had to test our catapults before doing so. However, with people forward, we could not test the catapults. The whole ship witnessed this development because everything was being recorded on video.

I sent the Minority Affairs Advisor and the Chief Master-at-Arms forward to see what they could learn, and then I placed my Marine Detachment on Standby in case I needed them. After this, I talked with the XO and discussed going to General Quarters which would have directed all hands to go to their assigned battle stations. I dropped this idea because I believed that any dissident group would disobey the command. In my opinion, the blacks wanted a confrontation and would work to get one, one way or another. I wanted to avoid any sort of physical confrontation because resultant injuries and allegations of brutality or inhumane treatment would serve only to blacken the Navy's eye and strengthen the positions of those critics who said the Navy was racist.

The air became electric with emotion. As my Minority Affairs Advisor walked forward, I looked down below me where I saw two plane captains - one black, one white - from the A7 squadrons leaning on the tail surface of their aircraft watching the scene forward. I thought to myself, "that young black man is not part of this problem, but, if I lost control of the situation,

CATALYTIC PHASE

"THIS PHASE CONSISTS OF TWO VERY DISTINCT EVENTS WHICH ARE INTIMATELY COUPLED TOGETHER. THESE EVENTS ARE A "TRIGGERING EVENT" FOR THE MAJOR RACIAL EVENT... THE TRIGGERING INCIDENT MAY BE MANIFESTED AS THE APPREHENSION OF SOME BLACK LEADER, A BLATANT ACT OF RACIAL DISCRIMINATION ON THE PART OF WHITES OR BLACKS, OR SOME OTHER ACT. SUBSEQUENT TO THIS TRIGGERING EVENT THE BLACKS WILL THEN GATHER TO PROTEST THE EVENT, AND FROM THIS POINT THE BLACKS MAY CARRY OUT A SET OF RACIAL ASSAULTS OR SOME OTHER MAJOR INCIDENTS.

IT IS THE CONSENSUS THAT THE TRIGGERING EVENT MAY ONLY BE AN EXCUSE TO CARRY OUT THE MAJOR RACIAL INCIDENT AND NOT THE ROOT CAUSE.... IT IS PROBABLE THAT A TRIGGERING EVENT MAY BE SIMPLY CREATED TO MEET THE REQUIREMENTS OF THE BLACKS TO SUPPLY AN EXCUSE TO CARRY OUT THE MAJOR INCIDENT."

TURBULENCE PHASE

"THIS PHASE BEGINS AT THE GATHERING POINT AND CONTINUES UNTIL THE TERMINATION OF THE INCIDENT. DURING THE TURBULENCE PHASE A VARIETY OF ACTIONS MAY BE CARRIED OUT BY THE DISSIDENTS. THESE ACTIONS COULD INCLUDE BUT ARE NOT LIMITED TO THE FOLLOWING: A SET OF ASSAULTS CARRIED OUT WITH OR WITHOUT WEAPONS, GROUP CONFRONTATION WITH COMMAND PERSONNEL, THE DUTY STRUCTURE AND SECURITY FORCES, THE DESTRUCTION OF GOVERNMENT PROPERTY, AND OTHER UNLAWFUL ACTS.

A NUMBER OF DISPERSALS AND REGATHERINGS MAY OCCUR FOR THE ORIGINAL GROUP AND/OR OTHER GROUPS. ADDITIONALLY THE ORIGINAL GROUP MAY FRAGMENT IN SEVERAL SMALLER GROUPS, THEN REAPPEAR AS AN EVEN LARGER GROUP."

HASC REPORT

I told them they had five minutes to get back to work. With that, I turned and headed for the bridge.

people are going to get hurt and he could be injured just because of the color of his skin." I knew that my last course of action – the use of my Master-at-Arms' force and the Marines – probably could not contain the large group forward and that some of the black sailors would escape into the ship where they could hurt others, sabotage equipment, begin fires, and perhaps trigger violence because some of my white sailors wanted revenge for the injury of the six white shipmates. It could have gotten nasty.

My Minority Affairs Advisor returned and told me that the group was not going to move. I asked if it would help it I went down to them. He said that it would not hurt. I went down to the flight deck and as I walked towards the bow, the sailors on the flight deck and in the catwalks yelled "Shoot the bastards" and "Throw them over the side!" At this time, I did not need any cheering section.

When I got to the group the black sailors were sitting or lying on the deck. No one stood up, no one said "Yes, sir" or "No, sir" when I asked questions. When I asked them what this was all about, I got:

1 "There's no OJT (On the Job Training) program;"

2. "The lines of communications aren't open;"

3. "A black man can't get promoted in this man's Navy;"

4. "We want black pilots in the squadrons now."

I noticed something unusual in the make-up of the group. The senior person was a Third Class Petty Officer and a goodly number of them were from the air squadrons. I told them that we had discussed all these issues a few nights before, and that they were leaving me with no other option. I told them they had five minutes to get back to work. With that, I turned and headed for the bridge. I had not taken ten steps before I heard someone in the group say, "Let's give the Captain a chance." I kept heading for the bridge. By the time I got there, the group had broken up, ending the immediate problem.

The Air Department ran its catapult checks and launched at noon, on time.

Ψ

That afternoon, I spoke on secure radio with ADM Frederick Michaelis, COMNAVAIRLANT, about the overall situation.[8] He felt that I was not moving fast enough investigating the incident. I also sent another message to all the Commanders up the line on the status of the incident. I met later in the day with all the officers and chiefs on the *ROSIE* and explained why I did what I did.

On succeeding days, I met with various divisions. A percentage of those I spoke with felt that I should have cracked heads. That might have satisfied their feelings of frustration towards this small, militant group, but I had the safety of the ship and my crew in mind. I had the ultimate responsibility.

I did not know it at the time but the flight deck demonstration ended incidents on the *ROSIE*. However, I kept all the available khaki walking the ship at night to prevent any further incidents, and we continued with the exercise which ended on 25 July. Then, we headed back to Mayport.

The Long Green Table

RADM Lawrence Geis, Commander, Fleet Air Jacksonville (COMFAIRJAX), spoke with me by radio on the 23rd and informed me that he had been appointed an Investigating Officer, and that he would fly aboard before we reached Mayport to conduct a Judge Advocate General Investigation (an Article 32 investigation) of the incident. [9] He arrived on the 25th.

Geis told me that I was an "interested party" to the investigation, which in Navy parlance meant that you had better not have made any mistakes in the way you acted as Commanding Officer during the incident or else you will be in serious trouble. As an interested party, I could have a Legal Counsel. My ship's

ARTICLE 32
UNIFORM CODE OF MILITARY JUSTICE

INVESTIGATION

(A) NO CHARGE OR SPECIFICATION SHALL BE REFERRED TO A GENERAL COURT-MARTIAL FOR TRIAL UNTIL A THOROUGH AND IMPARTIAL INVESTIGATION OF ALL THE MATTERS SET FORTH THEREIN HAS BEEN MADE. THIS INVESTIGATION SHALL INCLUDE INQUIRIES AS TO THE TRUTH OF THE MATTERS SET FORTH IN THE CHARGES, FORM OF CHARGES, AND THE DISPOSITION WHICH SHOULD BE MADE OF THE CASE IN THE INTEREST OF JUSTICE AND DISCIPLINE.

(B) THE ACCUSED SHALL BE ADVISED OF THE CHARGES AGAINST HIM AND OF HIS RIGHT TO BE REPRESENTED AT SUCH INVESTIGATION BY COUNSEL. UPON HIS OWN REQUEST HE SHALL BE REPRESENTED BY CIVILIAN COUNSEL, IF PROVIDED BY HIM, OR MILITARY COUNSEL OF HIS OWN SELECTION IF SUCH COUNSEL BE REASONABLY AVAILABLE, OR BY COUNSEL APPOINTED BY THE OFFICER EXERCISING GENERAL COURT-MARTIAL JURISDICTION OVER THE COMMAND...

lawyer, LT Dick King, USN recommended that I ask for the Captain lawyer at Naval Air Station Jacksonville (NASJAX) and the head of the Legal Center there. He agreed to represent me.

Before he left, he told me that he would recommend that I be given a "Letter of Reprimand" for my handling of the incident. He faulted me for failing to use force immediately when the sit-in started and said that I should never have agreed to meet with the dissidents on the bow.

On 26 July 1973, we docked in Mayport, and Geis began the investigation. The admiral took sworn statements from the assault victims and everyone else involved in the incident. None of the victims would identify their assailants. He reviewed the video of the sit-in.

As all this took place, two other actions developed. First, the head of the Naval Investigation Service for the Sixth Naval District who was not part of RADM Geis' investigation team, visited me. He asked me if it would be okay for him to talk with the victims because he knew how to get them to talk. I approved his request, and he got the victimized sailors to identify twelve assaulters. Second, a major television network interviewed me concerning the racial incident. I could not say much because of the ongoing investigation.

RADM Geis completed his investigation in a week and departed the ship to prepare his report. Before he left, he told me that he would recommend that I be given a "Letter of Reprimand" for my handling of the incident. He faulted me for failing to use force immediately when the sit-in started and said that I should never have agreed to meet with the dissidents on the bow. I told him that I did what I had to do; that I made my decisions on the spot based on available information during a highly emotional and tense period of time while we participated in a demanding Fleet Exercise. I told him that I ended the incident with no further injury or damage to my ship and that it was easy to play Saturday's ballgame on Monday morning. After RADM Geis completed his report, I added my comments to the report, and the admiral forwarded his report and my comments up the chain of command. I had no idea what would result.

Ψ

I charged the twelve sailors who had been identified as the

assailants with violations of the Uniform Code of Military Justice and transferred them to NASJAX for trial by Special Court-Martial. For simple assault or for simple assault and battery, a Special Court Martial could adjudge a sailor confinement for three to six months and forfeiture of two-thirds pay for up to six months. But a Special Court could not order a punitive discharge - Bad Conduct or Dishonorable Discharge - for a sailor.

I TOLD THEM THAT THE REGRETTABLE INCIDENT WAS BEHIND US, THAT WE WERE NAVY MEN, BLACK AND WHITE, WITH A JOB TO DO, SO WE NEEDED TO GET ON WITH OUR JOBS.

If I had chosen to seek a General Court-Martial for some of the sailors, the sailors would have faced much harsher penalties. For Aggravated Assault "With a dangerous weapon or other means or force likely to produce death or grievous bodily harm," a General Court-Martial could have increased the punishment to confinement at hard labor for three years and ordered a Dishonorable Discharge. If the aggravated assault intentionally inflicted "grievous bodily harm with or without a weapon," a General Court could have upped the penalty to confinement at hard labor for five years and also ordered a Dishonorable Discharge.[10] As severe as the punishments for Aggravated Assault charges were, the allowable punishments for charges of "Riot" or "Mutiny" were much more severe.[11]

My Marine Orderly, a black, told me that many of my black sailors believed that I had railroaded their shipmates. Consequently, that afternoon, I mustered the entire crew on the flight deck and briefed them on the situation.

I told them that the regrettable incident was behind us, that we were Navy men, black and white, with a job to do, so we needed to get on with our jobs. I told them that I had charged the twelve sailors with assault, and that they were to be removed from the ship and sent to NASJAX for trial by court-martial in a totally unbiased location. I explained that I was transferring them to protect their rights and that such action was in their best interests. That seemed to calm everyone down.

Then, I arranged for the transfer of the victims of the assaults to other ships because they feared for their safety if they remained on the *ROSIE*. The victimized sailors eventually testified at the courts-martial. At the end of the trials, the Special

Courts-Martial found eight of the twelve assailants guilty as charged.

This racial incident knocked my feet from under me. I had always believed that if I had a strong suit in the Navy, it was my success in handling people. The incident caused lots of personal trauma, and I dreamed about it for six months.

This racial incident knocked my feet from under me. I had always believed that if I had a strong suit in the Navy, it was my success in handling people. The incident caused lots of personal trauma and I dreamed about it for six months.

Aftermath

On 7 August 1973, I met with the President and other officers of the Southern Christian Leadership Council who asked to be briefed on the racial incident. They wanted to know how it happened and what I did. The SCL leaders did not seem to have an axe to grind and appeared honestly interested in helping to improve race relations and dismiss the notion that the Navy had a racist attitude.

FDR's racial incident seemed to finally get many heads out of the sand and made people realize real problems existed and that the Navy differed from days gone by.

FDR's racial incident seemed to finally get many heads out of the sand and made people realize real problems existed and that the Navy differed from days gone by.

ADM Michaelis sent two Norfolk-based carrier captains down to Mayport so that I could brief them on the incident and answer any of their questions. Later, I briefed all the commanding officers of the ships based at the Naval Station Mayport so that they could ponder how they would react to a similar incident, and they could meet with their officers to brainstorm solutions to this real threat. Very belatedly, ADM Michaelis sent me a copy of the report of the Congressional Sub-Committee Investigations of the racial incidents on the *Constellation* and the *Kitty Hawk*. These investigations listed the chronology of each incident. The first seventeen steps of their incidents followed in sequence.[12] That, I concluded, proved too coincidental to be spontaneous.

The leaders of the troublemakers on the two carriers were very visible. Consequently, CAPT Townsend and CAPT Ward isolated them and had them removed from their ships. That explained to me why I did not see or identify the leaders on the

ROSIE. They stayed behind the scenes and gave orders. Am I talking about a conspiracy of "get a carrier-a-year club?" I have no other explanation. The incident on the *ROSIE* was a well-organized plot, not a sand lot pick-up game. I recommended to ADM Michaelis that he not keep the report of the Congressional Sub-Committee Investigation[13] hidden away but distribute it to all Carrier Captains.

THE INCIDENT ON THE ROSIE WAS A WELL-ORGANIZED PLOT, NOT A SAND LOT PICK-UP GAME.

My Minority Affairs Advisor had orders to shore duty, but I asked him to delay his detachment date so that he could help monitor the progress of our affirmative action programs for OJT, help ensure that the lines of communication were in fact open, and help ensure that all the programs affecting minorities were in full swing and healthy. I asked him to stay aboard until we reached our first port in the Mediterranean. He agreed to stay.

Ψ

To close out this part of the story, a year later, in August or September 1974 when we returned to Mayport after four months in dry dock in Philadelphia, the NIS Agent from Charleston came aboard to introduce the new NIS Agent who was being assigned to Mayport. We had coffee in my in-port cabin and he asked me, "It's been over a year now; have you ever figured out who the leader was?" I smiled and nodded yes, and said I had a chief suspect. I asked him if he had a suspect. He said, "Yes, your Minority Affairs Advisor." I said, "I AGREE."

"IT'S BEEN OVER A YEAR NOW; HAVE YOU EVER FIGURED OUT WHO THE LEADER WAS?" I SMILED AND NODDED YES, AND SAID I HAD A CHIEF SUSPECT. I ASKED HIM IF HE HAD A SUSPECT. HE SAID, "YES, YOUR MINORITY AFFAIRS ADVISOR." I SAID, "I AGREE."

And years later I learned that during my consideration for Flag Rank, VADM Thomas Moorer, a distinguished aviator in World War II and a future Chief of Naval Operations and Chairman of the JCS, who at the time of the incident was my reporting senior and who was a member of the Board, supported me. When asked by other promotion board members if he would have handled the situation differently, Moorer said that he definitely would have handled the situation on the *ROSIE* differently, but he would not have been as successful as CAPT Morin was in defusing a very bad situation.[14]

Sidebar Notes

CAPT James Morin
Roosevelt 1973-1974, a cruise book, photo, 6.

1973 Racial Breakout of Crew, USS Franklin D. Roosevelt
Roosevelt 1973-1974. Statistics developed by John Bradley, 2010.

Precedents for Racial Incidents
HASC Hearings-1972, 148, 180.

Generalized Incident Model
HASC Hearings-1972, 189-90.

Article 32, Uniform Code of Military Justice, Investigation
Manual for Courts-Martial, United States, 1951, 424.

End Notes

[1] U. S. Navy photo. *Naval Historical Collection* (NHC) #97636, c. 1970-1973.

[2] Commander, Naval Air, Atlantic.

[3] Ralph Wayne Cousins retired in grade as an Admiral on 31 May 1975. He earned a Navy Cross, Distinguished Service Medal, and Legion of Merit during his service. *1993 USNA Register of Alumni*, Class of 1937, #12549, 243.

[4] Charles John Youngblade retired as a Rear Admiral on 1 July 1979 and died in 1990 at age 64. He earned a Legion of Merit during his service. *1993 USNA Register of Alumni*, Class of 1949, #20542, 297. A photo of the captain shows that he had earned a Bronze Star for Valor and what appears to be a second Bronze Star. He also earned several Air Medals, and a Navy Commendation Medal for Valor with a probable second award. He had served in Heavy Attack Squadrons and Commanded RVAH9 which flew Vigilantes. A graduate of the Naval War College, Youngblade served as the Operations Officer and the Executive Officer on the USS *Constellation* from December 1967 to January 1969 and then commanded an oiler. Prior to taking command of the *ROSIE*, he served as the Executive Assistant and Senior Aide to the Deputy Chief of Naval Operations for Operations and Readiness. *Change of Command Ceremony Pamphlet*, 12 April 1973.

[5] Joseph Park Moorer retired as a Vice Admiral on 1 October 1980. He earned a Distinguished Service Medal, Legion of Merit, Bronze Star, Purple Heart, and a Meritorious Service Medal during his service. *1993 USNA Register of Alumni*, Class of 1945, #16827, 273.

[6] Bill Kish, *Email to John Bradley*, 17 April 2005.

[7] Named for French Marshal Ferdinand Foch, hero of the Battle of the Marne at the start of World War I. A corps commander, army commander, and a group commander during the war, Foch became the Commander in Chief of all Allied armies in 1918 and was designated a Marshal of France. Britain appointed him a British Field Marshal; Poland appointed him a Marshal of Poland. *Wikipedia* (23 Nov 2010).

[8] Frederick Hayes Michaelis retired as an Admiral on 1 August 1978 and died in 1992. He earned a Navy Cross, Distinguished Service Medal, Silver Star, Legion of Merit, and a Distinguished Flying Cross Medal during his service. *1993 USNA Register of Alumni*, Class of 1940, #13900, 252.

[9] Lawrence Raymond Geis retired as a Rear Admiral on 31 July 1974 and died in 1980 at age 64. *1993 USNA Register of Alumni*, Class of 1939, #13452, 243.

[10] *Manual for Courts-Martial, United States, 1951*, 224.

[11] *Ibid*, 221 and 223.

[12] They are not listed the available *HASC Hearings-1972*. report that I downloaded from the internet. The pages are listed in the index, but no pages were included.

[13] *HASC Hearings-1972*.

[14] RADM James B. Morin, *Email to John Bradley*, 29 July 2008.

Twelve

MORE HARD WORK

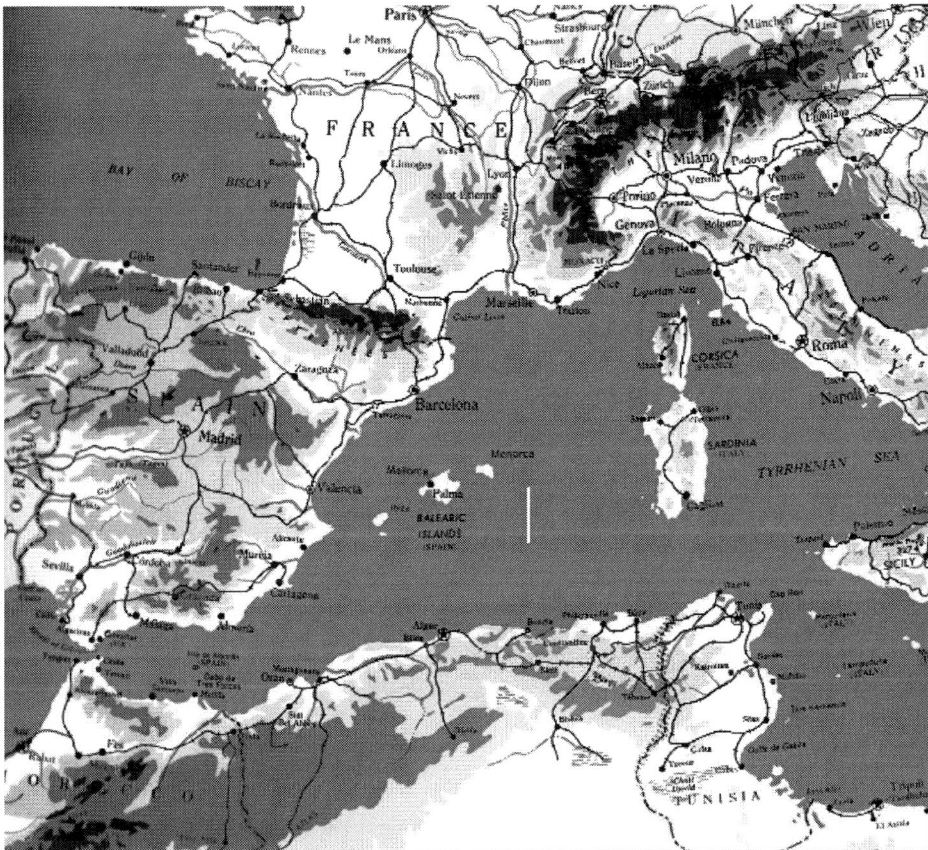

Western Mediterranean[1]

Cold War and the Mediterranean 1973-1974

Important Events		James Morin
Paris Peace Agreement signed.	**Jan 1973**	
Army Lt. Col. became last American KIA. Last U. S. troops left SVN.	Mar 1973	
	Apr 1973	**Morin took command of USS *Franklin D. Roosevelt*.**
U. S. House of Representative rejected request for more military aid for SVN.	Apr 1973	
U. S. Congress barred further U. S. combat actions after 15 Aug.	**Jun 1973**	
	Jul 1973	**Morin dealt with racial crisis on the *ROSIE*.**
President Richard M. Nixon resigned as President. Gerald Ford became President.	9 Aug 1973	
	Sep 1973	**Morin departed with *ROSIE* for MED. Began routine operations.**
Yom Kippur War.	**Oct 1973**	*Rosie* operated for 35 straight days during war.
1973 Oil Crisis.	Oct 1973	
	Dec 1973	*Rosie* in port in Spain for Christmas.
	Jan 1974	Morins vacationed in Greece.
	Feb 1974	*Rosie* operated in the Aegean Sea.
	3 Mar 1974	***Rosie* left Palma for Mayport, FL. Encountered heavy storm in the Atlantic.**
	Mar 1974	Morin and RADM Downey faced Article 32 Investigation.
	3 May 1974	*Rosie* arrived in Philadelphia to go into dry dock.
	28 Jul 1974	Dry dock repairs on *Rosie* completed.
	27 Aug 1974	Sarah Morin married Bruce Bishop.
Soviet leaning strongman deposed U. S. leaning Emperor Haile Selassie in Ethiopia.	Sep 1974	
NVN abrogated Paris Peace Agreement. Invaded SVN. President Ford did not react.	13 Dec 1974	
	20 Dec 1974	**Morin turned over command of the USS *Franklin D. Roosevelt* to CAPT Easterling.**

Cold War 1973-1974

The Mediterranean Sea proved as critical to the western Allies during the Cold War as it had been during World War I and World War II. Fortunately, after World War II, Great Britain still controlled Gibraltar and the western entrance to the sea, as well as the Suez Canal at the eastern end. Non-aligned Turkey controlled the vital Dardanelles and the Bosporus through which the Soviets had to sail their ships to get into the Mediterranean. And Greece, with Turkey, oversaw the Aegean Sea. But as the western Allies demobilized and the Soviet Union absorbed East Germany, Czechoslovakia, Austria, Hungary, Yugoslavia, Bulgaria, Romania, and Poland into her Communist bloc, the Mediterranean became the soft-underbelly of the western Allies. Any Soviet penetration of the Mediterranean countries would have been disastrous to the West.

Because of Soviet movements against Turkey and Iran in 1946, President Harry Truman sent the battleship USS *Missouri* into the Mediterranean with escort ships. The ships that remained formed a Mediterranean Naval Force which by 1950 had become the U. S. Sixth Fleet. Soon the fleet contained a carrier task force of generally two carriers which rotated in and out from stateside bases. By that time, the western Allies - the U. S., Great Britain, France, Belgium, the Netherlands, Luxembourg, Western Germany, and Italy - had formed the North Atlantic Treaty Organization (NATO) to defend Western Europe and the Atlantic from Soviet military actions. Moreover, the western Allies had stationed a substantial number of army divisions in France and Western Germany, and the U. S. Seventh Army with five divisions provided the core of the ground forces. Deployed U. S. air forces, the U. S. based Strategic Air Command, the Seventh Army, and the Sixth Fleet became the primary forces for the defense of Western Europe.

The first Soviet threat to the Mediterranean countries developed when Greek Communists supported from Yugoslavia tried to overthrow the Greek government, and when the Soviets supported actions against Turkey. As mentioned earlier, the Communist actions triggered the Truman Doctrine and American military support for Greece and Turkey. With those issues settled, Greece and Turkey joined NATO in 1952, and they completed the virtual encirclement of the Soviet satellites in Europe and increased NATO's control over the vital Dardanelles and Bosporus. Only Communist Yugoslavia and Soviet satellite, Albania, had direct access to the Mediterranean. The U. S. S. R. meanwhile continued to build up a presence in the Mediterranean.

However, nothing remained static in the Mediterranean. In 1956, Soviet-leaning President Abdul Nasser of Egypt national-

Exotic Mixture

"Even before the Cold War was fully underway, France sought to establish a linkage between the international Communist movement and anti-colonial struggles.

Under pressure from the United States to relinquish overseas colonies, the French Fourth Republic tried to make Washington believe that any surrender to nationalist movements in the colonies was automatically an invitation for Moscow to take over. Time and again, Lenin's old adage was trotted out in support of this view: from the East, he had said, the road to Paris passed through Algeria.

This strange mixture of the exotic and the familiar, of Africa and Communism, did not really coalesce until a vacuum emerged after the American defeat in Vietnam, when pro-Soviet regimes took hold in what had been the Portuguese part of Africa and in Ethiopia.

For the first time it seemed that these countries were not merely pawns in a geopolitical game. Although the constant threat of socialist control over sectors judged vital by the West still existed, what seemed even more worrying was Communism's apparent success in offering the Third World an illusory remedy for underdevelopment.

On top of all the suffering that these countries had already experienced, their leaders seemed to believe that the force of history was to make them the final inheritors of the glorious October Revolution."

The situation in the Mediterranean basin became critical again in 1967 after several years of border incidents in Israel.

ized the Suez Canal. A joint British-French-Israeli attack that year failed to regain the canal, in great part because President Dwight Eisenhower and the Soviet Union pressured Britain and France to stop their operations. The Israelis, however, gained the Gaza strip and access to the Indian Ocean at the Gulf of Aqaba. The war, however, highlighted Soviet interest in the Mediterranean.[2]

But the West had a strong ally, Israel, in the eastern Mediterranean and could count on most of the Arab and Moslem states that ringed the sea to oppose Soviet incursions into their countries.

In 1958, political problems developed in Lebanon, and the U. S. reacted to secure the area and stabilize the Christian-led government. The U. S. Sixth Fleet landed its marines to form a peacekeeping force with Army paratroopers who flew in from Germany. While not directly involving the U. S. S. R., the incursion affected Soviet-leaning Syria.

During the years that followed, there were frequent border clashes between Israel and Syria, Egypt, and Jordan. In addition, Palestinian guerrilla groups, supported by Syria, attacked Israeli outposts and the Israelis responded.

Δ

The situation in the Mediterranean basin became critical again in 1967 after several years of border incidents in Israel. President Nasser of Egypt "his prestige much eroded through his inaction in the face of Israeli raids, requested the withdrawal of UN forces from Egyptian territory, mobilized units in the Sinai, and closed the Gulf of Aqaba to Israel."[3]

Israel mobilized and on 5 June launched a massive, surprise air attack on the Egyptian Air Force and crippled it. Tank-heavy Israeli forces overran the Sinai in three days, and then attacked east against Jordan. The Israelis captured and then annexed the Old City of Jerusalem and occupied the strategic Golan Heights overlooking Syria. Following the end of the Six-Day War, Israelis clashed with Palestinian guerrillas frequently and retaliated when Arab forces fired artillery into Israel.

The war brought the Soviets into the Mediterranean in force.

> *The first significant appearance of Soviet sea power in the Mediterranean came in 1967 during the Arab-Israeli conflict. During the Six-Day War (June 1967) the Soviets increased their force in a show of support for the Arab states. That was the first opportunity taken by the Soviets to demonstrate their willingness to influence major events in the area by the use of military power. During that crisis the Soviet Mediterranean Squadron numbered up to some 70*

units, some of which were in Port Said and Alexandria, to prevent Israeli attacks against those ports.[4]

By 1974, the Cold War had expanded to Africa and made the U. S. presence in the Mediterranean even more important.

In the years that followed, the Soviets increased their naval forces in the Mediterranean: "Such was the rate of Soviet increase that by the October War in 1973 Soviet units outnumbered the Sixth. Also in 1973, unlike in 1967, the Soviets were able to deploy amphibious ships in large numbers and also used anti-carrier tactics." [5]

Consequently, the Mediterranean Sea was no longer an American or a NATO lake.

Δ

By 1974, the Cold War had expanded to Africa and made the U. S. presence in the Mediterranean even more important.

On 12 September 1974, Haile Selassie lost power in Ethiopia. The pro-western government quickly disintegrated and by December the new strongman, Mengitu Haile Mariam, announced that "the country would henceforth be a socialist regime." Marxist-Leninist groups soon grew, armed confrontations began, and eventually Mengistu launched a "red terror" campaign against opponents and destroyed them in 1977. That year, Ethiopia broke off military relations with the United States and, aided by Cuba and the U. S. S. R., began Sovietizing the country. By 1979, a full-fledged Ethiopian Communist Party had organized, and in 1984 the Soviet Union integrated the party "into the system of world-wide Communism..."

During this ten-year period of upheaval, Mengistu killed thousands and most likely murdered the deposed emperor. To control the people and eliminate competitors, East German state security forces, the STASI, helped him as did Soviet security forces. To combat armed enemies, he drew on "the firepower and naval and aerial logistical support of the Soviet Army, as well as the Cuban expeditionary force." [6]

In July 1974, the Portuguese government withdrew its army from Angola and turmoil began. The next year, Cuban and Soviet advisors landed in the country. In August 1975, several hundred Cuban soldiers landed, and by October, a total of 7,200 Russians and Cubans had deployed to the country. The deployment of Communist troops triggered intervention by South Africa. Eventually, the new Communist-backed regime survived because of Soviet and Cuban air support.[7]

In December 1974, a new group took charge in Portuguese Mozambique. As in Angola, the political situation deteriorated and by 1977, new leaders "had openly proclaimed their allegiance to the Bolshevik ideal... calling for closer links with the international Communist movement."[8]

When CAPT Morin entered the Mediterranean, he faced a changed and changing situation.

The tired ROSIE was the biggest ship in the world at the time.

Ψ

MED Cruise

After "being at sea for the three months I had been in command, we remained in port for five weeks. During that time, I had to prepare the ship for our upcoming deployment to the Mediterranean. The crew performed substantial maintenance on the old and tiring carrier and the sailors loaded and stowed all our needed supplies for our extended cruise.

The tired ROSIE was the biggest ship in the world at the time. Originally, she had been built to "withstand the hail of kamikazes from a desperate Japan which had torn ships and men from the Philippines to Okinawa."[9] Having steel decks made her stronger than World War II carriers. To deal with a new operational environment, the Navy modernized her from 1954 to 1956, and when she rejoined the fleet, *ROSIE* had a new ten degree angled flight deck for landing aircraft, three new steam catapults to launch larger jet aircraft, a new centerline elevator, a hurricane bow, and no guns. In 1960, the Fresnel Lens System was installed, and in 1969, the Navy overhauled her once again. Before I took command, *ROSIE* had many achievements:

> In 1966, she deployed on her first and only combat tour to the Tonkin Gulf;
>
> By 1971, she had handled over 200,000 arrested landings;
>
> Prior to 1973, she had completed seventeen deployments to the MED.[10]

A young sailor wrote in 1973, "... the ship is in the twilight of its long career. Newer carriers are being built to replace her. But the Roosevelt steams on, a grand old lady with plenty of punch."[11]

Ψ

Our punch came from our embarked Air Wing, CAW-6, led by Silver Star winner, CDR G. Cane. Air Wing 6 had served aboard the ROSIE since 1970. CAG Kane commanded several squadrons.

> VF-41, "Black Aces," a fighter squadron flying F4B Phantom IIs. Commanded by CDR N. Dyer. It had several combat-experienced pilots.
>
> VA-15, "Valiants," an attack squadron flying A7B Corsair IIs. Led by CDR R. McClary. VA-15 also had a few combat-tested pilots.
>
> VA-87, "Golden Warriors," another attack squadron flying A7Bs. CDR E. Rice, the CO, led several combat-experienced pilots.
>
> VA-176, "Thunderbolts," a third attack squadron had A6A Intruders and KA6D tankers. Commanded by CDR J. McKenzie, it had additional Vietnam veterans.
>
> The Air Wing included other units.
>
> Carrier Airborne Early Warning Squadron 121 (VAW-121), known then as "The Griffins," provided us combat intelligence against possible enemy aircraft and missiles. CDR R. Rentschler commanded the valued group.
>
> The pilots in HC-2, "Angels of the Fleet," manned our rescue helos. LCDR J. Walker, a full-bearded Bronze Star holder, led them.
>
> LCDR W. Coltrin commanded a detachment of VFP-63, "Hackers," a photo reconnaissance unit.[12]

Ψ

Complicating life during these strange years, I found that the *ROSIE* was not a taut ship. Since taking command, we had worked such a demanding schedule that I had not had a chance to do much to improve the discipline aboard. I had found that when I walked down a passageway, a young sailor coming the

1973-1974
Department Heads
USS *Franklin D. Roosevelt*

Med Cruise

Executive Officer
CDR Eugene M. Clemens

AIM Department
LCDR M. G. O' Conner

Air Department
CDR G. E. Sheldon

Communications Department
LCDR J. A. Thompson

Deck Department
LCDR E. W. Richardson

Dental Department
CDR R. C. Terhune

Engineering Department
CDR H. O. Bullock, Jr.

Executive Department
LCDR L. R. Jacobs

Medical Department
CDR R. J. Coyle

Navigation Department
CDR W. D. Yonke

Operations Department
CDR R. W. McKay

Supply Department
CDR H. R. Croeber

Weapons Department
CDR B. S. Stevenson

At other times when I walked the ship and entered a space, the sailors, who were not used to seeing their captain all over the ship, would look at me as if saying "Yeah, what do you want?"

other way would race me for the hatch instead of halting, coming to attention, and allowing me to pass. My Marine Orderly, who was about five-foot-six and was a weight-lifter with shoulders a yard wide, often would nab a sailor who came flying through the hatchway by the neck of his shirt, push him back through the hatchway, and make him brace up against the bulkhead as he should have done.

At other times when I walked the ship and entered a space, the sailors, who were not used to seeing their captain all over the ship, would look at me as if saying "Yeah, what do you want?" To make sure that this did not happen again, I would get them on their feet, explain the correct military courtesy required in this situation, and ask who was the senior man? The sailors would compare enlistment dates and figure out who was in fact senior. I put that person smartly on report. Amazingly, the word quickly spread around the ship.

The crew had also gotten into the habit of not wearing their hats in port. In the Navy you do not salute without a hat on, so when I would walk down the hangar deck, the uncovered sailors would not render the proper hand salute to me. Well, I got these and other courtesies and traditions reinstated. I did not believe these young sailors were deliberately discourteous: they simply did not know any better because to save money Secretary McNamara had reduced the time young sailors spent in recruit training, and the senior Navy leaders had cut shipboard military courtesy from the training syllabus.

I also began to conduct personnel inspections... I personally inspected each sailor. For most of them, that was the first time they had met me eyeball to eyeball.

I also began to conduct personnel inspections. Rather than conducting an inspection of the entire crew at one time, a very time-consuming and tiring action, I scheduled inspections of individual departments. That way, I personally inspected each sailor. For most of them, that was the first time they had met me eyeball to eyeball.

Ψ

During this in-port period, the President of the *ROSIE*'s Enlisted Wives' Club asked me to be the guest speaker at their next meeting. They expected a good turnout when they knew they would get a shot at the Captain. In turn, I expected to get a lot of heat at the meeting. These were wonderful women: they kept families together and homes running while their men were gone. We had been at sea a lot since I had assumed command. Too much was asked of these women who got no credit for the difficult roles they played, and many were frustrated, tired, and somewhat angry at the fate life had dealt them. Their men had been at sea and after returning to the U. S. often had to work twelve to eighteen hours a day repairing machinery on the ship that could only be fixed in port and not while steaming at sea. Now, they faced another six-month separation as we prepared to deploy to the MED.

I represented their cause of unhappiness. Once with them, I made some opening remarks and then asked for q uestions and comments. They vented on me for about an hour.

I represented their cause of unhappiness. After joining them, I made some opening remarks and then asked for questions and comments. They vented at me for about an hour. After that, they became much friendlier because they knew that I did not have much control of our operating and deployment schedules. I could not solve their problems, but they at least got a chance to state them. I thought that it was important for each agitated wife to have me hear her out and know that I cared.

Before we deployed, I met one more time with the women at a Pre-Cruise Meeting. There, a Chaplain and representatives of the Personnel Support Center and the Red Cross briefed them about whom to contact in case of emergencies, family problems, medical care, pay matters, etc. The briefings were particularly valuable for the wives and young mothers who faced their first long separation. This meeting was much friendlier. One outcome developed as it always did: the Navy wives learned that they had to support each other.

Ψ

On 7 September 1973, we got underway for four days to carrier refresh the Air Wing. Before entering port, almost all the

SINCE THE ADMIRAL WAS NOT A PILOT, I WAS TO BE THE SENIOR NAVAL AVIATOR ON BOARD AND COULD MAKE THE DECISION TO CANCEL FLIGHT OPERATIONS WHEN SEA AND WEATHER CONDITIONS MADE FLYING TOUGH AND DANGEROUS...

Air Wing landed aboard, saving us from having to use the pier-side crane to lift the aircraft on deck. Once pier-side, RADM D. J. J Downey,[13] Commander, Cruiser Destroyer Flotilla Eight, his Chief of Staff, CAPT Harry Southworth, with whom I had gone through flight training, and the rest of the admiral's staff reported aboard. RADM Downey served as our Flag Officer for the MED cruise. Since the admiral was not a pilot, I was to be the senior naval aviator on board and could make the decision to cancel flight operations when sea and weather conditions made flying tough and dangerous without having to contend with an Aviator Admiral who might say "Launch 'em; It's good training!" I found RADM Downey to be a tremendous boss and enjoyed a spectacular relationship with him during the deployment. He invariably designated me to be the OTC - Officer in Tactical Command - of his entire formation while we steamed in the MED. I grumbled, worked hard, and realized that the Admiral was training me and giving me more command experience.

Just before we left, a Jacksonville TV station reporter called me and said that a racial demonstration was planned on the pier at the time of our sailing. I told him that I knew nothing about it and then immediately alerted Naval Station Security. Sailings are hard times for families, and we did not need a bunch of demonstrators mixing in with a large crowd of women and children who were saying good-bye to their husbands and dads for six months. As it turned out, no demonstration took place.

WE HAD AN UNEVENTFUL CROSSING ALTHOUGH THE USUAL RUSSIAN BEAR AIRCRAFT FLEW BY AS WE NEARED EUROPE.

We had an uneventful crossing although the usual Russian Bear aircraft flew by as we neared Europe. We entered the Naval Station Rota, in Spain, for our turnover with the carrier, USS *John F. Kennedy*, which my old friend CAPT John Dixon commanded.[14] Once we completed turnover, we entered the MED where we planned to spend fifty percent of our time at sea and fifty percent of our time in various ports. The 1973 oil crunch handicapped our operations. The Sixth Fleet Commander gave me my fuel allowance each quarter and as a result I had to modify operations to save fuel. To do that, I would steam on eight boilers rather than on ten or twelve and could only make twenty-eight knots. As long as I had a wind for flight operations, we could operate at that speed. However, during calm wind conditions which often occurred in the MED, I operated on the edge with eight boilers. The *ROSIE's* old catapult and arresting gear engines required more wind across the deck that the newer Forrestal Class carriers. With no wind, I would have the ship steaming at twenty-eight knots, but that was too slow to launch the F4s. Steaming at twenty-eight knots also ate up fuel at an alarming rate because the boilers operated at their maximum firing rate.

WE SETTLED INTO THE ROUTINE OF MED OPERATIONS AND PORT VISITS OR ANCHORAGES FOLLOWED BY FLIGHT OPERATIONS. FOR ME, HOWEVER, LIFE WAS NEVER HUMDRUM.

Ψ

We settled into the routine of MED operations and port visits or anchorages followed by flight operations. For me, however, life was never humdrum. At sea, my officers and I faced numerous challenges. At all times, we had to protect the ship against possible collision situations with merchant ships because their captains were unreliable. During flight operations, particularly at night, we had to be aware of sea and weather conditions; moreover, we had to identify an available airfield ashore in case we had an emergency. We also had to conduct routine underway replenishments for fuel, food stores, and ordnance around our flight operations.

When entering port, we faced other tasks. Normally, we had to pull alongside a tanker and top off the night before we entered. Once in the harbor, we had to ensure that our anchor had been well set before we could start reducing the number of boilers on line.

To train my ship and air wing officers as I had been trained, I delegated the ship handling evolutions such as replenishing, anchoring, and getting underway to them. As much as I wanted some "stick" time, I did not need any. It took a lot of patience (and guts sometimes) to stand by and not step in while one of these officers made his first approach to an oiler at night. To

IF THE OFFICER BEING TRAINED GOT TOO CLOSE, I WOULD LET HIM TRY TO WORK HIS WAY BACK OUT. IF HE CLOSED TO ABOUT SIXTY FEET AND HIS CORRECTIONS WERE NOT WORKING, I WOULD STEP IN BECAUSE HE WAS GETTING TO A POINT WHERE A SERIOUS PROBLEM COULD DEVELOP.

appear casual and unconcerned, I would sit in my chair or nonchalantly walk over to the wing of the bridge to observe the maneuvering. Once alongside an oiler, a conning officer had to maintain a proper position on the oiler and hold a desired distance abeam. If the officer being trained got too close, I would let him try to work his way back out. If he closed to about sixty feet and his corrections were not working, I would step in because he was getting to a point where a serious problem could develop. Some of the aviators had a good "seaman's eye" for relative motion, but a few of them just could not get the knack of the ship's motion or a feel for the time it took for such a big ship to respond to rudder or shaft RPM changes.

In port, my officers and I had to deal periodically with sailors returning to the ship who had drunk too much on liberty. On occasion fights broke out between some sailors or there was pushing and shoving between several of them. We all knew how to handle these relatively common situations. If, however, black and white sailors were involved, we reacted in a totally different way because we remained sensitive to such clashes after our recent racial incident. In most cases, I found that these incidents were not racially motivated, but I thoroughly investigated each incident to make sure there were no racial problems involved.

Yom Kippur War

In the fall of 1973, the Arabs and Israelis clashed once again in the Yom Kippur War. We were in Barcelona when hostilities broke out. Sixth Fleet immediately put us on a four-hour alert. The U. S. Government shipped war supplies and equipment to Israel and wanted to fly F4s and A4s to them. Our "friends" in Europe, afraid that the Arabs would shut off their oil, refused to permit U. S. aircraft to stage through bases in Spain, Italy, and Greece. The Air Force, however, responsible for the delivery of the F4s, deployed a massive number of aerial tankers and with in-flight refueling flew the fighters to the

Azores. After crew rest, the Air Force pilots flew the F4s non-stop to Israel. The Navy also used in-flight refueling to fly the A4s to the Azores. From there, Navy pilots flew to the A4s to the *ROSIE,* which by that time had taken a position about 100 miles southeast of Sicily. From our ship, the pilots flew the fighter-bombers directly to Israel. I saw many familiar faces among the attack pilots who landed on the *ROSIE.* Most currently served as instructors in the A4 Training Command Squadrons and the TAG RAG Squadrons. This deployment demonstrated the value of in-flight refueling. Air Force and Navy aircraft could refuel anywhere over the ocean. Additionally, it demonstrated the utility and capability of aircraft carriers over air bases in foreign countries. On the high seas, we did not have to worry about someone cancelling our landing rights.

Four contingents of A4s landed on four consecutive days. We then headed for the waters around Crete and joined with Commander, U. S. Sixth Fleet in USS *Little Rock,* a cruiser, and with the carrier, USS *Independence.* CAPT Charlie Smith,[15] an old shipmate, commanded the *Independence.* Soon twenty-eight U. S. warships stood off Crete. The Russians joined our group with thirty-three ships, creating a big gaggle of warships.

Through all this, we conducted flight operations twenty-four hours a day. We flew twelve hours; the *Independence* flew twelve hours. In my first experience with Russian ships when I was the CAG in the *Independence* in 1968-69, the Russian destroyer captains pulled some foolhardy stunts; for example, they cut across our bow coming too close to us on purpose. One day, while I was sitting on the catapult as the ship came into the wind, a Russian destroyer came up very close on our starboard side and cut in front of us. He did not miss our bow by much. I yelled at the Captain on the radio, "Launch me; launch me!" I wanted to get airborne and dump all my fuel on that guy. The Captain held up the launch. I am not sure I would have cleared the Russian's mast anyway.

In the years 1968-69, the Russian Captains had realized the foolhardiness of their capers and had mellowed quite a bit,

1973 YOM KIPPUR WAR

"DURING 1973 THE ARAB STATES, BELIEVING THAT THEIR COMPLAINTS AGAINST ISRAEL WERE GOING UNHEEDED (DESPITE THE MOUNTING USE BY THE ARABS OF THREATS TO CUT OFF OIL SUPPLIES IN AN ATTEMPT TO SOFTEN THE PRO-ISRAEL STANCE OF THE UNITED STATES), QUIETLY PREPARED FOR WAR, LED BY EGYPT'S PRESIDENT ANWAR SADAT.

ON OCT. 6, 1973, THE JEWISH HOLY DAY YOM KIPPUR, A TWO-PRONGED ASSAULT ON ISRAEL WAS LAUNCHED. EGYPTIAN FORCES STRUCK EASTWARD ACROSS THE SUEZ CANAL AND PUSHED THE ISRAELIS BACK, WHILE THE SYRIANS ADVANCED FROM THE NORTH. IRAQI FORCES JOINED THE WAR AND, IN ADDITION, SYRIA RECEIVED SOME SUPPORT FROM JORDAN, LIBYA, AND THE SMALLER ARAB STATES.

THE ATTACKS CAUGHT ISRAEL OFF GUARD, AND IT WAS SEVERAL DAYS BEFORE THE COUNTRY WAS FULLY MOBILIZED; ISRAEL THEN FORCED THE SYRIANS AND EGYPTIANS BACK AND, IN THE LAST HOURS OF THE WAR, ESTABLISHED A SALIENT ON THE WEST BANK OF THE SUEZ CANAL, BUT THESE ADVANCES WERE ACHIEVED AT A HIGH COST IN SOLDIERS AND EQUIPMENT."

COLUMBIA ELECTRONIC ENCYCLOPEDIA, 6TH EDITION, 2007

particularly after a number of sideswipes between destroyers playing "chicken" in the North Pacific. The United States and the Soviet Union had gotten together to establish common signals to prevent dangerous situations.

When *ROSIE* conducted flight operations and I exercised tactical command, I would raise a flag signal in the air that announced our intention to turn to a new course in order to launch. The Russians never acknowledged our signal, but when we hauled the flags down smartly to signal "EXECUTE," the Russian ships would turn with us and maintain their relative positions as if they were part of our formation.

We operated this way for thirty-five straight days until the war started slowing down.[16] Finally, Sixth Fleet ordered us to Athens for some liberty. Unfortunately while in port, student demonstrations against the Greek Government forced me to cancel liberty on a number of days.

1973 Oil Crisis

"In October of 1973 Middle-Eastern OPEC nations stopped exports to the US and other western nations. They meant to punish the western nations that supported Israel, their foe, in the Yom Kippur War ..."

Sarah Horton

Ψ

We headed back to the operating area south of Crete to start flying again as the *Independence* headed to Athens for some port time. We visited Athens one more time before we steamed to the western MED so that we could spend Christmas in Barcelona. During the transit to Spain I faced two situations - one in the Straits of Elefonisou and one off Barcelona - that clearly demonstrated that I was grossly underpaid for my position.

After leaving Athens, we sailed south through the Straits of Elefonisou the first evening. It was a crystal clear, moonlight night. I sat in my chair on the bridge doing some paperwork when I noticed the ship turning. I yelled at the Officer of the Deck who should have been aware that something was amiss. Immediately, I shifted control to After Steering, and we got ourselves squared away before the bridge reacquired steering control. By the time we returned to our base course, the ship had turned sixty degrees.

After that incident, we passed through the Straits of

Messina between Sicily and Italy where we had to be extremely careful to avoid the ferry and small boat traffic that transited between the two countries, and then we steamed through the Straits of Bonifacio where we had to be careful of the dangerous rocks.

The night before we entered Barcelona, we began topping off from the Oiler, Neosho, as we sailed about twelve miles south of Palma, Majorca. On that cold and dark night, the lights of Palma looked inviting.

The night before we entered Barcelona, we began topping off from the Oiler, *Neosho*, as we sailed about twelve miles south of Palma, Majorca. On that cold and dark night, the lights of Palma looked inviting. As we topped off, the Officer of the Deck reported a ship leaving Palma on a course and at a speed that would put us in the same piece of ocean about eight to ten miles ahead at about the same time. With my binoculars, I saw a passenger liner running towards us with all its lights on. I assumed that the captain was probably in the main dining salon having dinner with all the passengers on the first night out, so I had my signal bridge put a light on the liner to challenge him. The ship responded with a good flashing light. They probably had a former Royal Navy signalman as part of their crew. I immediately had a signal sent saying, "We are refueling. Request you stand clear." At that time, vessels refueling underway had no special privileges even though they had severely restricted maneuverability. The liner "ROGERED" our signal.

Shortly thereafter, I observed the ship turning port, and I concluded that he would pass astern of us and forgot about him. In a few minutes, the Officer of the Deck reported to me that the contact had resumed course and, if something was not done, we would meet sooner than planned. During this time, we had kept the admiral's Staff Watch Officer informed of the situation. Soon ADM Downey appeared on my bridge and asked, "Well, what do you think?" I told him that it would mean a lot of work for a lot of people to stop pumping, disconnect and retrieve hoses, and then make a new approach to reconnect, but considering the situation, we had no other option. He agreed. I called the Captain of *Neosho* to tell him to cease pumping and start disconnecting. This took some time, and meanwhile the liner converged closer and closer, seemingly oblivious

After everything calmed down, we re-rendezvoused with the Neosho, made an approach, re-connected hoses, and began refueling again. We had a long night.

to the dangerous situation developing. As soon as the oiler had retrieved its hoses and we cast off all lines, I took the CONN and ordered: "All stop, all back full!" I told the *Neosho's* Captain that I would tell him when his stern would clear my bow so he could maneuver as necessary. I signaled the emergency signal with *ROSIE's* horn. As we slowed rapidly, the *Neosho* came hard right. The liner passed less than 1,000 yards ahead of us, way too close for ships of our size.

Evidently, the liner's Captain must have gone to his bridge when he heard our emergency horn. Seeing two large ships bearing down on him must have startled him so much that, after passing ahead of us, he turned starboard and crossed my bow a second time. This near disaster probably happened because some Third Mate on the liner did not want to disturb his captain's dinner and failed to brief his skipper on the converging situation, thus jeopardizing the liner and its passengers and the *ROSIE* and *Neosho*. After everything calmed down, we re-rendezvoused with the *Neosho*, made an approach, re-connected hoses, and began refueling again. We had a long night.

Ψ

We anchored off Barcelona on 19 December for two weeks in port for the holidays. We found the winter weather to be treacherous in the MED as low pressure systems after low pressure systems marched across Europe. We cancelled boating on a number of days because of rough sea conditions. Many a MED-sailor returned aboard his ship by way of a cargo net over the fantail and discovered such a return to be tricky and difficult if he had been shopping and had carried a camera with him.

On Christmas Eve, Harry Southworth and I went ashore for a few hours.

On Christmas Eve, Harry Southworth and I went ashore for a few hours. When we returned to the Fleet Landing, the Beach Guard Officer told me that all boating had been cancelled, and that he had sent my gig crew to the USS *Yarnell* for the night. He told me that high winds were expected and that the seas were very rough. With my ship anchored a mile from the mole guarding the harbor, I did not want to be ashore as she dragged her anchor and ended up on the rocks. I recalled my gig crew and we got underway. Inside the breakwater, we were fine, but once we turned the corner - WOW! We bounced up and down and this way and that. As we neared the ship, crewmen lowered the accommodation ladder. On the first approach, it was too rough to jump for. On the second approach, I made my play and landed on the platform as a wave about three feet higher than the platform soaked my legs. The coxswain made another approach and one more got off. Rather than risk another transfer because the situation was so hairy, I told the coxswain to head back for the Fleet Landing. I climbed up to the bridge to check our bearings to make sure we were not dragging. I had two more boilers lighted just in case I needed steam to put some turns on the screws. After that was done, I spent Christmas Eve sitting in my chair, and I remained on the bridge until morning. Fortunately, I saw a lot of Barcelona and Costa Brava during the next few days and enjoyed the lovely city and the surrounding area.

I CLIMBED UP TO THE BRIDGE TO CHECK OUR BEARINGS TO MAKE SURE WE WERE NOT DRAGGING. I HAD TWO MORE BOILERS LIGHTED JUST IN CASE I NEEDED STEAM TO PUT SOME TURNS ON THE SCREWS. THAT DONE, I SPENT CHRISTMAS EVE SITTING IN MY CHAIR, AND I REMAINED ON THE BRIDGE UNTIL MORNING.

Many wives arrived in Barcelona to spend the holidays with their husbands and break up the six-month long separation. I wished more could have come. During their visits, however, my officers and I found it rather frightening getting some of the women on and off the boats on bad boating days.

We got underway on 4 January 1974 and returned to the area around Crete. Unfortunately, we had to leave a few sailors behind because the Spanish Guardia Civil caught them selling marijuana. The Spanish routinely threw drug peddlers in jail where they generally waited a year before standing trial. The incident got the attention of many of my other sailors. I never learned what happened to the sailors.

Everyone seemed rusty after our long stay in Spain, so it was good to get flying again. We worked our way to the eastern MED before returning to Athens on 14 January. The night before we entered the port, we faced another incident while refu-

THESSALONIKI

"NEXT LIBERTY WAS THESSALONIKI WHERE I EXPERIENCED OUZO, ZORBA THE GREEK "BOUZOUKI" MUSIC , AND LEARNED TO DANCE WITH PLATES ON MY HEAD AND THEN SMASH THEM ON THE FLOOR.

WE RETURNED TO ATHENS JUST IN TIME TO BE PULLED OUT OF PORT EARLY, THIS TIME BECAUSE OF A MILITARY COUP THAT OVERTHREW THE POPODOPULOS REGIME."

ENS STEVE LOCKWOOD

eling: we had to break away from the oiler USS *Savannah* when an Israeli vessel failed to give way.

My bride joined me in Athens for the in-port period and we did a lot of sightseeing. We enjoyed our nice break in the long cruise. After ten days, we left Athens for a week of operations in the Aegean Sea. Then, we sailed to Thessaloniki in northern Greece. Betty and many other wives travelled there and met us when we docked. The U. S. Consul General and his wife and the senior Greek Army and Naval Officers proved most hospitable to all of us. Not many of our ships visited the small port.

On 3 February 1974, we got underway again and Betty returned to the United States. We found operating in the Aegean to be beautiful with ever-present blue skies and water and rocky, barren islands with monasteries perched on their tops. While the area was very scenic, we never had sufficient wind to work with because the sea was a sheet of glass most of the time. One evening, while landing aircraft with what wind there was, I had the *ROSIE* heading toward an island at twenty-eight knots. I only had about ten miles to work with. Even though the Navigator assured me we had deep water right up to the island, I did not have a good feeling running that fast toward a mountain top sticking out of the Aegean. Consequently, I got on the radio and told all the pilots I wanted them to land on their first passes. We had no sea room for wave-offs or foul deck wave-offs which would have extended their recovery.

National Week became our next big evolution. Our force joined with the USS *America* (CVA-66) captained by CAPT Tom Replogle, a shipmate from *Coral Sea* days off Vietnam.[17] Commander Task Force 60, aboard *America*, ran a demanding and tiring exercise.

After a post-exercise brief in Suda Bay, Crete, we headed for the western MED. En route, we conducted flight operations and then entered port at Palma, Majorca on 24 February 1974. In port, VADM Michaelis, Commander Naval Air Forces, Atlantic Fleet, and his staff inspected us. We did well in the inspection: the crew impressed the staff officers with their smartness and military courtesy and the inspectors reported that fact. We had made progress. At the conclusion of the command inspection, the Admiral hosted a dinner for us.

Ψ

A month earlier, the Navy had considered me for Flag selection but did not select me for promotion to rear admiral. VADM Michaelis, who had sat on the selection board, told me that the Board did not select me because I had accrued too much command time! My mouth popped open on that one, and I said, "Admiral, when I was a young officer, it was people like you who kept stressing the importance of going to sea and being in command. I've had plenty of both and now you're telling me that that wasn't the correct path." Making Flag had never been a goal: I felt it was just too tough. Getting command of the *ROSIE*, however, meant I still was running strongly, but I also believed that the racial incident had wiped me out of any small chance I might have had to be promoted to rear admiral.

"ADMIRAL, WHEN I WAS A YOUNG OFFICER, IT WAS PEOPLE LIKE YOU WHO KEPT STRESSING THE IMPORTANCE OF GOING TO SEA AND BEING IN COMMAND. I'VE HAD PLENTY OF BOTH AND NOW YOU'RE TELLING ME THAT THAT WASN'T THE CORRECT PATH."

Ψ

Palma was our last port in the MED. On 3 March 1974, we got underway for Rota where we planned to drop off some material and then leave for the States as soon as we offloaded the cargo. Headed west, my trials and tribulations would continue because during the winter months low pressure cells marched across the higher latitudes of the Atlantic and brought extremely unpleasant weather. For the crossing, I briefed the crew to secure everything throughout the ship that could come loose during heavy seas and injure people. I also made sure that all the Department Heads inspected their respective areas to ensure we were prepared for any bad weather. Before leaving I was confident that we had tied down everything extremely well.

Storm

"It was during our return to the U.S. that I observed Captain Morin's amazing courage and leadership in what must have been one of the biggest challenges of his Naval career.

"Crossing the Atlantic in the Spring is usually a non-event, but that year a freak end-of-winter storm caught us without warning and we were stuck in a maelstrom... I was on the bridge every ten hours...I doubt if the skipper ever left the Bridge. Sitting in his Captain's chair or pacing the deck, I didn't envy him the crucial decisions he had to make.

For [3 1/2] days I ... personally witnessed a scene I'll never forget. The flight deck was thirty five feet above the waterline, the bridge sixty five, huge waves were breaking over the flight deck, blue water exploding with foam and spray, the one thousand foot long ship shuddering with each wave, the bridge windows completely obscured by water and spray, the wipers going furiously to little effect, the bridge windows clearing just in time to watch the pitch of the bow into the next wave to repeat the process...

The ship was taking a beating all this time, the steel catwalks that ran the length of flight deck had been twisted or torn off for more than a hundred feet from the bow. The O-2 level forward of the JBD's was flooded and damage control teams were working around the clock, the forecastle had been evacuated. Aircraft had been moved to the hanger deck closely together to accommodate as many as possible and secured with tie-down chains."

ENS Steve Lockwood

Going Home

Out in the Atlantic, we experienced some deep swells caused by a low that passed well north of us. The swells caused us to roll uncomfortably for a carrier. About halfway across, Fleet Weather Center in Norfolk notified us of a low dead ahead of us. The weathermen predicted winds of forty knots and seas from eight to twelve feet. I asked my Aerologist to brief RADM Downey and me about the conditions. As stated, the conditions would not present problems to us. About midnight, however, the wind had freshened noticeably, the seas started to increase, and the barometer began dropping rapidly. Soon, we began pitching quite badly and I kept reducing speed to improve the ride. By dawn, we had thirty-foot walls of water coming at us on the bow and we had green water washing over the flight deck. As we came shuddering out of a big one, I expected the wave to wash away the two F4s on the forward catapults. I could not turn the ship to head down wind-down sea because I believed that, if caught in the troughs during the turn, I would start breaking loose equipment and aircraft. A loose aircraft or tow tractor on the hangar deck could cause tremendous damage. I slowed the ship to five knots and barely maintained steerage.

By this time, lots of water had entered the ship through the ventilators and the hangar deck had a foot of water on it constantly. We lost the whip antennas forward. The cascading water bent the forward catwalks upward and flooded the forward catapult machinery rooms and made the catapults inoperative. Many seasick sailors performed damage control beneath decks. The wind velocity peaked at one knot less than a hurricane. Fortunately, the seas remained on our bow the entire day. I was concerned that the constant pitching would loosen the tie downs on the aircraft on the flight deck, but I did not dare allow anyone out on the flight deck to check or adjust them.

Toward late afternoon, the winds began to abate somewhat and the distance between the wave crests increased. But the storm had tired everyone. With improved weather, I allowed the

Air Department and the Air Wing to send people out on the flight deck to check security, but required that each man wear a harness with a line attached to it that shipmates secured. Sitting in my chair on the bridge, I noticed a rogue wave coming down on us about thirty or forty degrees to starboard rather than dead-on as the large waves of had done all day. The rogue wave hit near the forward starboard elevator door which had been closed to keep out water and pushed the door into the hangar deck and pinned a Chief Petty Officer up against the tail of an aircraft. We rushed the injured CPO into surgery where he died.

Ψ

About 4:00 p.m., a U. S. destroyer steaming about 200 miles southeast of us sent us a message asking us if we could help one of her sailors who had appendicitis. RADM Downey called me and the senior helo pilot to his bridge to discuss the request. The helo pilot wanted to go. We would have to send the helo out to the destroyer without navigation aids and the pilot would have to find the destroyer, pick up the sick sailor, and return to the *ROSIE*. The pilot would not have much fuel to deal with a navigation error during the long trip. Therefore, I told the admiral that I thought I would be able to turn the ship soon and then I would steer to the destroyer to reduce the return distance for the helo. We launched the helo with a doctor aboard. Then, I announced to the crew that we would be turning and to standby for heavy rolls, and immediately called the Chief Engineer, told him about my plan. Furthermore, I told him that when I asked for increased revolutions to give them to me fast. I took the CONN and came left in ten degree increments. After thirty degrees, I rang up all ahead full and left full rudder. We came about smartly and spent very little time in the troughs. Once we began steering with the wind and had the seas behind us, I allowed people to go on the flight deck and we licked our wounds.

ROGUE WAVE

"ON THE THIRD DAY OF THE STORM A ROGUE WAVE HIT THE SHIP ON THE STARBOARD SIDE BY THE ELEVATOR. THE GIANT WAVE PUSHED IN THE MASSIVE HANGAR BAY DOORS OFF THEIR TRACKS AND SET UP A CHAIN REACTION WITH LETHAL EFFECTS. SEVERAL AIRCRAFT WERE PUSHED INTO EACH OTHER AND TWO CPO'S FROM THE AIR WING WHO HAD BEEN INSPECTING TIE-DOWNS WERE CRUSHED...

NOW CAPTAIN MORIN WAS FACED WITH A TERRIBLE DECISION, THE STORM CONTINUED TO RAGE WITH NO END IN SIGHT, TO TURN THE SHIP OUT OF THE STORM WOULD REQUIRE TAKING AT LEAST ONE WAVE AMIDSHIPS. THE HANDLING CHARACTERISTICS OF ANY AIRCRAFT CARRIER MADE THIS A RISKY PROPOSITION, BUT THE FDR WAS EVEN MORE AT RISK...[BECAUSE OF MODIFICATIONS, THE FDR] WAS MORE TOP-HEAVY THAN ANY OTHER AFLOAT. THE SHIP SIMPLY COULDN'T TAKE MUCH MORE PUNISHMENT ....,

CAPTAIN MORIN ADDRESSED THE CREW ON THE 1-MC, EXPLAINED THE MANEUVER WE WERE ABOUT TO MAKE AND ENSURED THAT THE SHIP HAD EVERYTHING STOWED AND ALL THE HATCHES BATTENED DOWN. WITH THE ISLAND TO THE INSIDE OF THE TURN TO ACT AS A COUNTERWEIGHT, HE TURNED THE SHIP.

THE FDR TOOK ONE NINETEEN DEGREE ROLL, RIGHTED HERSELF, AND PROCEEDED OUT OF THE STORM.

NOT MANY MEN IN LIFE ARE CALLED UPON TO MAKE THAT DECISION."

ENS STEVE LOCKWOOD

Article 110
Uniform Code of Military Justice

Improper Hazarding of a Vessel

Any person subject to this code who willfully and wrongly hazards or suffers to be hazarded any vessel of the armed forces shall suffer death or such punishment as a court-martial may direct.

Any person subject to this code who negligently hazards or suffers to be hazarded any vessel of the armed forces shall be punished as a court-martial may direct.

***Manual for Courts-Martial, United States, 1951*, 445.**

The helo returned safely with the patient just after dark. That was pretty good flying when you consider that we were out in the middle of the Atlantic. Once we secured the helo, we proceeded on a southerly heading to get away from the effects of the storm before heading toward Mayport again. Changing course delayed our arrival by a day.

Ψ

Before we arrived, Atlantic Fleet informed RADM Downey, the Transit Commander, and me that we would see the end of the long green table again, and that VADM John Finneran, Commander Second Fleet, would be the investigating officer.[18] Finneran was to determine responsibility for the damage to the ship and for the death of my crewman. I thought to myself: "Here we go again!" As I said before, this was not a fun tour.

The investigation lasted a week after we entered Mayport. Finneran did not fault us for not being ready because we had prudently taken every precaution for heavy weather before we entered the Atlantic. He also did not fault us for how we handled the ship and fought the damage. The Admiral concluded that we had been victimized by an Act of God, but faulted the Fleet Weather Center staff for failing to notify us in a timely manner when they observed the low pressure center rapidly intensifying. If I had been found culpable for the losses and the fatality, I would have been "awarded a court-martial" (as I would have been after the racial incident).[19]

Ψ

We spent one month in Mayport during which time we let people go on leave and contracted with the local shipyard to repair our damage. On 18 April 1974, we got underway and sailed into the JAX operating area for Fleet CARQUALS. The days and nights proved tedious because many of the pilots made their first night landings on a carrier. The Marines out of Cherry

Point, NC also came out to get some carrier work. One A6 squadron commander flew so badly at night that I sent him ashore. His Wing Commander, a colonel, came up to the bridge and I told him that I did not want to see that guy again on my ship. Toward the end of this ten-day period, the student aviators from the Training Command came out for their initial CQ. When they finished, we returned to Mayport for one week before leaving for a four-month period at the Philadelphia Naval Shipyard for dry docking. So after a six-month cruise, my married personnel got one month in port, ten days at sea, one week in port, and then deployed to Philly for four months. It was a brutal profession.

So after a six-month cruise, my married personnel got one month in port, ten days at sea, one week in port, and then deployed to Philly for four months. It was a brutal profession.

The week before we left for Philly, we off-loaded all our ordnance and much of our fuel. When a ship is dry docked, she settles onto blocks which are shaped and spaced to conform to the shape of the hull and to support the tremendous weight. Lightening the ship meant she would draw less water and would ride higher. Each day I conferred with my Chief Engineer and Navigator to go over their calculations to determine how much of my mast we had to cut off to give me a four-foot clearance under the Delaware Memorial Bridge. This calculation seems relatively simple once you determine how high above the waterline the top of the mast will be after offloading and taking in account the weight of the fuel that will burn off during transit. But, we had to compute the distance between the water surface and the bridge span, and tides would affect that measurement. To add to our planning, we had to be ready at plus or minus fifteen minutes of slack water time when there was no current. On our scheduled day, slack water time was 8 a.m. I wanted to be off Philly at 7 a.m. so that there would be time for the tugs to get in position and be made up. My Navigator then had to determine the time we would pass under the bridge and the level of the tide at that time. We ran the numbers time after time until finally we decided that the shipyard people in Mayport needed to cut off twelve feet of the top of my mast to give me a four-foot clearance.

Delaware Memorial Bridge

After sailing up the east coast, we arrived at the mouth of the Delaware River at 11:30 p.m. on 3 May 1974. The pilot came aboard for the all-night trip up the river. Because we took up so much space, my primary concern was meeting a ship coming downstream at a turn in the river, and my secondary concern was losing steering control in the narrow channel. Just as when I flew, I had to think out emergency procedures ahead of time.

We approached the Delaware Bridge just after dawn on the 4th and as we neared the bridge everyone on the ship looked at our sawed off mast and the bottom of the span. From a distance and looking ahead at an angle, the four-foot clearance just was not there. All I could do was take a deep breath and watch like everyone else. I had a terrible feeling as we approached, but our calculations turned out to be correct and we glided safely under the bridge.

By 7 a.m., we stood off the shipyard and the shipyard pilot flew aboard and the river pilot departed. I asked the new pilot for his tug make-up plan. Four Shipyard tugs approached us and he had arranged for four commercial tugs to help. There had not been an aircraft carrier at Philly in eight years; consequently, we faced some unexpected problems. My Officer of the Deck got my attention and pointed to the pilot's shaking knees. His nervousness did not give me much confidence in him. By 7:45 a.m., we should have been starting our approach to the dry dock gate, but the four commercial tugs were nowhere in sight. The pilot called the dock master on the radio to find out where the tugs were. We learned that the commercial tugs would not be available until 10:00 a.m. I thought, "Great," and asked the pilot, "OK, what's your plan?" He told me that we could not anchor in the area and that he would have to put me pier side. I asked him where and checked the chart with my Navigator. He told me that the mean low tide at the designated pier was forty-two feet deep. The *ROSIE* drew only thirty-three feet at the time, so that would not be a problem. However, we did face a problem: the "L" shaped pier was only 500 feet long and that meant that about half of my ship would stick out into the river and be in the current which would slowly increase to two-and-one-half knots at ebb tide. In addition, an Iranian destroyer was docked at the base of the "L."

No one in the port had planned for us to dock at a pier so no line handlers had been sent to the pier and no camel, a barge-like float, had been put in place to serve as a buffer between ROSIE and the pier and the pier pilings. We had to wait.

No one in the port had planned for us to dock at a pier, so no line handlers had been sent to the pier and no camel, a barge-like float, had been put in place to serve as a buffer between *ROSIE* and the pier and the pier pilings. We had to wait. Then the ebb started. With only four tugs, I knew we would be hard pressed to get pier side and so I prepared to use the ship's power to help. Meanwhile, the base command must have gone to the Navy Exchange and everywhere else, to corral sailors to be line handlers. The Commandant of the Fourth Naval District, RADM Joe Coleman,[20] had arrived to greet the ship. Soon, dressed in his whites, he began hauling lines as we neared the pier.

We tried to work forward as much as possible to reduce the amount my ship would stick out into the river. As we pulled in, I knew we were nearing the destroyer because the Irani sailors kept looking at the *ROSIE*. Somehow, we could not close on the pier. When the tugs pushed my bow, the stern would come out; when they pushed on the stern, my bow would come out.

During one of the evolutions, the forward part of my flight deck struck and knocked over a whip antenna on the docked destroyer and forced the Iranian sailors to run for safety. I rang "back two-thirds" to ease away when my Chief Engineer called me to report that he would have to wrap up some machinery because we were sucking in mud. I told him that I could not afford to lose power at this point. The fact that we were sucking in silt explained my problem: we were bottoming amidships and pivoting around that point. So much for forty-two feet of water there at mean low tide. We finally got our lines over so we could control the bow and the stern; then, the tugs positioned themselves amidships and pushed us close enough for the remaining hours at pier side.

Of course, I immediately notified the Fleet Commander and COMNAVAIRLANT that I had nudged the Iranian destroyer. I felt that they would receive my message and say, "Oh no! Not

OF COURSE, I IMMEDIATELY NOTIFIED THE FLEET COMMANDER AND COMNAVAIRLANT THAT I HAD NUDGED THE IRANI. I FELT THAT THEY WOULD RECEIVE MY MESSAGE AND SAY, "OH NO! NOT MORIN AGAIN!"

Morin again!" Miffed at my misfortune, I left the ship to call on the District Commandant. RADM Coleman, a former carrier commander, turned out to be very sympathetic and said that the Shipyard - because of the lack of tugs and the lack of a plan to go pier side - had handled the whole episode poorly. In addition, he said that the Shipyard would take care of the bill for the damage to the Iranian destroyer. When I told him that I seemed to be constantly sending out Morin "Delta Sierra's" to the Fleet Commander, he replied that every carrier commander was in the same fix. It was just the nature of the beast when you command a capital ship.[21]

Near-mid afternoon, the Shipyard prepared to move us into dry dock. With eight tugs, the move from the pier into dry dock went smoothly. We lined the *ROSIE* up so that she would settle on the blocks perfectly when the water was pumped out of the dry dock.

The next morning I inspected the hull in the dry dock. An aircraft carrier's size, viewed from directly underneath it, is very impressive. I then met with the Shipyard Commander to go over our repair package. Once the work started, we met weekly to discuss progress, snags, priorities, parts problems, etc. Although a necessity, no ship's captain enjoys a shipyard period. Shipyard workers trash the ship. Fire is a constant threat because of the amount of cutting and welding done daily. Crewmen can be injured.

MY WIFE TOLD ME THAT IF THE ROSIE DID NOT RETURN ON TIME FROM PHILLY AND I WAS AT SEA ON SARAH'S WEDDING DAY I MIGHT AS WELL KEEP GOING AND STAY AT SEA.

I had one special concern: completing our yard period and returning to Mayport on time on 27 August so that I could attend the wedding of our eldest daughter, Sarah, to Mr. Bruce T. Bishop of Norfolk in Norfolk on 31 August. My wife told me that if the *ROSIE* did not return on time from Philly and I was at sea on Sarah's wedding day I might as well keep going and stay at sea.

Ψ

We completed the dry dock work and refloated the ship on

IN OCTOBER, BOB DUNN, MY SHIPMATE FROM WASP AND LEXINGTON CRUISES IN THE 1950S AND AN A4 SQUADRON COMMANDER AT LEMOORE, WAS SCHEDULED TO RELIEVE ME OF COMMAND OF THE ROSIE.

28 July 1974. The detailed inspection that followed showed no leakage, so we moved to the carrier pier for the final work. The Engineering Department then prepared for the strict and difficult Light Off Examination which the Fleet Commander's Engineering Staff would conduct. My Chief Engineer and I had met at breakfast for weeks so he could prepare me for my oral examination by the Chief Inspector. I had to be able to describe my entire steam plant and answer any questions put to me regarding my machinery, auxiliary machinery, electrical power generation, the production of fresh water from sea water, etc.

We did well on the inspection. Soon we began taking on fuel and then conducted dock trials. On 24 August 1974, on schedule, we sailed for Mayport. We still had work to complete, but would complete the repairs and reinstall the top of our mast during the next five weeks in Mayport. I enjoyed the run down the Delaware River in daylight. Car traffic stopped on the bridge and people got out to watch our passage. I'm sure that many didn't think that our mast would clear the bridge. All cheered when we passed under the bridge.

After we arrived in Mayport, I made it to Norfolk and gave the bride away. It was a grand wedding.

In October, Bob Dunn, my shipmate from *Wasp* and *Lexington* cruises in the 1950s and an A4 squadron commander at Lemoore, was scheduled to relieve me of command of the *ROSIE*.[22] The Navy informed me that due to problems on the USS *Saratoga* (CVA-60) that was also home ported in Mayport, Bob Dunn would relieve the Captain of the *Saratoga* ASAP, and I would remain in command indefinitely because there was no other carrier CO available in the pipeline at the time. That meant that I had to take *ROSIE* through all the hard times - work-ups, Fleet Training group times, an Operational Readiness Inspection, etc. - all over again. I was not too motivated to do all that again. Moreover, I faced the possibility of sailing for the MED again in early January 1975.

In mid-December, NAVAIRLANT Staffers inspected us on Damage Control Preventive Maintenance, Personnel Qualifica-

With all my lines still over and the brows still in place, the ROSIE began moving toward the Saratoga, only a few hundred feet away. I caught the motion out of the corner of my eye and yelled, "What the hell is going on?"

tions Standards, Nuclear Weapons Training, and much more. In turn, I started another full round of department personnel inspections. The ship began looking good, as did the crew. Many new sailors with little sea time had joined the crew since we had returned in March. Consequently, we conducted extensive training to prepare them for their new jobs and responsibilities and integrate them into the crew.

On 7 October 1974, we planned to get underway to begin our training. The *ROSIE* stood astern of the *Saratoga* at pier side that morning, and as the Officer of the Deck (OOD) ran through his check-off list preparing for our departure, I sat in my chair doing some paperwork. Nearby, a young petty officer worked to correct a problem with the bridge's engine order telegraph and its repeater in Main Control. As he talked on sound-powered phones to his counterpart in Main Control and without the OOD's permission, he rang up "All two thirds" to see if the repeater in Main Control would receive the command. The Engineering Officer of the Watch saw and heard the "ahead bell" and incorrectly assumed that I wanted power even though we had not sounded the traditional alert order, "Standby to Answer All Bells." With all my lines still over and the brows still in place, the *ROSIE* began moving toward the *Saratoga*, only a few hundred feet away. I caught the motion out of the corner of my eye and yelled, "What the hell is going on?" I saw the "all ahead two third's" rung up and immediately ordered "all stop, all back two third's," and then called Main Control on the public address system and repeated the orders. The Navigator backed me up on another intercom system and the OOD used the phone. As we moved towards the *Saratoga*, I ordered the OOD to announce over the ship's loudspeaker system for all hands to stand clear of all lines and the brows. My lines stretched like rubber bands and one parted. Both brows fell to the pier but did not hurt anyone. Just another routine day for a carrier Captain!

For the next two months, we conducted strenuous training at sea with little in-port time. We prepared for our Operational Readiness Evaluation and got through everything in fine order.

At this time, the new captain should have been aboard so that he could learn his ship and his people. I was getting tired. During this period, the Navy notified me that Pete Easterling[23] would relieve me and ordered me to report to Norfolk to become the Chief of Staff for COMNAVAIRLANT.

Before the change of command took place, I faced a new and unique seamanship challenge. While conducting our training in the JAX operating areas under the command of RADM B. B. Forbes,[24] Commander, Carrier Division 6, the Fleet Commander asked him if the *FDR* could assist the USS *Basilone,* a destroyer,[25] which was about 100 miles north of us and could not make feed water for its boilers. Dangerously short on feed water, its captain did not think he could make it to Charleston, SC. RADM Forbes cancelled flight operations and ordered me to proceed to the *Basilone* to help. Aboard the *ROSIE* we discussed and then developed our plan. My Chief Engineer and First Lieutenant (Deck Department Head) came up with a method of rigging fire hoses so that we could transfer our own feed water to the destroyer. The seamanship problem, however, remained mine: I would have to maneuver *ROSIE* alongside *Basilone,* maintain station on her to rig hoses, and then transfer water.

When we arrived, I found that the destroyer had slowed to four knots and that created a big problem because under five knots, I would not have much control of the *ROSIE.* Forbes, a former commander of the *ROSIE,* knew this, but left things up to me. After evaluating the situation, I decided that I would approach the *Basilone,* stay astern of her at 500 yards at co-speed (four knots), and see how *ROSIE* handled.

Luckily, we had excellent sea conditions.

I rendezvoused with the destroyer in the late afternoon, and I took the CONN because I could not trust anyone else with the evolution. I noticed that the COMCARDIV's Surface Operations Officer, a former destroyer captain and one of the few non-aviators on the Admiral's staff, had joined me on the wing of my bridge. I am sure that he wondered how I would handle this task because many surface officers did not believe that airmen

1945-1979
RISE OF NAVAL AVIATORS
CHIEFS OF NAVAL OPERATIONS

	Air Officer	Line Officer	Sub Officer
Nimitz, C.W.			X
Denfeld, L. E.		X	
Sherman, F. P.	X		
Fechtler, W. M.		X	
Carney, R. B.		X	
Burke, A. A.		X	
Anderson, G. W.	X		
McDonald, D. L.	X		
Moorer, T. H.	X		
Zumwalt, E. R.		X	
Holloway, J. L. III	X		
Hayward, T. R.	X		

USS *BASILONE*

I RENDEZVOUSED WITH THE DESTROYER IN THE LATE AFTERNOON, AND I TOOK THE CONN BECAUSE I COULD NOT TRUST ANYONE ELSE WITH THE EVOLUTION.

such as me - without a lifetime of ship handling experience under our belts - were qualified to command big ships.

With my best helmsman on the wheel and after a short period of maintaining station, I felt comfortable enough to start my approach. I eased up alongside of *Basilone*, got our lines over quickly, and then got the fire hoses over. Once the destroyer's crew secured and emplaced the fire hoses, we began transferring water. Meanwhile, my helmsman worked hard to hold the ordered heading. I kept watching the rudder indicator and he would tell me when the ship did not respond to her rudders. If that happened, I could drag the screws on one side to help ease our head around. When the CO of *Basilone* reported that he had sufficient water so that he could bring up his speed, I had him do so. I slowly increased my speed to match his. That delighted the helmsman because his job became less harrowing. We soon steadied at ten knots and after transferring 9,400 gallons of water, I sent the *Basilone* merrily on its way to Charleston. A pleased RADM Forbes announced to the Fleet Commander that the *ROSIE* had come through. And coincidently, we impressed Forbes' Surface Operations Officer.

About this time, CDR Gene Clemens, my XO, departed for a new assignment. He had served with me through all the hard times, and I owed him a lot for his leadership and hard work. Subsequently, the Navy gave him the command of an oiler. He would retire as a Captain.

Turnover

ON 9 DECEMBER 1974, WE RETURNED TO MAYPORT. I ENTERED THE ST. JOHN'S RIVER FOR THE LAST TIME ABOARD THE ROSIE.

On 9 December 1974, we returned to Mayport. I entered the St. John's River for the last time aboard the *ROSIE* and requested that Mayport's senior harbor pilot do the honors for me. I did not want anything to go wrong at this stage of the game. CAPT Easterling reported aboard and we spent a week together before turnover. We inspected every part of the ship and held numerous inspections of the personnel. I signed all the required items over to him, and on 20 December, he relieved me as Captain of the USS *Franklin D. Roosevelt.*

I heaved a sigh of relief when I left. The ship had taken a lot out of me but I left a lot of me behind.

I HEAVED A SIGH OF RELIEF WHEN I LEFT. THE SHIP HAD TAKEN A LOT OUT OF ME BUT I LEFT A LOT OF ME BEHIND.

My next stop was a desk.

Ψ

In two weeks, the *ROSIE* left for the MED. The first time CAPT Easterling took her to sea, he deployed for six months. About two months after leaving Mayport, he sent me a letter and thanked me for giving him a well-trained crew and a good ship. He said, I made his job easy.

Sidebar Notes

Exotic Mixture

Courtois, *Black Book of Communism, 683.*

1973-1974 Department Heads, USS *Franklin D. Roosevelt*

Roosevelt 1973-1974, a cruise book, photo, 8 and 10-11.

1973 Yom Kippur War

"Yom Kippur War," *Columbia Electronic Encyclopedia*, 6th Edition, 2007.

1973 Oil Crisis

Sarah Horton, "The 1973 Oil Crisis," ttp://www.envirothonpa.org/documents/ The1973 OilCrisis.pdf (15 June 2010).

Thessaloniki, Storm, and Rogue Wave

Steve Lockwood, *Paper*, 26 November 2003.

Article 110, Uniform Code of Military Justice, Improper Hazarding of a Vessel

Manual for Courts-Martial, United States 1951, 445.

Delaware River Bridge

Wikipedia photo (19 July 2010).

1945-1979 Rise of Naval Aviators Chiefs of Naval Operations

Various biographical sources.

USS *Basilone*

Wikipedia photo (19 July 2010).

End Notes

[1] Department of Military Art and Engineering, U. S. Military Academy, *Terrain Study Maps* (West Point, NYE: n.d.), Map 42.

[2] "The 1956 War," *The Columbia Electronic Encyclopedia*, 6th Edition, c. 2007, hotdog://wraw. Inoculates. com/ ce6/history/A0856667.html (16 July 2010).

[3] "The 1967 War (The Six-Day War)," *The Columbia Electronic Encyclopedia*, 6th Edition, c. 2007, http://www.infoplease. com/ ce6/history/A0935692.html (16 July 2010).

[4] "Mediterranean *Eskadra*," *WMD Around the World*, http://www.fas.org/nuke/guide/russia/ agency/mf-med.htm (17 Jul 2010).

[5] *The 1967 June War: Soviet Naval Diplomacy and the Sixth Fleet.* Corporate Author: CENTER FOR NAVAL ANALYSES ARLINGTON VA. Personal Author(s): Wells, Anthony R. Report Date : OCT 1977. Accession Number, ADA047236, Defense Technical Information Center http://oai.dtic.mil/oai/oai?verb=getRecord &metadataPrefix=html&identifier=ADA047236 (22 June 2010).

[6] Courtois, et. al., *Black Book of Communism*, 68-69.

[7] *Ibid,* 695-700.

[8] *Ibid*, 701-04. In the 1980s, a total of 8,850 Soviet advisors, 53,900 Cubans, and a large number of East Germans operated in Africa (685).

[9] AMS2 Cal Crego, "A Grand Old Lady," *USS Franklin D. Roosevelt Cruise Book, 1973-1974 MED Cruise*, 3.

[10] *Ibid*, for all ship details.

[11] *Ibid.*

[12] *Roosevelt, Cruise Book, 1973-1974 MED Cruise*, 215-300 for details of CVW-6 and its squadrons.

[13] Denis-James Joseph Downey retired as a Vice Admiral in October 1976. He graduated from the U. S. Naval Academy in 1946. *1993 USNA Register of Alumni*, #18016, 281.

[14] Dixon retired as a Rear Admiral. Probably John Curtis Dixon, Jr., Class of 1949, U. S. Naval Academy, who retired in October 1981. Dixon had earned a Legion of Merit, a Distinguished Flying Cross, and a Navy Commendation Medal. *1993 USNA Register of Alumni*, #20412, 296.

[15] Smith retired as a Captain, USN.

[16] During this critical period, the Military Airlift Command delivered 22,395 tons of military cargo - including tanks, howitzers, and helicopters - to Israel. Walter J. Boyne, "Nickel Grass," *AIR FORCE* Magazine, December 1998, 54-59.

[17] Replogle retired a Rear Admiral. After he retired from the Navy, he joined Browning-Ferris Industries, a waste management company, and managed a large collection company located in Springfield, VA. Eventually, he transferred to Dallas to manage a large collection company there. As the District Manager in Dallas, he worked for Jim Mattly, the Regional Vice President of the Southwest Region, headquartered in Houston. Mattly had been a lieutenant, junior grade on the *Coral Sea* when Replogle had been a lieutenant commander. Personal knowledge. JHB.

[18] John Glennon Finneran retired as a Vice Admiral in September 1977. He graduated from the U. S. Naval Academy in 1947. *1993 USNA Register of Alumni*, #19045, 287.

[19] Morin, *Email to Bradley*, 18 July 2010.

[20] Coleman retired as a Rear Admiral.
[21] A capital ship is one that displaces 10,000 tons of water or mounts eight-inch or larger guns.
[22] Dunn retired as a Vice Admiral. Probably Robert Francis Dunn who retired as a Vice Admiral in June 1989 after serving as the Deputy Chief of Naval Operations-Air. Dunn graduated from the U. S. Naval Academy in 1951. He earned a Distinguished Service Medal, two Silver Stars, four Legions of Merit, and three Distinguished Flying Crosses. *1993 USNA Register of Alumni*, #22177, 308.
[23] Easterling retired as a Vice Admiral.
[24] Bernard Brown Forbes, Jr. retired as a Vice Admiral in March 1977. He graduated from the U. S. Naval Academy in 1945. Forbes earned a Distinguished Service Medal, two Legions of Merit, a Bronze Star, Naval Unit Commendation Ribbon, a Meritorious Service Medal, a Navy Achievement Medal, and a Combat Action Ribbon. *1993 USNA Register of Alumni*, #316831, 273.
[25] Named for Medal of Honor winner, Sergeant John Basilone, U. S. Marine Corps. His citation reads:

> *For extraordinary heroism and conspicuous gallantry in action against enemy Japanese forces, above and beyond the call of duty, while serving with the 1st Battalion, 7th Marines, 1st Marine Division in the Lunga Area, Guadalcanal, Solomon Islands, on 24 and 25 October 1942. While the enemy was hammering at the Marines' defensive positions, Sgt. Basilone, in charge of 2 sections of heavy machineguns, fought valiantly to check the savage and determined assault. In a fierce frontal attack with the Japanese blasting his guns with grenades and mortar fire, one of Sgt. Basilone's sections, with its guncrews, was put out of action, leaving only 2 men able to carry on. Moving an extra gun into position, he placed it in action, then, under continual fire, repaired another and personally manned it, gallantly holding his line until replacements arrived. A little later, with ammunition critically low and the supply lines cut off, Sgt. Basilone, at great risk of his life and in the face of continued enemy attack, battled his way through hostile lines with urgently needed shells for his gunners, thereby contributing in large measure to the virtual annihilation of a Japanese regiment.*

In 1945, Gunnery Sergeant Basilone was KIA on Iwo Jima and the Navy awarded him a Navy Cross. Basilone was the only enlisted marine to win a Medal of Honor, a Navy Cross, and Purple Heart.
[26] The Captain of the *Basilone* was either CDR Noel Harper Petrie, Jr. (Nov 1972 – 1974) or CDR Charles A. McClellan (1974 - Mar 10 1976). http://www.navsource.org/ archives/ 05/ 824. htm (19 July 2010).

Thirteen

IN THE CLOUDS

Office of the Joint Chiefs of Staff

Cold War and Washington, D. C. 1975-1979[1]

Important Times		**James Morin**
NVN Politburo approved plan for the assault on Saigon in 1976	**Jan 1975**	
NVN overwhelmed SVN. President Ford refused to assist as per the Paris Peace Agreement.	Feb-Apr 1975	**CAPT Morin selected for promotion to Rear Admiral.**
1st Cuban and Soviet advisors landed in Angola.	Mar 1975	
Republic of Vietnam (SVN) ceased to exist.	**30 Apr 1975**	Morin appointed Deputy Director of Operations at the National Military Command Center, JCS in the Pentagon.
Cambodia fell to the Communist Khmer Rouge.	Apr 1975	
Khmer Rouge seized the Mayaguez; U. S. Marine force recovered the ship.	May 1975	
Communist Pathet Lao took over Laos.	Dec 1975	
	c. Apr 1976	Morin appointed Director, Studies and Gaming Agency of the Joint Staff.
North Korean soldiers murdered two U. S. officers and wounded additional U. S. and ROK soldiers in a tree-trimming incident at Panmunjom in the Demilitarized Zone.	18 Aug 1976	
Mao Tse-dung died.	**Sep 1976**	
New leaders in Mozambique turned to the U.S.S.R. and China for military aid. Proclaimed "allegiance to the Bolshevik idea."	1977	
Jimmy Carter inaugurated as President.	Jan 1977	
President Carter relieved MG John Singlaub for publicly criticizing Carter's plan to withdraw U. S. forces from Korea.	21 Mar 1977	
New government in Ethiopia broke off military relations with the U. S. after deposing the emperor.	14 Apr 1977	Morin appointed JCS representative at the Law of Sea Negotiations.
North Koreans shot down U. S. CH47 helicopter. Killed 3 and wounded 2 Americans.	14 Jul 1977	
President Carter and Soviet Premier Leonid Brezhnev began negotiations for new arms limitation treaty - SALT II.	Dec 1977	
President Carter brokered Camp David Accords between Egypt and Israel. Accords led to a peace treaty.	Sep 1978	
U. S. formally recognized the Reople's Republic of China (Red China).	1 Jan 1979	
Israeli-Egyptian Peace Treaty signed.	Mar 1979	Morin went on terminal leave.
	Apr 1979	**RADM Morin retired after completing 33 years of service.**

Cold War 1975-1979

DÉTENTE, HOWEVER, HAD NOT WORKED AS SOME AMERICANS HAD HOPED.

The Russians now refer to the long rule of Leonid Brezhnev (1964-1982) as the "period of stagnation."

In the early 1970s, the Soviet stance toward the United States became less overtly hostile and negotiations between the United States and the Soviet Union resulted in summit meetings and the signing of strategic arms limitation agreements. Brezhnev proclaimed in 1973 that "peaceful coexistence" was the normal, permanent, and irreversible state of relations between imperialist and Communist countries, although he warned that conflict might continue in the Third World. In the late 1970s, growing internal repression and the Soviet invasion of Afghanistan led to a renewal of Cold War hostility"[2]

Détente, however, had not worked as some Americans had hoped. On the face, the Cold War, seemed to have warmed a bit, but by 1975, the Soviet Union had built up its armed forces and narrowed the arms gap with the United States. The Soviets had:

- "Fifty percent (50%) more ICBMs;"
- "Three times as many army personnel;"
- "Three times as many attack submarines;" and
- "Four times as many tanks."

The United States maintained a strategic deterrent of 9,000 "deliverable nuclear bombs and warheads" compared to the 3,200 the Soviets had stockpiled.[3] While the intent might have been "peaceful coexistence," in reality the Soviet capabilities for worldwide war had grown dramatically and increased the Soviet threat to the United States and the West.

When the U. S. S. R. invaded Afghanistan, the United States moved to provide arms and support for the Afghans.

Ψ

Ashore Again

I took a month's leave. It was good to be home. I was pooped!

I did a lot of nothing while on leave. Then, I reported in at NAVAIRLANT to replace Jack Youngblade as Chief of Staff. I found that I had little to do until Jack left. I attended meetings, got briefed by each staff division, and handled several special projects.

The selection ended my duty in Norfolk because the Chief of Staff's position was a Captain's billet. It also ended my service with the U. S. Navy. The Navy immediately ordered me to join the Joint Staff in Washington, DC.

Ψ

One evening in early February 1975, Jack Youngblade called and told me that both of our names were on the Rear Admiral Selection List.

Over the years, when I found out that the Navy had selected me for promotion, I would be elated and simultaneously relieved. This time, I reacted differently. The selection overwhelmed me. Suddenly, all the hard work, family separations, sea duty, moves, the sweat, and frustrations proved worthwhile. Heady times followed: in the next few weeks, old friends and shipmates called and wrote letters of congratulations.

The selection ended my duty in Norfolk because the Chief of Staff's position was a Captain's billet. It also ended my service with the U. S. Navy. The Navy immediately ordered me to join the Joint Staff in Washington, D. C. Going to Washington after a command tour - as I did in 1966 - was anti-climactic. I considered that my position as Commanding Officer of the *FDR* was more prestigious than being another desk-bound Flag Officer in Washington.

Earlier, VADM Michaelis had told me that the selection board did not pick me up for promotion to rear admiral because I did not have any joint staff experience. During my years of service, I had darn little time to serve on a joint staff while I flew and commanded squadrons and ships. But, by this time, having such experience became a necessity for selection to Flag/General Officer rank. The Air Force and the Army had ramrodded this requirement because their career patterns were totally different than the Navy's. Their officers served on more staffs than Navy officers, shared a common staff heritage, and worked more closely with each other than they did with the Navy. Moreover, at least in the case of the Army, the successful officers attended branch schools as captains and the Command & General Staff College (C&GSC) or the Armed Forces Staff College (AFSC) as majors and lieutenant colonels. C&GSC trained the student officers to be battalion, brigade, and division staff officers in the Army, and the AFSC trained the officers to be joint staff officers. The Air Force, Army, Marine, and Naval officers who attended the Air Force, Army, Navy, or National War College gained added joint staff education. They also studied national and international military and political issues. Because of my extensive sea duty, I did not attend the AFSC or one of the war colleges.

Ψ

In Washington, I assumed the position as Deputy Director of Operations at the National Military Command Center, Joint Chiefs of Staff. In the command center, an admiral or general served as the World Duty Officer, twenty-four hours a day, seven days a week. I had a fascinating job because when on duty I knew what was going on everywhere in the world. During any crisis - for example, the evacuation of Saigon, *Mayaguez*, Lebanon, the rescue attempt in Iran - the Secretary of Defense and the Chairman of the Joint Chiefs of Staff came to the Command Center and talked, as needed, to the on-scene commander. The miracle of worldwide secure communications helped the senior leaders handle critical issues and allowed civilians such as the Secretary of Defense and even the President to get involved in the details of an operation and make the minutest decisions. This "blessing" of instant communications often did not help the on-scene commander and in a time-critical situation could be costly.

The best way to handle crisis is simple: give the on-scene commander his mission; then let him figure out the "how." If the senior civilian or military leaders believe that the commander cannot handle the mission, they should not have given

1976
National Leaders

President
Gerald Ford

Secretary of Defense
Donald Rumsfeld

Joint Chiefs of Staff

Chairman
GEN George S. Brown

Chief of Staff, Air Force
GEN David C. Jones

Chief of Staff, Army
GEN Frederick C. Weyand

Chief of Naval Operations
ADM James L. Holloway, III

him the job in the first place. In my opinion, we have started growing an officer corps (particularly in the Air Force which has enjoyed for years a remarkable communications tie-in all over the world) which will not take action without checking with the next guy up the chain of command. The aborted rescue of the hostages in Tehran confirmed my opinion. President Jimmy Carter and other Washington officials managed that debacle minute-by-minute.

I HEADED A STAFF OF BRAINY INDIVIDUALS WHO CONDUCTED THE STUDIES AND OPERATION ANALYSES OF VARIOUS MILITARY AND POLITICAL PROBLEMS. WE CONDUCTED A NUCLEAR EXCHANGE WAR GAME ANNUALLY.

Ψ

After a year in the Command Center, I moved to a new job for a year: Director, Studies and Gaming Agency of the Joint Staff. The incumbent had retired early, and I filled in to complete the Navy's turn in the position. I headed a staff of brainy individuals who conducted the studies and operation analyses of various military and political problems. We conducted a nuclear exchange war game annually. High ranking civilians from industry and government joined to war game a political-military crisis such as the North Korea invading South Korea again or a NATO-Warsaw Pact encounter. I found it surprising that the "Doves" became quite sanguine when they had to make tough decisions in a crisis situation.

I WAS NOT HAPPY IN THIS JOB BECAUSE I HAD A BASIC SUSPICION OF OPERATION ANALYSTS.

I was not happy in this job because I had a basic suspicion of Operation Analysts. In my view, they can prove anything someone wants proved. The analysts dealt in the world of probabilities knowing that no one will ever hold them accountable for their decisions. The government had spent – read "wasted" – lots of money on myriad studies over many years. This JCS agency, however, produced studies cheaply compared to those the government contracted out to private companies in the business. During my time in the job, the JCS routinely tasked us to conduct a study for someone in the Department of Defense or other department. Project officer after project officer met with individual requesters and determined the assumptions and other parameters for each study and then charged off and produced a paper. I always felt that our finished products served no purpose. An official or several officials might have read our papers and then filed them away.

I became the JCS Representative at the Law of the Sea Negotiations where I served as the Military Advisor to Ambassador Eliot Richardson.

Ψ

After a year, I became the JCS Representative at the Law of the Sea Negotiations where I served as the Military Advisor to Ambassador Eliot Richardson. Richardson impressed me and my job interested me. I concluded that the U. S. State Department types lived a good lifestyle.

During this assignment, we worked at the United Nations in New York for two months and then worked in Washington for two months. Next, we worked at the United Nations in Geneva for two months and then returned to Washington for two months.

The negotiations aimed at creating a treaty that would encompass all the rules that had evolved over the years in international law and common usage relating to:

- High seas rights;
- Overflight rights;
- Straits passage;
- Territorial waters; and
- Freedom of navigation.

In addition, the conferees worked to develop new agreements concerning:

- Fishing rights;
- Mineral rights beneath the sea off a nation's coast;
- Deep sea bed mining;
- Right of innocent passages; and
- Other related issues.

Lawyers from the State Department and other U. S. agencies staffed most of the large American delegation. Other delegates came from a variety of interest groups, including, mining, oil, fisheries, and environmental.

Land-locked nations such as Afghanistan and Ecuador, with no coasts, wanted to be part of the negotiations on articles dealing with fishing and deep sea mining!

The negotiations had started in 1973, and the delegates had agreed to many articles. In 1977, about 150 countries made up the United Nations and each country had one vote. The small nations ganged up and used their votes to frustrate and delay progress. Land-locked nations such as Afghanistan and Ecuador, with no coasts, wanted to be part of the negotiations on articles dealing with fishing and deep sea mining! Third World nations that did not have the trained people or the capital to conduct deep sea bed mining wanted, nonetheless, to share the profits derived from it. The negotiators also faced language problems when they tried to get several nations to agree on the wording of an article because in the translation from one language to another, the intent and nuances had to be stated correctly to satisfy everyone. From the perspective of this pragmatic sailor, the negotiations progressed at a snail's pace and became a lawyer's paradise.

Ambassador Richardson worked hard to hammer out the treaty, but I had to keep an eye on him to protect our Defense interests.

Ambassador Richardson worked hard to hammer out the treaty, but I had to keep an eye on him to protect our Defense interests. For example, in an attempt to reach a compromise on some issue, he might offer to give up or modify U. S. rights in "Straits Passage." Nations bordering the straits might have said, "We respect your rights to use the straits but we want advanced notice if a warship is going to go through." I would have to tell the Ambassador, "Like hell" to prevent such an agreement. During the negotiations, Peru tried to claim territorial seas out to 200 miles and stated further that no nuclear-powered vessel could sail in their territorial waters. At the time, the United States remained one of the few countries that recognized the three-mile territorial water limit. Almost every other country recognized twelve miles, so we gave way on that issue. Throughout the negotiations, we had to put up with asinine recommendations from other countries, but for the first time, we saw eye-to-eye with the Soviets on nearly everything. Now that they had a blue-water Navy, they did not want some little country dictating the rules that would affect where and when they would sail.I had counterparts – the Military Representatives of the other nations. From time to time, we lunched together. I found several of the foreign officers to be interesting people, especially Major General Peter Barabolya of the Soviet Union. I got along very well with him. The first day our group spent in Moscow, he introduced me to Stolichnaya Vodka at lunch. Our waitress served us the vodka in small fruit glasses. I sipped mine, but he told me, "No, no, Admiral Morin, we drink!" Then, he knocked his drink back. Not to be outdone, I did too. The waitress immediately filled our glasses, and the general proposed a toast. I knocked that one back too, but he sipped his. I figured he wanted to put me away early, so I did not knock any more vodkas down at lunch. I always found Barabolya fun to talk to, but I always talked with him through his interpreter, whom I figured to be KGB.

I always found Barabolya fun to talk to, but I always talked with him through his interpreter, whom I figured to be KGB.

In Geneva, the Japanese sought our help to declare the water between their Central and Northern Islands as territorial seas instead of a strait. At points, the strait is only three miles wide. Under existing law, the center of any strait is considered high seas. The Russians would sail or fly reconnaissance missions right down the center without permission and that drove the Japanese bonkers.

In Geneva, the Japanese sought our help to declare the water between their Central and Northern Islands as territorial seas instead of a strait.

The Japanese Ambassador invited Ambassador Richardson and a few of us to dinner one evening. I sat on the left of the Japanese Ambassador and one of his staff sat to my left. During the dinner, the two Japanese jabbered away with each other. At a break, I blurted out a few comments in the little Japanese I remembered from my two years of language study at Berkeley. Stunned, the Japanese Ambassador said, "Oh, Admiral Morin speaks Japanese."

The two did not speak to each other for the rest of the evening at dinner. I most likely spoiled their meal because they probably assumed that I understood all that they talked about, and if they had discussed sensitive issues, my apparent knowledge of their issues and comments must have perplexed them. I did not have the foggiest of what they had said, but I probably

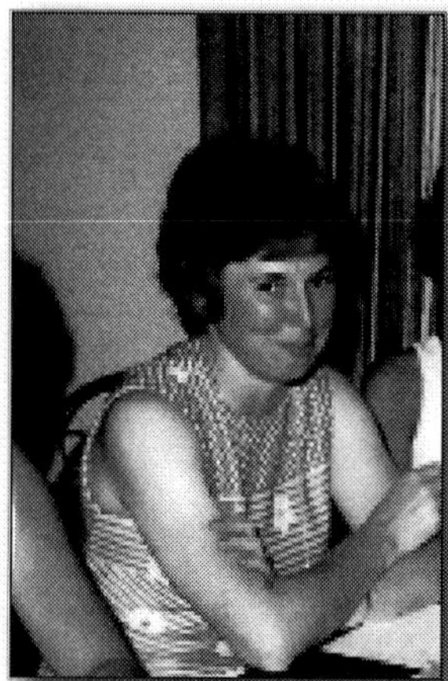

Betty Morin

scored a coup for the United States. We did not support the Japanese proposal concerning their strait. We did not want to set a precedent and close off other narrow straits such as Hormuz and Malacca which we needed to use.

Ψ

In the summer of 1978, my wife had another serious operation. As a result, I decided to retire the following year with four years in grade. I did not need any more sea duty, and she did not deserve any more. VADM Bob Baldwin[4] the Chief of Naval Personnel, surprised by my action, called me over to his office to discuss my decision. He wanted me to stay on, but he understood the circumstances. I have often said that a successful career in the Navy required one's own good health as well as the good health of your family members.

In the late summer of 1978, Congress again pinged the military to reduce the number of Flag/General Officer billets. The Secretary of Defense decided to give up the Flag billet assigned to the Law of the Sea negotiations and fill it by using a retired admiral. Shortly after the decision, VADM Shannon Cramer,[5] a submariner of some repute, relieved me, and the Navy assigned me as the Chief of Staff of the Defense Intelligence Agency (DIA).

On my last trip to the United Nations in September-October 1978, I flew into LaGuardia and saw the Butler Aviation hangar. That reminded me that Jim Mattly, a Silver Fox of some repute, worked for Butler. I called him and Jim invited me to dinner at his home in New Jersey. That evening, I mentioned that I intended to retire in 1979 and he asked me what I was going to do. I told him "medicine, dentistry, law, I can't do. Anything else, I'll try." He asked, "What about Butler?" He told me about the company and arranged for me to visit the operation in LaGuardia. I said that I was interested and he arranged for me to meet the Company President.

In the meantime, I checked into the Defense Intelligence Agency as one of the few people there without an intelligence background. In early 1979, Jim Mattly called to tell me that Butler was holding open a position for me a Baltimore-Washington International Airport. He asked how soon I could leave the service. I checked with my boss at DIA who told me that I could go as soon as the Navy ordered in a replacement. Navy said a replacement would be in by the end of February.

On 1 March 1979, I began my terminal leave and on 30 April 1979, I officially retired.[6] I opted not to have a retirement ceremony because not having worked for the Navy in four years, I felt that a ceremony would be rather meaningless. Four years in the Joint Staff in four different jobs did not give me a feeling of belonging. I cannot say that I enjoyed those four years. I saw a tremendous waste of effort in the Pentagon. I used to say facetiously that if 100 represented the Pentagon's output of work at the time and if you cut out half of the civilians in the Department of Defense and half of the military officers that output would drop to ninety-five.

Legion of Merit

**Awarded to
Rear Admiral James B. Morin**

for

"Exceptionally Meritorious Conduct in the Performance of Outstanding Services from July 1976 to July 1978."

Ψ

I look back on my thirty-three years of service with a great deal of satisfaction. The Navy trained me, educated me, and gave me all the challenge, competition, and responsibility I could have asked for. With six commands, I had more than my fair share. I flew operationally right up to the time I took command of the *LaSalle*. I served with great squadron mates and shipmates. The people made everything worthwhile. I worked with some dunderheads who should not have been naval officers or naval aviators, but overall I worked with a group of men whom I admired as a special breed of American. God bless them.

Our moves to new places always proved to be exciting. The children benefitted by them. To them, each move was an adventure, and they handled the disruptions as well as I could have expected. The long cruises – especially the wartime cruises – challenged my wife and the children. Betty and other Navy wives bore the burdens well while their men sailed off to do their duty

AS I LEFT THE NAVY, I HAD ONE MORE TASK TO COMPLETE.

in peace and war. These unique women took on the full responsibility of raising the children, running the house, managing the household budget, getting the children to school, maintaining the car, etc., and they got darn little credit for their contributions to their families and the Navy. The vast majority of Americans never knew about their sacrifices and most likely would never have cared anyway.

Ψ

I owe a great deal to my wife, my squadron mates and shipmates, and my sailors because they helped me day-in and day-out to perform my various jobs, kept my spirits up, and contributed to my successes that led to Flag rank.

As I left the Navy, I had one more task to complete.

End Notes

[1] Gutzman, *Vietnam,* 16-17. Courtois, *Black Book of Communism,* African references. "Korea Demilitarized Zone Incidents, 1967-2001," http:// www. global security .org /military/dmz-lits.htm 2 of 5. (16 July 2010).
[2] http://www.biblio.org/expo/soviet.exhibit/coldwar.html, paraphrased and quoted.
[3] Martin, *America and Its Peoples,* II, 881.
[4] Robert Bemus Baldwin retired as a Vice Admiral in August 1980. He graduated from the U. S. Naval Academy in 1945. Baldwin earned a Distinguished Service Medal, Legion of Merit, Bronze Star, and Purple Heart. *1993 USNA Register of Alumni,* #16675, 273.
[5] Shannon Davenport Cramer, Jr., retired as a Vice Admiral in August 1977. He graduated from the U. S. Naval Academy in 1944. He earned a Distinguished Service Medal. *1993 USNA Register of Alumni,* #16137, 268.
[6] Ironically, Jim Mattly left Butler Aviation and joined Browning-Ferris Industries in Houston in April 1979.

Sidebar Notes

1976 National Leaders
Various sources.

Legion of Merit
Official Morin citation.

Fourteen

THE CEREMONY

Silver Foxes, 2000

Standing:
Kish (DFC), Hyde (Gold Star), Floreani (DFC), Hiebert (DFC), Holthaus (DFC), Chasko (Gold Star), Rivers (Gold Star),

Kneeling:
Blosser (DFC), Morin, Mattly (DFC), Starling (Gold Star)

Chief of Naval Operations

Blosser, George
Chasko, Gerald
Floreani, Richard
Hiebert, Edwin
Holthaus, Kenneth
Hyde, Albert
Kish, William
Mares, Ernest
Mattly, James
Rivers, Wendall
Jay Starling

The President of the United States
takes pleasure in
presenting the
DISTINGUISHED FLYING CROSS
to

____

FOR SERVICE AS SET FORTH IN THE FOLLOWING

CITATION

For extraordinary achievement while participating in aerial flight as a pilot with Attack Squadron 155 aboard the USS CORAL SEA (CV 43) in combat operations in the Tonkin Gulf from February to November 1965. In the successful completion of over 100 combat missions __________________________ contributed materially to the success of United States effort. He was involved in numerous major air strikes against bridges, power stations, military facilities, and fuel depots in the Red River Delta, an area of the most intense air defenses, including surface to air missiles. He also flew day and night road reconnaissance missions to slow the flow of supplies from North Vietnam to South Vietnam, attacking choke points at bridges and rivers where the enemy was quick to strengthen its defenses. He unflinchingly pressed home his attacks in the face of a determined and resolute foe. The strain of daily combat operations for periods of four to five weeks over an extended deployment never kept him from performing his duty as a naval officer and naval aviator. ________________________'s personal bravery, superb airmanship, and loyal devotion to duty reflected great credit upon him and were in keeping with the highest traditions of the United States Naval Service.

For the President,

/s/

J. L. Johnson
Admiral, United States Navy
Chief of Naval Operations

One More Task

THE GRAYING SILVER FOXES ARRIVED AT PENSACOLA NAS IN MAY 2000 AND DISCOVERED THAT THEY HAD BEEN SCHEDULED TO ATTEND A LUNCH AT THE NAVAL AVIATION MUSEUM AND FOLLOWING THAT TO ATTEND A CEREMONY IN THE ATRIUM OF THE MUSEUM TO WATCH NEW NAVY AVIATORS RECEIVE THEIR GOLD WINGS.

The graying Silver Foxes arrived at Pensacola NAS in May 2000 and discovered that they had been scheduled to attend a lunch at the Naval Aviation Museum and following that to attend a ceremony in the atrium of the Museum to watch new Navy aviators receive their gold wings. They did not know why.

Δ

Jim Morin had retired and joined Butler Aviation, Jim Mattly's former company, in 1979. Soon after he left the Navy, Morin lost his beloved Betty - described as the "cutest" girl in his high school class and "as merry as the day is long" - who had been with him from his first days in the Navy.

Morin worked at Butler until he finally retired in 1991 and moved to Tallahassee, FL. While in his new home, he met an old friend, Rita Johnson, the widow of Jack K. Johnson with whom he had gone through flight school. Jim and Rita had first met in the early 1950s when the Morins and Johnsons had lived in Pensacola and when Morin and Johnson served in the same Carrier Air Group during their first tour of sea duty. Jim and Rita have been partners ever since, and Rita has become a valued and popular addition to the Silver Fox family.

In the years after he left active duty, Morin believed that he had failed as a leader to gain just recognition for the Silver Foxes who had flown and fought under his command in 1965. He stated in 2000:

> *For 35 years I have felt I hadn't done my job as CO because my guys hadn't gotten the recognition they richly deserved.*[1]

Morin got Distinguished Flying Crosses awarded to Jerry Chasko, Albie Hyde, Jay Starling, and Wendy Rivers while in Southeast Asia, but not for his other officers. And, he also got the Navy to award a Silver Star to his second Marine pilot, CPT Ed Huff, during his short second tour when he flew from the USS *Constellation.*

CDR MORIN PRESENTED AN AIR MEDAL TO LT HIEBERT.

He remembered that, "Most of my junior officers completed their obligated service in 1966 and were discharged from the Navy," and that "After 8 months of combat operations, each of the Silver Foxes had surpassed 100 missions over the North. Each had been awarded anywhere from eighteen to twenty-two Air Medals (including Strike/Flight awards.) Moreover, after a

Morin had been unable to get any higher combat awards for his pilots during their tours even though he had submitted recommendations for Distinguished Flying Crosses for his pilots in 1965.

cruise of 11 months, we were finally going home and medals weren't as important as that." Morin had been unable to get any higher combat awards for his pilots during their tours even though he had submitted recommendations for Distinguished Flying Crosses for his pilots in 1965. The Navy rejected the recommendations for DFCs; therefore, most of his veteran pilots returned home having earned only multiple Air Medals, sometimes perfunctory awards, for their combat service.[2]

Morin, however, managed to get non-combat decorations awarded to many of his valuable and productive support personnel. He got the Navy Commendation Medal for his Ordnance/Avionics Officer and his Assistant Maintenance Officer and the Navy Achievement Medal for several of his Maintenance Chiefs (CPOs) and Petty Officers and for a "bunch of ... ordnance Petty Officers."[3]

The Silver Fox Skipper continued to fret about his failure during the intervening years and especially at the reunions that the Silver Foxes organized around the country where they relived the hard days of the long cruises," griped about the "poorly managed air war," laughed at "the funny stories," and honored their "many fallen squadron mates." Because of the tough missions and the closeness of the Silver Foxes to Morin and each other, a special comradeship developed between most of the former officers that continues to this day. But that comradeship increased Morin's concern that he had not gained the proper recognition for his pilots that he deeply believed they richly deserved.

After one reunion, the Silver Foxes considered holding their next reunion at Pensacola NAS in 2000. Pensacola was a special place for the Silver Foxes because all had trained there. Morin had served as an instructor there, and he lived nearby in Tallahassee.

Δ

Before the reunion, Jim Morin discussed the upcoming event with Diane Mattly who had helped organize earlier get-togethers. Then, according to several people, he called and/or wrote the Silver Foxes and advised them to be there. Some believed it to be a command performance because their Skipper had never done anything like this for a previous reunion.[4] Bill Kish reported that Morin said:

> *I want you guys to bring a sport coat and don't spill anything on it at lunch because we will be guests at a winging ceremony in the early afternoon.*

He added: "I am not sure if any of us were all that wild about attending the ceremony but after all, this was the skipper's deal and one could wear a bib if needed."[5]

Δ

Months prior to the Pensacola reunion, Morin began working to get DFCs for his twelve surviving pilots. He contacted Wes McDonald who had served as the CAG on the USS *Coral Sea* in 1965 and who had completed his distinguished career as a four-star admiral. Morin told ADM McDonald that he had carried around guilt for thirty-five years because he had failed as a leader to get proper recognition for his pilots, and then he asked the admiral for someone to contact in Washington so that he could submit the belated recommendations for awards. McDonald supported the proposal and advised Morin to call the Awards Board in the Office of the Chief of Naval Operations. ADM McDonald also said that he would have someone from the Awards Board contact Morin. A CDR Maher did call, and he told Morin what he had to do to push the paperwork forward.[6]

Morin collected the information needed from his pilots and prepared the recommendations for the award of a DFC to each surviving Silver Fox. He then submitted the paperwork with McDonald's endorsement to the Awards Board.[7]

Having McDonald on his side most likely assured a positive outcome. As the CAG, who had flown with the Silver Foxes and knew them personally, his endorsement carried extra weight. Moreover, McDonald had been a heavyweight during his last years of service: he was the last U. S. Navy admiral to hold three important naval commands simultaneously: Supreme Commander, Allied Atlantic Command, a NATO command; Commander, U. S. Atlantic Command; and Commander, U. S. Atlantic Fleet. He had also served as the Deputy Chief of Naval Personnel. In addition, after retiring McDonald remained active in Naval Aviation affairs. [8]

Also, while not directly involved in the process, the CNO, ADM Jay Johnson, was a naval aviator and had flown combat tours in Crusaders and also had worked in Naval Personnel. If the CNO had been a submariner or a destroyer commander, the results might have been different.

To his extreme pleasure, the Navy notified Morin in April 2000, a month before the scheduled reunion, that the Awards Board had approved the awards of the DFCs to his pilots.[9]

Once notified that the awards had been approved, Morin kept the information secret from everyone as he planned an awards ceremony. Toward that end, he contacted the Training Wing Commander at Pensacola and requested he be allocated ten minutes at the conclusion of a Winging Ceremony to present the decorations to his aviators.

During this time, apparently unknown to Morin, a handful of admirals from around the country called Diane Mattly inquiring about several of the Silver Foxes. She did not know why they

Attendees

George "Gorgeous" & Joan Blosser

Jerry & Jeanne Chasko

Dorothy Fidelibus

Richard "D'vinch" Floreani

Evan "Padre" Greco

Ed "Hoobie" & Marilyn Hiebert

Ken "King" & Joan Holthaus

Al "Albie" & Zeb Hyde

Jim "Crunch" & Diane Mattly

Bill "Razor" Kish

Jim "Skipper" Morin & Rita Johnson

Chris Morss

Les Parmenter VA-153

Wendy & Betty Rivers

Jay & Kay Starling

Jim & Mae Tillis

BLUE ANGEL A4S
NAVAL AIR MUSEUM

called, and they never told her the reason for their calls. Most likely, they were gathering or confirming information related to the evaluation and approval of Morin's recommendations.

"The Navy has come a long way since Beeville"[10]

Before lunch, the Silver Foxes gathered in the Museum. Wendy Rivers was viewing the Vietnam War Exhibit and telling war stories. Several Silver Foxes gathered around him, and soon other visitors drew close to Wendy to listen to his fascinating stories. Chris Morss noted, "Just outside the exhibit, after we had started to leave it, a woman came up to Wendy and said, "Thank you for serving." Wendy had to turn away to hide his tears. In 1973 so many prisoners had returned only to be reviled."[11] That began a happy and emotional day.

The officers gathered for lunch at the Cubi Club in the Museum. Filled with photographs of naval fliers and the insignia of the many squadrons, air wings, carriers, and naval units that served in Vietnam, the club was the natural setting for recounting past experiences and enjoying the company of comrades and the wives who had joined the group. There were lots of laughs and many smiles, but none of the Foxes devined what lay in store for them. Morin, of course, maintained his secrecy.

When the veterans and their wives gathered for the winging ceremony beneath the four "historic" A4s flown by the Navy's Blue Angel demonstration team that were suspended high above them, they found that they had reserved seats and that their Skipper had a place on the stage with an Admiral, a Rear Admiral, the Base Captain, and a Chaplain. Moreover, media representatives had set up cameras and video equipment to cover the ceremony. George Blosser thought that he should have figured out that something important was going to happen for the Silver Foxes.[12] Bill Kish remembered his "austere" winging ceremony at Beeville with only the Captain of the Base and a Navy photographer in attendance with three new pilots.[13] And Morin must have thought the ceremony better than his when he expected Hollywood starlets would present his wings.

The ceremony began on time, and the senior officers presented each of the twenty new pilots with their Wings of Gold for satisfactorily completing Navy Flight Training, bringing back memories to many about their graduations years before. After completing the presentation of the wings, the Master of Ceremonies introduced the Silver Foxes, saying:

> *Before adjourning we have one more item of business: We would like to recognize and acknowledge the pilots of VA-155 from the USS Coral Sea deployment in 1965 to Vietnam who are here enjoying a reunion.*

He then quickly added before anyone left the area, "I would now like their commanding officer from that period, retired RADM Morin, to come up and say a few words." Bill Kish thought to himself, "Who cares about what happened 35 years ago...," but commented: "Admiral Morin got up and the place took on a different atmosphere."[1]

AS THE SKIPPER CONTINUED WITH HIS COMMENTS, HIS VOICE BROKE AS HIS FEELINGS TOOK OVER AND HE BEGAN TO SPEAK ABOUT HOW HE BELIEVED THAT HE HAD FAILED US AS OUR LEADER BECAUSE WE HAD NEVER BEEN RECOGNIZED FOR WHAT WE DID...

Δ

Jim Morin recalled:

> *At the microphone, I reviewed the asinine war in which we had fought, the problems we had in getting awards approved, and said that an injustice had been corrected. I read a generic citation and I must admit that emotion overcame me so that I had to stop and catch my breath.*
>
> *I continued, "The President of the United States takes pleasure in awarding the Distinguished Flying Cross to...."*

One pilot recalled more detail:

> *[ADM Morin] began with a few stock phrases about young aviators, what Naval Aviation means, etc. He launched into a series of comments about how the Vietnam War had been handled by not only the civilian leadership of the country, but by news media, and the senior leadership of the Navy. I thought this is pretty powerful stuff and where is he going with this? As he got to his concluding comments, he began to speak personally about what our squadron had done together and how proud he was to be associated with us. ...*
>
> *As the Skipper continued with his comments, his voice broke as his feelings took over and he began to speak about how he believed that he had failed us as our leader because we had never been recognized for what we did...*
>
> *As he finished his comments, he said he wanted to have everyone acknowledge our contributions and that the Navy would do it officially by awarding us each a Distinguished Flying Cross.*

Another pilot recalled:

> *After the last student had received his wings, our squadron was introduced as a "reunion group" from 1965 and ADM Morin was given an opportunity to speak. This did not surprise any of us, as such courtesies are normal. The Skipper made a few remarks and then concluded with "Ladies and Gentlemen, I would*

My guys said that receiving the DFC after 35 years was more meaningful to each of them than had they received it in 1965-66.

like to tell you that the Navy will usually right a wrong sooner or later...this time it has taken 35 years. I would like to read the following citation." The Skipper then read a citation for "duty above and beyond" during 1965 while onboard the Coral Sea.

I remember thinking "That sounds like some of the flying we participated in during our long cruise." The gravity of what was about to happen did not hit me.[15]

ADM Charles S. (Steve) Abbot rose and moved in front of the podium to present the decorations, and Jim Morin, after reading the official citation, then said,

"Will the following officer please come forward to receive this award ... LTJG George Blosser."

ADM Abbot presented 2d award of DFC to Jerry Chasko

Morin then called the others forward in alphabetical order - Jerry Chasko, Dick Floreani, Ed Hiebert, Ken Holthaus, Albie Hyde, Bill Kish, Jim Mattly, Wendy Rivers (who received a standing ovation), and Jay Starling - to receive their DFCs, relieving George Blosser's fear that his squadron mates would question why Morin had called him forward first.[16] Chasko, Hyde, Starling, and Rivers received Gold Stars in lieu of second awards.

Morin concluded: "It was a very moving ceremony. My guys said that receiving the DFC after 35 years was more meaningful to each of them than had they received it in 1965-66."

As mentioned, Jim Hughes and Ernie Mares did not attend the reunion; Morin mailed their decorations to them.

ADM Abbot presented 2d award of DFC to Wendy Rivers

Real Awards

Tragic events highlighted the importance of the legitimate combat awards given to the Silver Foxes.

ADM Johnson's predecessor as CNO committed suicide apparently because he had been accused of wearing an unauthorized "V" device indicating an award for valor on his Navy Commendation Medal and his Navy Achievement Medal.

In 1998, B. G. Burkett and Glenna Whitley published *Stolen Valor*, and related myriad stories about fake Vietnam veterans who claimed high decorations for their wartime service and about Vietnam vets who claimed decorations they did not earn.

Typical was the story of Bill Kane, a Princeton honor graduate, a graduate of Yale Law School, and a Fulbright Scholar. Kane claimed that he "had been skirmishing in jet fighters above the jungles of Southeast Asia and had been awarded a Medal of Honor for his exploits." In 1991, Kane committed suicide. Three years later, the New York *Times* published a story about his

Vietnam service and his other incredible experiences. Kane turned out to be a draft dodger .

In 2002, Senator John Kerry triggered hostile political responses for his Vietnam service where he earned three dubious Purple Hearts and gained an early return from the combat zone because of his trivial wounds.

In 2010, a young man in Houston gained media and police attention when he attended a gala event for the new Mayor wearing an Army Mess Dress uniform of a general officer and a mind boggling array of medals including two Army DSCs and two Silver Stars. He was not a veteran.[17]

RITA JOHNSON AND JIM MORIN
PENSACOLA, 2000

"A most memorable day for us all"[18]

The surprised pilots soon faced the press, posed for many pictures, and talked about the ceremony. Several newly-minted naval aviators and their instructors also came over to "shake the hands of some real heroes."[19]

George Blosser commented that he was surprised that the Navy would award the DFCs, but he was not surprised that his Skipper had got them the awards because it was typical of the man "who does whatever it takes to complete a mission."[20]

Dick Floreani told the local newspaper, "I was absolutely surprised and baffled by it," adding that the "squadron members ...will forever share a bond forged in war."[21]

Wendy Rivers added, "I knew the skipper was trying to get medals for the younger guys and I told him not to do anything for me because I've got plenty."[22]

That evening, a Chief Petty Officer whose son had received his wings during the ceremony made two notable comments about Morin: "... please tell your Skipper that General Patton [had] been known to get all choked up when speaking about his troops;" and "tell him that we are very proud for what he has done for our Nation, the Navy, and his squadron."[23]

Δ

The next day, a lady approached one of the Silver Foxes, congratulated him on receiving a DFC, and asked for a favor. Her husband had passed away in Vietnam and her son had just gotten his wings, and she asked, "Would you mind speaking with him since I don't think he really realizes how special those Wings are, and you obviously do." The old A4 pilot and his wife spent about an hour with the new aviator and his fiancée.[24]

In July, Executive Editor Mike O'Callaghan, a former two-time Governor of Nevada, wrote a special article in the Las Vegas *Sun* and entitled it: "Morin: A Man of Honor." He reported:

> *Jim Morin developed a bond with the men he led into combat 35 years ago that has never weakened or been doubted by his fellow pilots,*

and he ended it with,

> *It took 35 years, but Jim Morin, a strong combat leader with undying loyalty to his men, finally brought a bit of justice for those who flew into danger with him.*[25]

Sometime after the ceremony, a young reporter from *Delta* Magazine who was preparing a story about the event called ADM Morin to get information for her story. After doing so - she had been reluctant to call "an admiral" - she declared; "what a great man, it was like speaking to my dad."[26]

Δ

Years later, most likely mirroring the thoughts of most of the pilots of VA-155, a reflective Silver Fox wrote:

> ***DOWNWARD LOYALTY IS ALMOST UNHEARD OF IN OUR SOCIETY AND SOMETHING LIKE THIS [CEREMONY] COULD ONLY OCCUR IN AN ORGANIZATION LIKE THE NAVY WHERE HONOR AND LOYALTY MEAN SO MUCH.***

He added:

> *It is all the more remarkable when you consider Jim Morin had been retired from the Navy for more than 20 years after a 33 year career when [the ceremony] took place." And he concluded that "the only possible reward was doing his duty..." and "his duty in the Navy made his life worthwhile."*[27]

Honored Leader

James Brendon Morin, dubbed in his high school yearbook as "most reliable," is one hell of a guy, and he became one hell of a leader. As a naval officer, he gained the admiration and trust of those who served with him. Moreover, he gained the deep respect of subordinates that bordered on love for some Silver Foxes and that stamped him as a special and honored leader.

End Notes

[1] James B. Morin, "Belated Honors," Sea Stories," http:// www.able-dogs.com/ Stories/honors.htm (20 November 2000).
[2] Comment by George Blosser, "Silver Fox Memories," 28 December 2003, 6.
[3] James B. Morin, *Email to John Bradley*, 18 May 2009.
[4] Conversations with Diane and Jim Mattly and Marilyn and Ed Hiebert, Houston, TX, 22 July 2010.
[5] Kish, *Email to Bradley*, 17 February 2005.
[6] James B. Morin, *Email to John Bradley*, 17 August 2010.
[7] *Ibid*, 7 for comments about the CNO.
[8] *Wikipedia*, 14 Aug 2010. ADM MacDonald retired in 1985 and died in 2009.
[9] Morin, "Belated Honors," 2.
[10] Kish, *Email to Bradley*, February 2005
[11] Chris Morss, "Notes for John Bradley," 15 July 1010.
[12] Blosser, "Silver Fox Memories," 7.
[13] Kish, *Email to Bradley*, February 2005, 1.
[14] *Ibid*, 2.
[15] Blosser, "Silver Fox Memories," 7.
[16] Kish, *Email to Bradley*, February 2005, 1, and Blosser, "Silver Fox Memories," 7.
[17] Fake Awards

The Department of the Navy Board for Correction of Naval Records 2 Navy Annex, Washington, DC, 20370-5100, AEG: Japha Docket No: 6956-98, 24 June 1999 denied the request from ADM Jeremy Boorda's children to correct the records to show that their deceased father had been entitled to wear a Combat "V" device on his Navy Commendation Medal and his Navy Achievement Medal.

B. G. Burkett and Glenna Whitley, *Stolen Valor, How the Vietnam Generation Was Robbed of Its Heroes and Its History* (Dallas: Verity Press, Inc. 1998), 194-5 (Kane).

John H. Bradley was interviewed by a Houston TV reporter and identified the U. S. medals that the young man wore. They included: two Army Distinguished Service Crosses, a Department of Defense Distinguished Service Medal, an Army Distinguished Service Medal, two Silver Stars, a DFC, a Legion of Merit, a Soldier's Medal, a Bronze Star, a Meritorious Service Medal, an Army Commendation Medal, and a Purple Heart. He also wore a Combat Infantryman's Badge with two stars indicating awards in three wars, a Master Parachutist Badge, a Pathfinder's Badge, and an Air Assault Badge as well as a high-level British decoration worn around his neck.

In March 2011, a federal judge ruled that lying about one's war record and awards was not a crime under current law because it was permitted under the free speech provisions of the 1st Amendment to the U. S. Constitution.

[18] Morin, "Belated Honors," 3.
[19] *Ibid.*
[20] Blosser, "Silver Fox Memories," 7.
[21] Beth McPherson, "*Silver Foxes* rewarded. 6 receive Distinguished Flying Cross," Pensacola NewsJournal.com, Saturday, May 27, 2000.
[22] *Ibid.*
[23] Blosser, "Silver Fox Memories," 8.

[24] *Ibid.*

[25] Mike O'Callaghan, "Where I Stand, Morin: A Man of Honor," Las Vegas *Sun*. July 2000.

[26] *Ibid.*

[27] Kish, *Email to Bradley*, February 2005, 2.

Epilogue

Officers who lead individual pilots into combat differ from most officers who lead ground troops into battle or command combat ships because they must perform as individual warriors as well as combat leaders.

In many ways, leaders such as Jim Morin resembled medieval knights who led and inspired their men in battle, often proved to be the bravest and most acclaimed warrior among them, and thus gained the deep fealty of their warriors.

John Bradley

APPENDICES

JIM MORIN AND HIS CHILDREN

Appendix A
Aviation-related Ranks and Positions on Aircraft Carriers

Rank	Aviation Status	Positions
Chief Warrant Officer (CWO)	Non-Aviator	Technical support position
Ensign, (ENS) 2nd LT, USMC (2LT)	Aviator	Administrative or technical support position
Lieutenant, junior grade (LTJG) 1st LT, USMC (1LT)	Aviator	Administrative or technical support position
Lieutenant (LT) Captain, USMC (CPT)	Aviator	Flight or section leader
Lieutenant Commander (LCDR) Major, USMC (MAJ)	Aviator	Squadron Executive Officer or Squadron or Wing staff officer
Commander (CDR) Lt. Colonel (USMC) (LTC)	Aviator	Squadron Commander; Squadron Commander designate; Wing Executive Officer or staff officer
Commander (CDR) Lt. Colonel (USMC) (LTC)	Aviator	Commander, Air Group (CAG), Commander Air Wing (CAW)
Captain (CAPT) Colonel, USMC (COL)	Aviator	Commander, Air Group (CAG), Commander Air Wing (CAW)
Captain (CAPT)	Non-flying Aviator	Captain, aircraft carrier
Rear Admiral (RADM)	Non-flying Aviator	Commander carrier task group
Vice Admiral (VADM)	Non-flying Aviator	Commander carrier task force

Captains

Army, Air Force, and Marine (USMC) captains rank with Navy lieutenants in Grade O3.
Navy captains rank with Army, Air Force, and Marine colonels in Grade O6.
To differentiate the ranks of captains in the text, CPT is the abbreviated title used for Army, Air Force, and Marine captains, and CAPT is the abbreviated title used for Navy captains.

Appendix B
U. S. Navy Decorations

Decoration	Reason
Medal of Honor	Actions above and beyond the call of duty in combat
Navy Cross	Distinguished actions in combat not meriting a Medal of Honor
Navy Distinguished Service Medal	Distinguished service at the highest level
Silver Star	Gallantry in Combat action
Distinguished Flying Cross	Distinguished action in flight or heroic action in aerial combat
Legion of Merit	Exceptionally meritorious conduct – generally not in combat
Sailors and Marine Medal	Lifesaving or peacetime heroic act
Bronze Star Medal	Meritorious service in a combat zone
Air Medal	Meritorious achievement in flight, normally in a combat zone
Navy Commendation Medal	Commendable service, generally in peacetime
Purple Heart Medal	Wounds in combat; death from wounds in combat

Valorous Action

The Legion of Merit, the Bronze Star Medal, the Air Medal, and the Navy Commendation Medal normally are not awarded for heroic combat actions. If one of the decorations is awarded for valorous actions in combat, a "V" device is awarded to signify heroic action, and it is worn on the medal or ribbon.

Additional Awards

The Navy awards a Gold Star for a second, third, and fourth award and a Silver Star for a fifth award. Additional Gold or Silver Stars are awarded for additional awards. The small stars are worn on the medals and ribbons.

During the Vietnamese War, the armed services awarded a large number of Air Medals to aviators and pilots. Because of this, the Navy replaced Gold and Silver Stars and the Army and Air Force replaced Oak Leaf Clusters with bronze numerals for two or more awards.

Appendix C
Sorties over North Vietnam
April 1965-March 1973

The Navy flew more sorties (52%) over North Vietnam than the Air Force; the Air Force flew substantially more over South Vietnam.[1]

Service	Sorties	Percent of Total Sorties
U. S. Air Force	226,000	43%
U. S. Marine Corps	27,000	5%
U. S. Navy	275,000	**52%**
Total	578,000	

Notes

[1]Nichols and Tillman, *On Yankee Station*, 163

Appendix D
Air Losses

Causes

Anti-aircraft fire (AAA) from large caliber guns caused the most losses of naval aircraft. Smaller caliber automatic and semi-automatic weapons caused the most losses for air force and marine aircraft, most likely during close air support missions that supported ground forces in South Vietnam.

Causes for In-Flight Combat Losses[1]						
Service	**AAA Fire**	**Semi-automatic & automatic fire**	**Surface to air Missiles**	**MiG Fighters**	**Unknown**	**Own Ordnance**
U. S. Air Force	26%	**47%**	7%	4%	16%	1%
U. S. Marine Corps	14%	**50%**	2%		33%	
U. S. Navy	**37%**	18%	15%	2%	25%	3%

A4 Aircraft

The naval service lost more A4 & TA4 aircraft (257) than any other type of aircraft during the Vietnamese War. Overall the Navy and Marine Corps lost 711 aircraft.

Losses of A4/TA4 "Skyhawk" Aircraft[2]						
Service	**North Vietnam**	**Percentage**	**South Vietnam**	**Percentage**	**Total A4s/TA4s**	**Percentage all Losses**
U. S. Marine Corps	7	11.3%	51	82%	62	8.7%
U. S. Navy	172	88%	4	2%	195	27.4%
U. S. Navy & U. S. Marine Corps	179	69.6%	55	21%	257	36%

VA-155

Richard Floreani reported that during the 1965-1966 cruise, "almost half, twenty-nine of the Coral Sea's sixty plus aircraft were shot down. Eleven pilots were killed and two became prisoners of war." VA-155 lost only one killed and one captured."[3]

VA-155 Losses, Jan-Nov 1965[4]

Date	Aviator	Target	Cause	Result
7 Feb 1965	LT Ed Dickson	Barracks and port at Dong Hoi, NVN	Ground fire	KIA
29 Mar 1965	CDR Jack Harris		Ground fire	Rescued by a submarine
Aug 1965	LT William Fidelibus	Armed road reconnaissance, NVN	Ground fire	Rescued by a helicopter
30 Sep 1965	LCDR Wendy Rivers	Rail marshalling yard, Vinh, NVN	Ground fire	POW
Last line period 1965	LCDR Paul Moore[5]	Reconnaissance	Ground fire	Rescued by a destroyer

Notes

[1]Nichols and Tilman, *On Yankee Station*, 164-5.
[2] *Ibid.*
[3] Floreani, "Silver Fox Squadron," 5.
[4] USS Coral Sea, "Navy Jet Pilot Escapes Injury," Sea Breeze, Jan 1966, 39. The rescue crewman, M. J. Austin "was lowered into the moderately choppy seas to help Lt. Fidelibus into the rescue harness." It was Austin's fifth rescue of the day. The rescue pilot was Lieutenant, j. g. B. G. Brandow.
[5] No specific date established. Morin, page 152-54, said during the last line period and implied it was after Wendy Rivers was shot down. Bill Kish wrote: "I do not remember the date but I am pretty sure it was before we lost Wendy and after the end of July.

Appendix E
USS Liberty Incident

Several aspects of the USS *Liberty* incident added to the drama of the Cold War.[1]

1. No clear reason for the Israeli attack has been discovered.

2. The incident may have resulted in a nuclear attack.

Joe Meador, wounded in the attack, was evacuated to Crete en route to Malta. On Crete, ground crewman asked him if he were from the *Liberty*. They reportedly said:

... they had handled the refueling of a US fighter which, to their amazement, had an atomic bomb underneath. They said it had been launched from the [USS] America to bomb Cairo as a result of the Liberty's distress call. After being recalled, it could not land on the ship's deck with the atomic bomb slung beneath it and had therefore been diverted to a land-based strip.[2]

CAPT Joseph Tully, the captain of the USS *Saratoga* at the time, however, in a talk to *Liberty* survivors years later mentioned only regular bombs. So did RADM Geiss.

Rear Admiral Geis in [USS] America recalled the aircraft with instructions not to jettison their 500 pound bombs. This created a problem for Saratoga, but fortunately there was a NATO Ammo Depot in Crete which was nearby, and so four of Saratoga's aircraft were diverted to NATO Ammo Depot and airfield.

Meanwhile, Vice Admiral [William] Martin ordered another strike sortie in support of Liberty at about 1500, which was accomplished, only again to be recalled by Rear Admiral Geis, with again some of Saratoga's aircraft being diverted to Crete for unloading.[3]

LT Jim Nanjo, a B52 crewman in the 741 Bomb Squadron at Beale Air Force Base, CA reported that on the morning of 8 June 1967, he and other bomber crewmen had been alerted and manned their bombers. Once aboard his aircraft, sealed orders were opened and this told him that this was not a "practice" alert. Moreover, his aircraft and other B52s moved to the end of the runway to await orders to take off. He recalled years later that the aircraft were "cocked" and that they had nuclear weapons aboard.[4]

3. Some American leaders feared that the Soviets or Egyptians had attacked the ship. The former could have precipitated a nuclear war; the latter would have justified an Israeli or American attack on Egypt.

4. When the Israelis promptly took responsibility for the attack, they removed any chance that the Egyptians or the Russians had attacked the ship. President Johnson immediately notified the Soviet Premier that the Israelis admitted the attack; the Soviet leader, in turn, notified Egypt.[5]

5. The U. S. Sixth Fleet launched two strike groups and recalled both.

Captain Tully, commanding *Saratoga*, reported that he launched 12 fighter bomber aircraft and four tankers to rescue the *Liberty*, and that "... Admiral Lawrence Geis, who commanded the Sixth Fleet carriers, radioed Tully and ordered the planes - -

which were still in view - to return." This apparently followed Secretary McNamara's order to recall the aircraft at 0824 Washington Time (WT).

Given permission to launch again at 0950 WT, Tully did so, but soon received an order from Geis, who had received an order from McNamara, to recall them.[6]

6. Early on, President Johnson decided not to blame Israel for a deliberate attack.

RADM Geis told LCDR David Lewis of the *Liberty*, that he had requested confirmation of McNamara's second order and "the Commander-in-Chief, Lyndon Johnson came on and ordered them recalled, with the comment, I will not embarras our ally." He also told Lewis that "he was sure it will all be hushed," and "he had to tell someone he tried to help us." Geis asked Lewis to keep his comments confidential.[7]

RADM Clarence A. Hill, Jr., in a speech to *Liberty* veterans in 2008 reported that Geis took stronger action:

The U.S. carriers Saratoga and America were in the region fully loaded with aircraft, which were dispatched to help. However, President Lyndon Johnson and Secretary of Defense Robert McNamara recalled the carriers' planes. "We were unable to do anything to rescue the Liberty," Hill lamented, "and it left a stain on the Navy."

Everyone acquiesced except for Rear Admiral Larry Geis, who commanded the carriers in the Sixth Fleet. "Geis got back to McNamara to ask why his carriers' planes were being recalled," Hill said. "Geis said he wouldn't turn around unless the commander in chief himself told him to. McNamara brought President Lyndon Johnson to the phone. 'Mr. President,' Admiral Geis said, 'They're killing Americans.' Johnson's answer was 'I don't care how many Americans are being killed. I don't want to embarrass one of our allies.'"[8]

7. The Navy threatened its sailors with courts-martial if they talked about the incident.

10 June 1967: Vice Admiral McCain orders Rear Admiral Isaac Kidd to convene Navy Court of Inquiry.

11 June 1967: Admiral Kidd boards Liberty with small staff to head Navy Court of Inquiry.

14 June 1967: Liberty arrives in Malta. Total news blackout imposed. Rear Admiral Kidd warns crew: "You are never, repeat never, to discuss this with anyone, not even your wives. If you do, you will be court-martialed and will end your lives in prison or worse."[9]

8. The fact that the United States took no action to support the *Liberty* was unconscionable. RADM Hill summed up the feelings of many when he said:

It's true that we might not have reached the Liberty in time to help prevent the first attack, but we could have prevented those torpedo deaths. Officers were willing and able to do anything they could to help but were prevented from doing so by our own country team.[10]

8. The United States Government and the Israeli Government covered up the facts surrounding the incident. Admiral Thomas Moorer, the Chief of Naval Operations at the time, always believed that the Israelis deliberately attacked the *Liberty*. RADM Hill commented:

> *Admiral Moorer confirmed these facts with the commanders of the America and Saratoga, and called this coverup "the most disgraceful act I witnessed in my entire military career."* [11]

9. The U. S. Government, the Israeli Government, and the U. S. Navy have not provided the survivors or the families of the casualties with a full and truthful accounting of the incident, maintaining that it was an unfortunate accident.

There remain many unanswered questions about the incident: Why was the USS *Liberty* deployed to the area? Was the ship deployed there to become a sacrificial target? Did the captain know about an impending attack? Did the Israelis notify the U. S. of the attack prior to conducting it? Why did U. S. aircraft not immediately come to the aid of the American ship?

But the most important questions concern the cover up.

1. Why did the Navy not conduct a detailed inquiry into the incident and call all the survivors to testify along with all the carrier officers and pilots who were involved in responses to the attack?

2. Why did Congress not conduct a detailed inquiry into the event and call all the survivors to testify along with all the carrier officers and pilots who were involved in responses to the attack and the political officials in the Johnson Administration who dealt with the incident?

3. Why did the Navy and the Johnson Administration (and following Administrations) fail to provide the full truth about the attack to the survivors and the families of those killed in the attack?

Δ

The lack of a full Congressional Investigation where sailors died in a combat action compares unfavorably to the Congressional Investigation of the racial incidents on carriers discussed earlier in this book. Politics apparently triggered the cover-up on the USS *Liberty* Incident as it did with the investigation of the racial incidents.

Notes

[1]For a general overview of the incident, with some good specifics, see *Wikipedia.* In particular see *THE USS LIBERTY , [Paper] #34 in AMEU's Public Affairs Series.* Americans for Middle East Understanding.

[2] Peter Houman, *Operation Cyanide, Why the Bombing of the USS Liberty Nearly Caused World War III* (London: Vision, 2003), 183 (Meadors). While it appears that Houman focused on the possible nuclear attack to make his book important, he provided much useful information from people he interviewed.

[3]Captain Joseph Tully, USN, Former Commanding Officer USS Saratoga, *Speech to USS Liberty survivors gathered in San Diego for 28th anniversary reunion . Wikipedia* (25 Dec 2010).

[4] Houman, *Operation Cyanide,* 8-9 (LT Nanjo).

[5] *Ibid,* 148-9.

[6] *Ibid,* 91.

[7] *Ibid,* 175.

[8] *Navy Memorial Service for Admiral Thomas Moorer,* The U.S. NAVY Memorial in Washington, D. C. March 4, 2008. *Wikipedia* (25 Dec 2010).

[9] *[Paper] #34 in AMEU's Public Affairs Series.*

[10] *Ibid.*

[11] *Ibid.*

APPENDIX F
Awards, James B. Morin

Distinguished Flying Cross
with Gold Star in Lieu of Second Award
Heroism
10 September 1965 & 5 October 1965

Legion of Merit
Exceptionally Meritorious Conduct
July 1976-July 1978

Bronze Star Medal with Combat "V"
Meritorious Service
22 June 1965 to 1 July 1965

Air Medal
with Strike/Flight Numeral 12
Meritorious Achievement
7 February 1965-22 June 1966

Navy Commendation Medal
with Gold Stars in Lieu of the Second and Third Awards with Combat "V"
Meritorious Service from 15 to 21 May 1965
Heroic Achievement on 12 August 1965
Meritorious Service from 5 to 15 October 1965

Appendix G
"Never Forgot You"

e-mail 5/11/2005 to RADM James B. Morin

Sir:

I'd thought I'd take a moment to try and express my gratitude to you for the tremendous impression you made on me early in my Naval career.

I checked into VA-155 fresh out of AME "A" school in October 1964. I met with disappointment on check-in when the Maintenance CPO informed me that I would not go directly to my shop as I had been told while training over at the RAG. All of my buddies were in Plane Captain School at the RAG while I was told that I would be needed in the shop so I was to learn all I could, and be prepared to hit the deck running. I did my level best to make that happen. When I checked in the Maintenance CPO, after realizing that I was the AME he anxiously awaited and not realizing I was a young Black kid, asked me "Boy, ain't you afraid of that Ejection seat and the liquid oxygen?" To which I replied, "NO Chief." He then said, "You're assigned to the line crew." I asked why then did you not send me to P/C school so that I'd at least be on the same pace as the other guys. He said "you can learn the job" and then turned away.

I was brought to your office later on and I remember you saying, "I feel old." You said that I was born the year you graduated from the Academy. I couldn't believe you sat me down in your office and talked to me like a real person.

I spent approximately three weeks on the Line doing whatever I was tasked to do. One day I was told to go TAD to the Galley and I asked what did I do wrong because everyone seemed pleased with my productivity. At the time the norm was to send "losers" to the Galley. I checked out of the Squadron on a Thursday and into the Galley that afternoon. That night I along with two squadron mates hitchhiked to Fresno and went U. A. I returned Saturday and reported back to the Galley.

Later I was called down to the hanger and you happened to see me topside and called me into your office. You then told me how impressed you were with me when we first met. You said that your first thought when you saw me was "that is a sharp young man who will be an asset to the squadron." You asked me why did I do it and I said "because I did not feel that I was wanted or appreciated and that I was a dirt bag." You asked me if anyone had explained the policy of sending people to the Galley and why. I said they only told me to go. You took time to explain to this young kid the facts and you told me that what had happened in my situation would not occur with another kid again. I returned to the Galley with the full understanding of my role for being there and did my job well.

Another time you sat me down was in Yokosuka, Japan when we got extended. I got myself into some trouble with the Shore Patrol subsequent to our squadron party. I had a report chit that was rather thick. You said, "Son, I don't want to send you to the ship's CO with this because he will hang you." You then gave me the choice of restricting myself to the ship the remainder of the in-port period and during our time in Subic where we were headed to replenish ammo. I had to think long and hard on that one as I loved the liberty in Subic. I finally made the correct choice and you said, "Son, that's a wise decision."

I remember the day you were leaving the Constellation. I was up on the fuselage of one of our birds installing an ejection seat. You walked by, stopped, and climbed up on the plane to say good-bye to me! And then you said you were going to Washington and told me to contact you if there was anything you could do for me.

I spent 30 years in the Navy. I retired in 1993 as a CWO4 and through the years I always remembered to try to emulate you and a few other fine gents that I chose as my role models.

Thank you sir, for setting an example I could follow and try to impress others.

Sincerely,

Bill Maynor

INDEX

A

B

C

D

E

F

G

H

M

T

U

NOTES

Asterisks - SILVER FOXES

Italicized names -KIA, DOW, died on active duty

NFN - no known first name

sb - sidebar